ABOUT ISLAND PRESS

Island Press is the only nonprofit organization in the United States whose principal purpose is the publication of books on environmental issues and natural resource management. We provide solutions-oriented information to professionals, public officials, business and community leaders, and concerned citizens who are shaping responses to environmental problems.

In 2003, Island Press celebrates its nineteenth anniversary as the leading provider of timely and practical books that take a multidisciplinary approach to critical environmental concerns. Our growing list of titles reflects our commitment to bringing the best of an expanding body of literature to the environmental community throughout North America and the world.

Support for Island Press is provided by The Nathan Cummings Foundation, Geraldine R. Dodge Foundation, Doris Duke Charitable Foundation, Educational Foundation of America, The Charles Engelhard Foundation, The Ford Foundation, The George Gund Foundation, The Vira I. Heinz Endowment, The William and Flora Hewlett Foundation, Henry Luce Foundation, The John D. and Catherine T. MacArthur Foundation, The Andrew W. Mellon Foundation, The Moriah Fund, The Curtis and Edith Munson Foundation, National Fish and Wildlife Foundation, The New-Land Foundation, Oak Foundation, The Overbrook Foundation, The David and Lucile Packard Foundation, The Pew Charitable Trusts, The Rockefeller Foundation, The Winslow Foundation, and other generous donors.

The opinions expressed in this book are those of the author(s) and do not necessarily reflect the views of these foundations.

EXPERIMENTS
IN
CONSILIENCE

EXPERIMENTS IN CONSILIENCE

*Integrating Social
and Scientific Responses
to Save Endangered Species*

Edited by
Frances R. Westley
and Philip S. Miller

ISLAND PRESS
Washington · Covelo · London

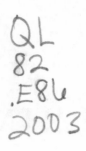

QL
82
.E86
2003

Library of Congress Cataloging-in-Publication Data

Experiments in consilience : integrating social and scientific responses to save endangered species / edited by Frances R. Westley and Philip S. Miller.
p. cm.
Includes bibliographical references and index.
ISBN 1-55963-993-8 (alk. paper) — ISBN 1-55963-994-6 (pbk. : alk. paper)
1. Endangered species. 2. Wildlife conservation. 3. Human ecology. I. Westley, Frances R. II. Miller, Philip S.
QL82.E86 2003
333.95'22—dc21

2003007224

British Cataloguing-in-Publication Data available

Book design by Teresa Bonner
Composition by Wilsted & Taylor Publishing Services

Printed on recycled, acid-free paper ✪

Manufactured in the United States of America

09 08 07 06 05 04 03   10 9 8 7 6 5 4 3 2 1

This book is dedicated to
Ulysses S. Seal
June 13, 1929–March 19, 2003

Contents

List of Acronyms

AEUB	Alberta Energy and Utility Board
AWAG	Algonquin Wolf Advisory Group
AZA	American Zoo and Aquarium Association
CAMP	Conservation Assessment and Management Plan
CBSG	Conservation Breeding Specialist Group
COSEWIC	Committee on the Status of Endangered Wildlife in Canada
CPR	common property regime
CRE	Central Rockies Ecosystem
DEC	Department of Environment and Conservation
DRC	Democratic Republic of the Congo
ESGBP	Eastern Slopes Grizzly Bear Project
GIS	Geographic Information System
GNWT	Government of the Northwest Territories
HTO	hunters' and trappers' organization
ICDP	Integrated Conservation and Development Project
IGCP	International Gorilla Conservation Program
ISIS	International Species Information System
IUCN	World Conservation Union (formerly the International Union for the Conservation of Nature and Natural Resources)
LDR	Less Developed Regions
LeastDR	Least Developed Regions
MDR	More Developed Regions

NGO	nongovernmental organization
NWT	Northwest Territories
OMNR	Ontario Ministry of Natural Resources
PASS	political-administrative system strength
PHVA	Population and Habitat Viability Assessment
PNG	Papua New Guinea
PRA	Participatory Rural Appraisal
PVA	Population Viability Analysis
RENEW	REcovery of Nationally Endangered Wildlife
RWED	Department of Resources, Wildlife, and Economic Development
SSC	Species Survival Commission
SSP	Species Survival Plan
TEK	traditional ecological knowledge
TEKS	traditional ecological knowledge systems
TWS	traditional Western science
USAID	U.S. Agency for International Development
USFWS	U.S. Fish and Wildlife Service
WCED	World Commission on Environment and Development

Preface

This book is the result of five years of walking a fine line: the line between theory and practice, the line between social and natural science, and the line between colleagues and friends. All people who were involved in any way in this volume were committed to trying to bridge these differences; ultimately we believed that such divides had to be crossed if we were to contribute to the survival of endangered species and spaces. But sometimes, each of us felt that we ourselves, or at least our disciplines, might be the endangered species. Consilience is a slow process and involves spending some time "far from land," in conceptual spaces where each and every one of us felt far from comfortable. This required patience and hard work and an ability to confront, but also to tolerate, our differences. We extend appreciation to our colleagues. Every person involved in the Network stayed with the process throughout. The result is this fascinating book, which documents our moments of convergence as well as the differences in perspective and approach that were maintained throughout.

There are many people to thank. We appreciate the insights and comments of Dr. Susie Ellis, Conservation International, who attended many of the meetings and enlivened them with her wit and wisdom. Colin Scott, Sally Walker, Sanjay Molur, Ruth Barretto, Oliver Coomes, Karen Peterson, and Mike Robinson also joined us for at least one meeting and gave us the benefit of their own experiences in consilience, both theoretical and practical. A very special expression of gratitude goes to Jenna Borovansky for the help she gave us in editing the final manuscript and for teaching us about megadocuments and why she hates Bill Gates.

Other much needed and valued assistance along the way came from Moriya McGovern and Tara Shaughnessy, who all helped with the diagrams and printing, and to Ronda Fisher and Emmanuel Raufflet, who both documented our rather intense meeting discussions.

We are grateful to Barbara Youngblood and Barbara Dean, from Island Press, for their enthusiasm, patience, insightful comments, and belief in this project.

We also would like to thank the Social Science and Humanities Research Council of Canada and the United States' National Science Foundation for their generous financial support of this Network research project.

To our families—Fred, Katie, Clara, and Jane, as well as Ellen, Sydney, and Jason—we of course owe a huge debt of gratitude for their support and patience, not only for this project but for all the projects that have taken us away from them through the years. No, Jason ... Daddy doesn't live at the airport!

Special thanks go to the participants in the various workshops that we attended as part of this project. One of the joys of being a part of this team has been witnessing this global band of conservationists in places as far-flung as Kampala, Canmore, Lae, Belo Horizonte, Yellowknife, and Dorset united in their determination against all odds to save the endangered species they love. We hope this book will be a tribute to their efforts.

Lastly, we owe the greatest debt of gratitude to one man, Ulysses S. Seal, who has inspired not only us, but thousands of conservationists and scientists around the world. Ulie's unique gifts fundamentally shaped CBSG and the workshop processes that are the subject of this book. His far-reaching vision, passion for conservation, and faith in the human species' ability to transform the world for the better gave us the energy and the determination to begin and complete this project.

Frances R. Westley
Philip S. Miller

PART ONE

Introduction

The Story of an Experiment: Integrating Social and Scientific Responses to Facilitate Conservation Action

FRANCES R. WESTLEY

Transdisciplinarity is a highly creative act; there are not formulas for reintegrating knowledge. However difficult the task, and however resistant it is to formalization, it is clear that the major failings of earth systems are due to the artificial fracturing of knowledge in the name of scholarship. The task ahead is to counter this tendency.

Rapport 2000

This is a story of an experiment. It centers on the problem of conserving the planet's endangered species, but it also tells the story of a new form of organizing for effective risk assessment, recommendation, and action. It focuses on the challenges of cross-disciplinary analysis as well as cross-functional, cross-disciplinary, and cross-sectoral action. Most centrally, it is the story of a sustained project in action research and the learnings that resulted.

In 1987, the Brundtland Commission published its influential report *Our Common Future,* which firmly established sustainable development on the international agenda for the coming decades. Among the priorities identified in the report was the conservation of species and ecosystems. "Species and their genetic materials," the authors argued, "promise to play an expanding role in development, and a powerful economic rationale is emerging to bolster the ethical, aesthetic, and scientific cases for preserving them" (World Commission on Environment and Development [WCED] 147). This imperative, in turn, became the focus of the World Resources Institute, the World Conservation Union, and the United Nations Environment Program's report, *Global Biodiversity Strategy.* In that document, a clear ethic of sustainable development, which implies a balance between social development and biological conservation, is presented.

Development has to be both people centered and conservation based. Unless we protect the structure, functions, and diversity of the world's natural systems—on which our species and all others depend—development will undermine itself and fail. Unless we use Earth's resources sustainably and prudently, we deny people their future. Development must not come at the expense of other groups or later generations, nor threaten other species' survival (WRI 1992, v).

This ethic has been widely endorsed internationally, as witnessed by the number of nations which have signed the Biodiversity Convention, established at Rio in 1992. Embedded in this overarching statement, are additional values: that of maintaining diversity, balancing human and nonhuman rights, and economic development and conservation. It also stresses the value of participation ... that stakeholders in the Earth's resources all have an equal right to participate in decisions concerning distribution of those resources. But all this raises the specter of despair: are such goals impossible to achieve? Many challenge the notion of sustainable development as oxymoronic: can we continue to reap an endless economic harvest from an increasingly depleted planet?

Certainly, the application of these principles is a difficult and challenging task, both scientifically and socially. The *Global Biodiversity Strategy* report urges that action is needed both to strengthen the tools and technologies of biodiversity conservation (in order to identify priorities and strengthen the capacity of on and off-site institutions to conserve species and habitats); and to expand the human capacity to conserve biodiversity (in order to increase awareness, disseminate information, promote research links between social and natural sciences, transfer technology and know-how, and build partnerships). While biologists disagree on the exact rate of extinction of species on the planet, it is widely recognized that it is not only rapid, but that it is accelerating. "Conservative" estimates place the current rate of extinction at around 1,000 species a year, but with the continued destruction of habitats around the world, this is anticipated to rise to over 10,000 species per year by the end of this decade (approximately one species per hour) (Wilson 1989, 1992).

Whose problem is this? In the broadest sense, it is all of humanity's, including the future generations who will be deprived of the biodiversity that their ancestors enjoyed. As a species, humans have relied on rich biodiversity for nourishment, medicine, aesthetic satisfaction, and even for psychological well-being (Kellert and Wilson 1993). Biodiversity has been the basis of trade and of much commerce. The loss of biodiversity challenges the very bases of human life on this planet.

In practical terms, therefore, implementing any strategy for maintaining biodiversity demands integrating both biological science and social science, expert and local knowledge, economic and conservation imperatives in actions designed to ensure stakeholder participation, equity, and justice, and even survival. The challenge is enormous and time is short. As the *Global Biodiversity Strategy* report states: "Irreplaceable genes, species and ecosystems are disappearing at a rate unprecedented in human history and essential development is at risk as a result. Immediate action is needed to defend these threatened living resources...." (WRI 1992,19).

The Biodiversity Research Network

In 1997, with the help of a grant from the Social Science and Humanities Research Council of Canada, a research network (hereafter referred to as the Network) was created to build interdisciplinary connections and stimulate an exchange of expertise among specialists concerned with the conservation of biodiversity. Team members shared a concern to (a) understand the ecology and population dynamics of key species in particular ecosystems; (b) understand the impact of local human populations on the survival of threatened ecosystems and species; and (c) develop tools and processes for securing the involvement, collaboration, and responsibility of a wider range of local stakeholders in conserving species in their habitats and the ecosystem management required to achieve this.

The first principle of this initiative was that this discourse should be multidisciplinary, due to the complexity and magnitude of the problem. A number of scholars, chief among them E. O. Wilson, have recently highlighted the need to find an integration between social and biological or natural sciences if we are to address the environmental concerns. Wilson terms this rapprochement "consilience" and argues that sound environmental policy can only be formed at the juncture of ethics, social science, and biology (Wilson 1998).

Such transdisciplinary teamwork is difficult to achieve, however, even in the exploration of the kinds of environmental problems where it is most necessary. As a society of specialists, we have a low level of interaction. We know how to separate into disciplines, but not to put the pieces back again: "Transdisciplinarity is not an automatic process that can be successfully carried out simply by bringing together people from different disciplines. Something more is required, although the 'magic ingredient' is difficult to pinpoint. Transdisciplinarity requires 'transcendence,' the giving up of sovereignty on the part of any one of the contributing disciplines, and the formation, out of the diverse mix, of new insight by way of emergent properties" (Somerville and Rapport 2000, xv).

Recent research on transdisciplinary projects suggests that success demands no less than a revolution in our knowledge institutions: the commitment of senior people in the field, funding and publication outlets, and the arduous process of building transdisciplinary communication and trust (Daily and Ehrlich 1999). Developing a sound base of trust and understanding is extremely time consuming and requires patience. Levels of commitment to this process will clearly vary, and bringing on new people after the process has started is always challenging (Naiman 1999). Part of the difficulty resides in the fundamental difference in discourse and dialects that have developed within each discipline, as well as the discipline-based nature of reward systems (Kostoff 2002). Therefore, a period of translation and mutual learning is always required (Wear 1999; Somerville and Rapport 2000), and not all researchers are willing and able to engage in this kind of collaboration (Nicolson et al. 2002).

With most collaborations, the period of translation and mutual learning is demarcated by several stages and phases, each with its own dynamic. The first stage is "problem definition/recognition" in which a statement of the problem or problems under consideration needs to be crafted so that all involved disciplines can relate it to their base of knowledge. Here, power dynamics make an early appearance, as different disciplinary groups jockey to have their "problem definition" dominate (Nicolsen et al. 2002). A second phase involves "defining direction." At the interdisciplinary level this is often a problem of methodology (Prickett et al. 1999). Here again, issues of dominance and power are critical. If more powerful or influential disciplines "hijack" this process, the less powerful will become disaffected and be prone to withdraw (Gray 1989; Westley 1999; Hardy and Phillips 1998). The development of mutual trust and commitment is fragile and easily reversed. However, concrete experiences (field trips, simulations, a specific research site) can provide shortcuts to this process (Prickett et al. 1999). Also, the use of analogy and sustained metaphor (e.g., the comparison between ecological patch and neighborhood; Grove and Birch 1977) can help build and facilitate interdisciplinary communication, as can the choice of "middle level perspectives/phenomena," such as a species or a habitat (Prickett et al. 1999). Finally, the critical role of "social interaction and long-term associations that allow friendships to develop" (Daily and Ehrlich 1999, 278) cannot be underestimated. This is the glue which allows the collaboration to hang together through frustrations, and ultimately allows constructive conflict to surface. Such conflicts, in turn, seem a central element of creative problem resolution (Brown and Ashman 1996).

Our research team faced the challenge of interdisciplinary research on two different levels. The first was at the level of the team itself. Members of the Network included American and Canadian experts in interorganizational collaboration, stakeholder processes, human demography and the environment, participative research, management and development, conservation biology and wildlife management, population genetics, reproductive biology, ecosystem dynamics, business and the environment, environmental management, and planning. Some of the Network members were located in university faculties, some in research labs, and still others in nongovernmental organizations (NGOs). Some members of the group had worked together intensively in other research or action settings and others had not collaborated previously.

The second challenge was at the level of the experiments that the Network undertook. The work plan was to bring Network members together at least twice a year. These meetings revolved around intensive discussions of issues involved and around plans to experiment with new, more integrative approaches to stakeholder inclusion and information intensification in conservation workshops. It soon became clear that these "experiments" would have to deal with three challenges to interdisciplinary integration:

1. *Integrating tools:* We were concerned with developing methods to allow some of the tools for analyzing human dimensions such as demography, economics, institutional and governance structures, and industry dynamics to interface with tools that assess a particular species' risk of extinction.

2. *Creating processes for integrating expertise and expanding inclusion:* We sought ways to link social scientists with expertise in such things as resource and agricultural economics, human demography, industrial geography, Indigenous cultures, and political and institutional processes, with biological scientists who understood conservation science. Our goal was to elucidate the dynamics of the social system that is the "human envelope" around endangered spaces and species.

3. *Exploring process:* We examined and monitored the ways in which experiments in the above two areas affected the process of conservation planning workshops and the implications for redesigning that process. We explored ways in which a wider group of stakeholders and their information could be incorporated into the workshop process, without reducing their ability to carry out effective risk assessment and to formulate helpful recommendations.

In order to ground this experiment in an ongoing stream of action, the experiment was designed to focus on a single type of conservation-planning workshop run by a single organization. The Conservation Breeding Specialist Group (CBSG) is one of more than 120 specialist groups comprising the Species Survival Commission (SSC) of the World Conservation Union (IUCN). Its small group of paid staff and extensive network of volunteer scientists and managers around the world are supported by annual voluntary donations from more than 150 institutions and organizations worldwide. The mission of CBSG is to facilitate endangered species survival through developing, testing, and applying scientifically based tools for risk assessment and decision making in the context of wild and captive species management. One of a number of tools employed by CBSG is the Population and Habitat Viability Assessment (PHVA) workshop. A PHVA workshop brings together stakeholders from the scientific, nongovernmental, and governmental communities in a highly interactive, participatory process designed to assist in the development of strategic recovery plans for threatened species and their habitats. Such processes are not unique to CBSG, but for our research project they provided a focal process in which the parameters had been relatively constant over the past ten years (to allow for comparison) and in which the dynamics were flexible enough to allow for an experimental increase in the variety of data and stakeholders introduced.

This book tells the story of this experiment. After this introduction, part II begins by describing the history of CBSG and putting the organization and the PHVA workshop in the context of larger conservation efforts currently underway. In Part III we describe the six workshops that were the focus of this experiment and that concerned the mountain gorilla in Uganda, Rwanda, and Democratic Republic of the Congo; the muriqui in Brazil; the Peary and Arctic Islands caribou in the Inuvialuit region; the tree kangaroo in Papua New Guinea; the Eastern Slopes grizzly bear in western Canada; and the Algonquin wolf in eastern Canada. Part IV explores the challenge of integrating social and biological data in risk assessment models, considering the role, in particular, of human demography, governance systems, and local stakeholders. In part V, the book concludes with a discussion of the lessons learned that have application to both theory and practice, including reflections on interdisciplinarity, integrated risk assessment, and future directions for research and action. We now look at each of these parts in greater detail.

Part II: Design for Consilience

The notion of action research informed the Network's research project, as well as the construction of the book itself. In part II we cover in some detail the background of the organization and ongoing workshop processes that formed both the subject of and the context for our research.

As noted earlier a key objective of the Network project was to experiment with notions of consilience in the context of ongoing conservation initiatives. In particular, the group decided to examine a set of workshops that have been designed and run by the CSBG. In the past ten years this group has pioneered new strategies to allow practical and effective conservation actions around endangered species all over the world. A small, scientifically based organization, CBSG's workshops facilitate planning meetings both to identify species and habitats deserving conservation and, more importantly, to assist stakeholders in producing practical research and management recommendations. With a staff consisting of three program officers, a voluntary chairman, and a large volunteer network of professionals, the CBSG has conducted or participated in more than 40 PHVAs in the last five years. CBSG has been described as "an endangered species fire brigade which goes from crisis to crisis with state-of-the-science advice on the emergency moves best calculated to avert calamity ... without the CBSG, there would (often) be no movement at all" (Alvarez 1993, 356). In chapter 2, Frances R. Westley and Harrie Vredenburg present an overview of CBSG's development, core competencies, and key strategies.

Central to these workshops is the PHVA process, which brings together biologists, wildlife managers, captive breeding specialists, and government officials in order to develop conservation objectives and management plans for the species in question. Because this kind of workshop process is central to the consilience experiment at the heart of our project, we have devoted several chapters to PHVAs. In chapter 3, Phil Miller and Bob Lacy explore how PHVAs relate to the more focused scientific process of Population Viability Analysis (PVA), and they look at some challenges of integrating the human dimension into such efforts.

A PHVA workshop uses a variety of tools, including a computer simulation called VORTEX (Miller and Lacy 1999), to model extinction scenarios and align stakeholders' research and action agendas around a common direction and plan. These processes build on the foundation of adaptive management approaches (Walters 1986), which use scientific simulations to generate dialogue and consensus among diverse groups concerned with ecosystems.

PHVAs embody the objectives outlined by the *Global Biodiversity Strategy* report. The workshops are grounded in a concern for the maintenance of biodiversity in general, and endangered species and their habitats in particular. They are based on cutting-edge science. They bring together a number of significant stakeholders to debate and design management plans for the species in question. The workshops take place in the range country (the country in which the endangered species population ranges) and so encourage the synthesis of local and international expertise. While the workshops are organized at the behest of the range countries' governments, they are designed to encourage equal participation among stakeholders and to minimize power differences. As their goal is a set of policy, action, and research recommendations, their output has implications for local governments.

PHVAs are more, however, than scientific analyses. In chapter 4, Frances Westley and Onnie Byers focus on the design side of PHVA workshops: the design of the flow of human and task interactions that makes such interdisciplinary and intersectoral collaborations possible. PHVA workshops, as developed by CBSG, are highly participative processes, deliberately designed to combine optimal precision with optimal inclusion. The overall design allows for groups of twenty to sixty people, generally wildlife managers and scientists concerned with a particular species, to explore the implication of population dynamics, genetics, and a variety of threats to habitat and species exploitation. Participants work in small groups to identify and analyze risks and, ideally, to provide specific measures of habitat fragmentation. Periods of small-group work alternate with plenary presentations that allow all groups to "vet" each other's analysis and recommendations. As more data is introduced and the complexity is increased, participants generally enter what is thought of as a "groan zone" in group dynamics, a period of maximum divergence and complexity when it feels as if no clarity or consensus is possible. Some of the tools that help to build consensus are the VORTEX model and the continual emphasis on prioritization and, ultimately, translation of analysis into specific plans to implement. The divergence allows for inclusion of a full range of data, views, and stakeholder needs; the convergence allows for precision of analysis, risk assessment, and focused recommendations.

PHVA workshops share these characteristics, and to some extent these dynamics, with a growing variety of participatory, multiparty stakeholder, whole-system initiatives, similar in intent to movements such as community-based resource planning, Participatory Rural Appraisal (PRA), the Campfire movement, and large-system change processes. Reviews of ecosystem management programs as well as collaborative initiatives of all kinds indicate that key success factors are the inclusion of significant stakeholders, and the trust and consensus the process builds (Yaffee 1996; Gray 1989; Brown and Ashman 1996). For initiatives that are science based, precision and soundness of the science are also critical (Yaffee 1996).

Lastly, in chapter 5, Harrie Vredenburg and Frances Westley present the results of an early longitudinal evaluation of the success of a sample of CBSG PHVA workshops in the years before the Network experiment was initiated. Based on questionnaire data administered to participants before and after workshops and a follow-up mail survey sent out two years after the workshops, this chapter explores the short, medium, and long-term indicators of workshop impact and success and the degree to which workshops have succeeded in specific countries. The success revealed by this analysis is one of the reasons the PHVA workshops were selected as the focus of the Network experiment. The chapter also uncovers the logic model that underlies the PHVA process design.

Part III: The Workshops

Part III provides this book's central focus. The Network project unfolded around six PHVA workshops among twice that number carried out by CBSG during the same period of our study. By focusing on a specific set of initiatives in which conservation principles were put into practice, the team hoped to both improve understanding of science-based collaboration and to practice designing and facilitating such processes. Our concern was to develop tools and processes that preserved the existing strengths of CBSG, but also applied an ethic of full stakeholder participation. In particular, our two central areas of concern were (1) how to create an interdisciplinary information base that would allow data and expertise from the social sciences to be integrated with that from the natural sciences; and (2) how to include a greater variety of stakeholders in the workshops, including at one end local, Indigenous groups and at the other industrial interests.

CBSG's PHVA workshops, all of which are designed along the same lines, but each of which takes place in a different country and around a different species, offered exciting possibilities for the researchers in this Network to explore the effective power of collaborative design in shaping outcome. As a group, these workshops embodied principles of collaborative theory, illustrated possible pitfalls, illuminated the impacts of success, and allowed for the evaluation of theory founded on single (different) cases in the context of multiple (similar) cases. To this end, six workshops (among those already scheduled by CBSG) were selected for study, on the basis that they offered a variety of institutional contexts, species and human population demographic trends, and possible stakeholders (table 1-1). Network members helped identify potential stakeholders and data sets to expand these workshop processes, attended the workshops, conducted interviews, took field notes, and analyzed results. Chapters 6 through 11 each describe a different workshop concerning the mountain gorilla of Uganda, Rwanda, and the Democratic Republic of Congo; the Brazilian muriqui; the Peary and Arctic Islands caribou; the Papua New Guinea tree kangaroo; the Eastern Slopes grizzly bear; and the Algonquin wolf.

Each case is presented as what it was: something of an experiment. The Network members hypothesized that in general (1) increased stakeholder participation leads to a richer result and a greater sense of ownership of the process and the product; (2) incorporation of human demographic information and other social science data into the modeling process would result in a more instructive picture of species viability and would lead to more useful management recommendations; and (3) a number of contextual factors would influence the success of conservation initiatives, in particular, issues of governance and economics. We therefore deliberately tried before each workshop to increase the variety and amount of information available at the workshop for risk analysis and to increase the variety of stakeholders present at the workshop itself. As the cases in part III reveal, success was mixed, and surprise and serendipity were our constant partners. An added bonus, which we will return to at the conclusion of this book, was that the workshops offered intensive interaction time for Network members—a key element in consilience.

Table 1-1.

General workshop characteristics of the six case studies presented in part III.

	Mountain Gorilla	Muriqui	Caribou	Tree Kangaroo	Grizzly Bear	Wolf
Location	Sheraton Kampala Hotel, Uganda	Hotel Grandville Del Rey, Belo Horizonte, Brazil	Yellowknife, Northwest Territories, Canada	University of Technology, Lae, Papua New Guinea	Camp Chief Hector, near Seebe, Alberta	Leslie M. Frost Natural Resources Center, near Dorset, Ontario
Total Participants	56	27	36	45	69	63
Network members	Byers Miller Seal Westley	Borovansky Raufflet Seal Vredenburg Westley	Francis Miller Seal	Borovansky Byers Miller Nyhus Williams	Byers Lacy Miller Raufflet Seal Vredenburg Westley	Byers Francis Miller
Scientists	16	14	3	9	30	17
Private stakeholders	5	0	1	12	6	6
NGOs	15	5	8	2	4	7
Wildlife managers	8	2	7	1	6	10
Government officials	3	0	13	1	8	9
Zoo representatives	2	0	1	7	1	1
Press/media	0	0	0	2	0	0
No survey	3	1	0	6	7	10

Part IV: Understanding and Integrating the Dynamics of Human Systems

One of the challenges that we address in this book is the problem of limits. While it is clear that the solution to complex problems, such as endangered species conservation, requires the integration of both biological and sociological data, the inclusion of both of these in risk assessments is problematic. As pointed out in chapter 3 and explored in the specific case studies (chapters 6–11), most PHVA-type workshops have had limited inclusion of social science data, such as demographic, land-use, cultural, and economic data, either in the modeling process or in the planning process. CBSG and the Network members were anxious to widen the breadth and variety of data used to inform decision making, but problems of translation and integration were evident throughout the experiment. While we knew, for example, that population growth or industrial development endangers species and their habitats, it was not straightforward to translate these threats into numbers that could be used in projecting probabilities of population extinction, nor was it a simple matter to bring stakeholders such as industry actors and Indigenous groups to the table. While the design of the CBSG workshops (see chapter 4) as well as the design of the simulation program VORTEX (see chapter 3) encourages a diversity of inputs, creating the bridges between disciplines and data as well as between diverse actors proved challenging.

Progress was made in the course of the Network project, however. At the project's beginning, Network members had little knowledge about how to make the link between proximate (biological) and ultimate (socioeconomic) causes (i.e., how to translate socioeconomic causes into biological data that can be entered into VORTEX). We recognized that human population dynamics, human governance structures and institutional arrangements, economic and industrial activity, and local participation and support appeared to be key variables in ecosystem degradation and species extinction. However, key elements and dynamics of these subsystems needed clarification and articulation to make them accessible to a transdisciplinary team. In the course of our study, we moved some distance in reconciling our understanding of these interlocking dynamics—surely a critical step in consilience.

In this fourth part of our book, Network members explore different social subsystems and their implications for conservation action and policy. Network members brought to bear years of previous experience within their own disciplines, which is reflected in these chapters. However, during the three years that Network members worked together, all confronted challenges to consilience, as they linked their theories and experiences to those of others, as well as to the action orientation of the PHVA workshops. The chapters in part IV therefore elaborate major themes relevant to both the workshops and Network, as well as present enduring concerns that preceded the project and will, undoubtedly, continue to preoccupy Network members in the future.

In chapter 12, George Francis explores the larger institutional contexts in which the PHVA workshops unfold. He argues that the governance system of the range country determines long-range outcomes for conservation and should therefore be analyzed in some detail. Governance systems are so complex, however, that few attempts have been made to explore the implications of such systems for decision making at the workshop level. Francis approaches this challenge of complexity by isolating key variables (an approach also recommended in chapter 13 by human demographer Gayl Ness).

On the level of formal systems, critical variables may be the regional or national system's political authority and whether a government can provide security and order. Furthermore, can a government pull together to focus on conservation issues? Is the government cohesive enough to act, or is it internally divided and fragmented? The range country occupied by the species in question may overlap with many different jurisdictions, and the institutional arrangements may vary from very simple to very complex. Finally, are the responsible authorities sufficiently committed to conservation of the particular ecosystem or species to support enhanced protection?

On the level of more informal systems, Francis points to the importance of a given social system's awareness of the potential threat of environmental damage. Individuals and organizations collaborate informally on social problems, first in the emergence and recognition of domains (which can be geographic areas, social or economic sectors, or certain kinds of problems and issues); then in the formation of regimes; and finally in the development of full-fledged institutions. The stage to which a certain social system "envelope" has evolved affects such variables as the degree of difficulty in mobilizing stakeholders and the amount of conflict and resistance to change among concerned organizations attending the workshop.

Francis suggests that these formal and informal elements often combine in distinctive configurations, each with particular challenges for conservation: protected areas in industrialized countries, protected areas in developing countries, protected areas on Indigenous peoples' lands in frontier regions. The ability of a workshop to create recommendations that will subsequently be implemented may vary considerably across these three "ideal type" scenarios. An awareness of the institutional context's configuration at the planning stages of a PHVA workshop, Francis argues, could help workshop planners to estimate the potential for the long-range success of workshop initiatives. Such an awareness could also potentially help to differentiate those workshops with a high likelihood of success from those with a low likelihood.

A similar concern for understanding human system dynamics and their impact on workshop success infuses Gayl Ness's chapter on human demography and the environment (chapter 13). Ness points out that considerable work has been underway for some time in building links between human population growth and its impact on the environment. Important observed patterns include, at a macro level, the distinction between population-growth patterns in the Western world (leveled off and dropping), the "tiger" economies of Southeast Asia (leveled off), and the underdeveloped economies of Africa and parts of South America (still on the rise). In terms of environmental impact, a lag effect is observed between the point when population size begins to level off and the point at which environmental destruction begins to wane. In particular, the number of young males (age 15–19) in a population is an important factor, as they constitute a major component of resource utilization (hunting, farming, forestry, gathering, etc.) and are an indicator of the likelihood of environmental damage induced by wars. The key, from Ness's perspective is to isolate minimum specifications—those critical variables that policy makers can manipulate to transform systems. Ness concludes by exploring a model for integrating human population data into species risk assessment and the relationship between that process, workshop success, and ultimately species conservation.

A slightly different tack is taken in chapters 14 and 15. In chapter 14, John Williams directly addresses the challenge of how to translate these macrolevel human demographic patterns that Ness mentions into local disturbances, which in turn can be included as wildlife population threats in the VORTEX model. Human demographers can easily perform projections; indeed interactive population dynamic models, which can project trends based on such data as mortality rates, births, immigration, and emigration, can produce estimates of composite impact for a given area.

However, other, more qualitative data-gathering and analysis approaches are required to understand the implications of these patterns in the human population for the species at risk and to capture local variation.

Williams draws heavily in this chapter on a project that began ten years before the Network experiment and that informed our approach to the Papua New Guinea (PNG) tree kangaroo workshop (see chapter 9). In discussing three workshops conducted by CBSG between 1993 and 1994 in India, Thailand, and Indonesia, Williams describes the use of Participatory Rural Appraisal (PRA), a popular methodology for data gathering that also has showed promise in integrating Indigenous groups into "science-driven" workshops. PRA was developed in large part by Robert Chambers of the Institute of Development Studies at the University of Sussex (Chambers 1994), and it quickly gained popularity as a method by which local Indigenous groups could gather and organize information about their own ecological or social system in a short, intense intervention. Much like the PHVA process, it requires specialists and locals working together, data gathering, sharing problem analysis, setting priorities, and building community support and capacities (i.e., empowerment and consensus building). Williams describes not only the challenges of collecting and integrating PRA-type data into a PHVA and involving local community groups in the workshop, but also the benefits of such local data gathering for understanding how the macrodynamics of human demography play out at the local level, which is where conservation action must start.

In chapter 15, David Lertzman zeros in with even greater precision on the problem of integrating traditional ecological knowledge (TEK) into deliberations such as the PHVA, which are founded on western scientific models. In numerous cases, such as those of the Peary and Arctic Islands caribou (chapter 8) and the PNG tree kangaroo (chapter 9), the participation of Indigenous stakeholders was a key to successful risk assessment. In both cases, however, such involvement required thought, openness, and patience to ensure full representation of diverse views. Drawing on previous research about a successful bridging initiative in Canada, Lertzman suggests a number of principles for working with Indigenous stakeholders.

Lastly, in chapter 16, Harrie Vredenburg explores another group of important stakeholders whose active involvement in PHVA processes seemed critical to the Network: representatives of private corporations, particularly in situations where industrial or resource extraction activity is having a major impact on the at-risk species and its habitat. There are two reasons to involve industrial interests: (1) to provide more information for the VORTEX model, including both direct and indirect effects of resource development (e.g., forestry brings human in-migration, which brings secondary service industries); and (2) to obtain participation in PHVA and, ultimately, to achieve buy-in to specific conservation initiatives (i.e., corporate and/or regulatory decisions to avoid or alter resource development in ecologically sensitive areas). While securing such stakeholders' participation was viewed as a challenge, it was felt that corporations might welcome such involvement, as it fosters goodwill locally and helps fulfill international regulatory requirements. Again, like Williams's, Vredenburg's chapter highlights the importance of understanding specific local relationships and initiatives when estimating the impact of industrial or other economic activities. When it comes to the human system dynamics, macroeconomic models only tell a partial story and may be of little assistance in understanding how sound policy recommendations can result in effective conservation action. Vredenburg also again underscores the fundamental challenges of consilience: bringing stakeholders to the table is only the first obstacle. Keeping them there is equally difficult, requiring active bridge building across divides of language, discipline, culture, and ideology.

Part V: Reflections on Consilience

The story of this experiment was, ultimately, a return journey. Along the way, Network members alternatively experienced excitement and demoralization, clarity and confusion, satisfaction and frustration. In the end, new understandings were nurtured in the context of old ones, and trust was built on a foundation of collegiality and hard work. It was less a journey to pure novelty than to surprised recognition, in which we discovered the familiar in new places and the new in what we thought was familiar. New understandings of old problems and old interpretations of new problems left us with the experience of progress. Most importantly, as is appropriate in experiments with consilience, new relationships were forged, between people, between concepts, and between processes. In this last part of the book, we offer these reflections on consilience.

In the first chapter in this concluding section (chapter 17), Gayl Ness draws on forty years of research and practice in the area of population, environment, and development to reflect on the current Network experiment. He proposes nine basic observations: (1) interdisciplinary work requires building bridges between specializations; (2) specialization is necessary and useful, but also requires bridges to other disciplines; (3) a focused task, such as the PHVA, offers both advantages and disadvantages to bridge building; (4) the PHVA provides the opportunity to test, or validate, hypotheses developed in interdisciplinary Network discussions; (5) resources are needed to create interdisciplinary teams; (6) even the most effective solutions drift or lose their power with time unless innovative work is constantly included in the process; (7) all problems and solutions are location specific, no solution works for all places or for all times; (8) communities affect habitat and species survival, but the effective communities are communities of interest, and not simply geographic communities; and (9) finally, there is a bottom line, which for the Network is species preservation. Interdisciplinary work, such as that proposed by the Network, ultimately must prove useful to the task at hand. Will such work enhance species survival? If it does, it will be useful and its costs justified. If not, it will not be useful, and its costs will not be justified.

In the second concluding chapter (chapter 18), Phil Miller and Bob Lacy return to the discussion of PHVAs as exercises in modeling. Models can be powerful tools for integrating knowledge across disciplines, but they can also be barriers to understanding because models often represent specialized knowledge from only one or a few disciplines. One of our realizations during the Network experiment was that much of what we struggled with involved the difficulty of getting disparate models to talk to each other. Within each discipline, the way in which we as specialists represent our knowledge is our model of the problem or issue at hand. In order to incorporate knowledge from another discipline into our understanding, we must find ways to get that discipline's model to speak to our model.

Miller and Lacy also discuss the conceptual and mechanical problem of quantitative model integration in detail, and they present a history of the Network's early attempts at stimulating a productive analytical synthesis. While these initial attempts were not as successful as originally anticipated, we learned extremely valuable lessons about the types of tools needed to facilitate an improved understanding across disciplines and, hence, an improved modeling environment. Our approach of using models to explore how the knowledge of diverse disciplines interacts has evolved from an original plan for merged "megamodels" to our current concept of developing "metamodels," which link systems that retain their original structure and integrity. Miller and Lacy describe this revised approach and present selected examples of its preliminary application.

We conclude with chapter 19 by Frances Westley, Phil Miller, and Bob Lacy. The authors highlight the learnings from the Network project, integrating the chapters and the cases, and speculate on directions for future research and practice.

Throughout the writing of this book, we have tried to be transparent about our failings as well as our successes. We seek to provide anyone interested in multiparty workshop processes based on science and on simulation the opportunity to (a) learn from our attempts to make these processes as powerful and effective as possible and (b) to understand the dynamics of effectiveness. We believe that as much, if not more, is learned from failure as from success.

No challenge in biocomplexity is as significant as strengthening the capacity to integrate knowledge across specialties and levels—the capacity for consilience. Much of the natural world is at risk. Species are being driven to extinction, ecological communities dismantled, and ecosystem processes disrupted. These environmental problems are the result of complex interactions among human and natural processes. Research efforts based solely within traditional scientific disciplines are not able to fully encompass and interpret these interactive processes.

But as we noted in our introduction, there is no magic formula for consilience. Our aim in writing this book is to share our learnings with those interested in pursuing transdisciplinary research and action. It is our hope that in doing so we will help enhance researchers' ability to conduct analyses of biocomplexity issues so that such analyses are sufficiently robust to inform reliable policy decisions.

PART TWO

Design for Consilience

Chapter 2

The Art of Walking through Walls: Strategy and Structure in the Conservation Breeding Specialist Group

FRANCES R. WESTLEY AND HARRIE VREDENBURG

It is one thing to talk about consilience in the context of interdisciplinary teaching and research and another to talk about it in action. Key to understanding consilience in action is understanding the dynamics of successful collaboration, across action/theory barriers, sectoral barriers, and interdisciplinary barriers. The Population and Habitat Viability Assessment (PHVA) workshops that were the applied focus of this project and this book are one type in a series of initiatives developed and delivered by an unusually collaborative organization: the Conservation Breeding Specialist Group (CBSG).

In this chapter, we will explore this organization, which has been the catalyst for much global conservation. CBSG is unusual in a number of respects. First, it is unusual in mission: science-based, but focused on changing social processes that act to transfer scientific knowledge into conservation action. Second, it is unusual in organization: a small group of staff supported by a much larger group of very active volunteers. Third, and related to the second point, it is unusual in its use of resources: remaining singularly lean in the interests of freedom of action and movement and defining resources largely as talent and motivation. Fourth, it is unusual in its scope and impact: having started small, it now has a presence in Latin America, Asia, and Africa and has been involved in conservation action all over the world. Last, it is unusual in its leadership: much

Reprinted by permission. Frances R. Westley and Harrie Vredenburg. 1997. "Interorganizational Collaboration and the Preservation of Global Biodiversity. *Organization Science* 8 (4). © 1997, the Institute for Operations Research and the Management Sciences, 901 Elkridge Landing Road, Suite 400, Linthicum, MD 21090 USA.

of its success can be attributed directly to Dr. Ulysses Seal, the scientist who built the CBSG network and created the vision of science-based action, which has infused the organization. In this chapter we will describe the origin and development of CBSG, exploring its strategies and operations, and place these in the context of theory on effective collaborative action.

The Conservation Breeding Specialist Group and Its Chairman

The CBSG is a division of the Species Survival Commission (SSC) of the World Conservation Union (IUCN) and is based in Minnesota, USA. The group is an informal nucleus for individuals both inside and outside of larger organizations, including international conservation organizations, agencies in national governments responsible for wildlife national parks, and related areas and zoos concerned with breeding endangered species. CBSG works through a network of relationships.

CBSG started life as the Zoo Liaison Committee, a small subcommittee of the Swiss-based IUCN. The IUCN had established a series of taxon-based specialist groups to advise various organizations as to the status of particular taxa in the wild. The Zoo Liaison Committee was set up to act as a liaison between these groups and the captive breeding community. It remained largely inactive until, in 1979, Ulysses Seal was appointed chairman of the Captive Breeding Specialist Group, as it was then called. Since then, it has been difficult to separate the actions of the "group" from those of its chairman.

Seal came at the problem of endangered species and the conservation of biodiversity from a rather unusual angle. He was trained initially in psychology (bachelor's and master's degrees), switched fields and, in 1957, he received his doctorate in biochemistry. He did postdoctoral work in chemical carcinogenesis and then accepted a position as research scientist at the Veterans Administration Hospital in Minneapolis, Minnesota, where he continued his biochemical research, funded by a succession of research grants. His productivity was so high (some three hundred single and co-authored scientific articles in the course of his career), that he was able to maintain himself, and his extracurricular work with animals, on "soft money" or research grants. Despite being sought after by several university faculties, he preferred the independence associated with his grant-funded research position.

While he had always had a personal interest in animals, his research led him to zoos as sources of exotic species' blood samples with which to work. In exchange, he helped zoos develop more effective immobilization drugs in order to facilitate sampling. In the process, Seal championed a scientific "medical" approach to the management of zoo animals, an approach he found sorely lacking. "What I did in the course of time I spent with [zoo people] was to hear about problems. These guys would all tell me: 'Here's a problem. Here's something we don't know, here's something we can't do.' Wherever I thought I could bring to bear what I knew about human medicine, I did. At one time I had over thirty projects going with field and zoo people on a variety of species all over North America" (Seal personal communication).

Through this ongoing work with captive animals, Seal became aware of a number of potentially dangerous anomalies in the management of captive species. First and foremost was the almost total absence of medical records for individual animals and genetic records for the groups of animals in captivity. As animals were moved between zoos, this often meant that few keepers had a good sense of how to manage the health of animals under their care, or how to ensure the genetic health of future generations through scientific breeding of selected pairs. This was particularly critical for endangered species; it was not enough to merely care for such animals, rather they must be genetically managed if the species were to survive. Seal recognized that to solve this problem, an intensification of the kind and quality of information available about individual animals was needed, as well as far greater collaboration between keepers and organizations where such information was "stored."

From 1979, when Seal took on the chairmanship of CBSG, until 1990, CBSG was a one-man operation, staffed by Seal himself on a part-time basis. In 1990, an executive officer was hired, and then a secretary in 1991. By 1995, the staff had grown to three program officers and three office staff, with Seal, having retired from his research position at the Veterans Administration Hospital, devoting full time to CBSG activities. Funding, like staffing, has also always been low. The group was and is supported by donations from its core donor base, composed largely of zoos.

With only 15 members by 1984, by 2002 CBSG had a formal membership of 1,010 from 95 countries. CBSG is one of the largest of over 120 specialist groups comprising the SSC, one of six IUCN commissions. One hundred and thirty-five donor institutions and organizations provide annual funding, enabling a CBSG operating budget of approximately $350,000 a year. Formally, CBSG "reports" to the chairman of the SSC. It also has a steering committee of some sixty individuals, selected by Seal, to act in an advisory capacity.

As a convener, Seal (and by extension, CBSG) was widely recognized as powerful and legitimate. His personal dedication and energy were described as "prodigious." As one colleague said: "He was an extraordinarily successful entrepreneur. He had no personal interest; he didn't even draw a salary. In fact from time to time he used his private resources to finance this activity. But he was extraordinarily successful at making the programs go. No one I have ever met worked as hard as he did, by a factor of two." Furthermore, CBSG's interventions were welcomed as scientific and largely nonpartisan. As one commentator noted: "If Seal and his team showed up to analyze panther problems, there would be no nonsense. They would not be influenced by factional interests or provincial views" (Alvarez 1993, 429).

Over a period of twenty-five years Seal and his associates were responsible for following four major initiatives, three specific and one a general operating approach, which have systematically addressed the "problem" as Seal defined it (figure 2-1).

1. The establishment in 1973 of an International Species Information System (ISIS) that systematically collects genetic, behavioral, and

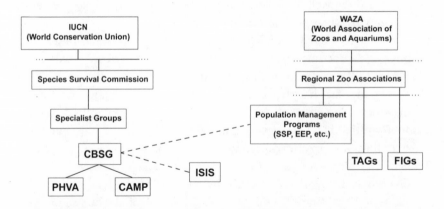

Figure 2-1. The relationship between the Conservation Breeding Specialist Group (CBSG) and the world's wild and captive population management communities. (Adapted from Westley and Vredenburg 1997.)

demographic information on individual animals in zoos worldwide. The information was pooled into a database, is continually updated, and is made available as a basis for captive breeding mating decisions.

2. The initiation of Species Survival Programs (SSPs), which called for collaboration between zoos designed to maximize the effectiveness of breeding programs for endangered species.

3. The development of a series of workshops including Population and Habitat Viability Assessments (PHVAs) and Conservation Assessment Management Plans (CAMPS), which bring together (on a cross-sectoral, transnational basis) all groups and individuals concerned with the survival of a given species. The cornerstone of the PHVA process is the use of VORTEX, a computer program for modeling the dynamics of species extinction risk under different scenarios. The results help to build consensus around appropriate strategies for managing the species.

4. The establishment, empowerment, and amplification of a far-flung, international network of professionals and activists, which spans all three of the previous initiatives. This has been an important and deliberate element of the CBSG strategy. Frequent telecommunications and infrequent meetings link the network, and members can be relied upon to act over time in the interests of problem solution and domain transformation.

In figure 2-1, we show the link between these initiatives as steps in addressing the challenge of rapid extinction scenarios. The PHVA workshops, which are the focus of this book, are explored at greater length in chapter 3. Here, however, we will explore the nature of the collaborative network that Seal forged and relate his strategies to the theory on building successful collaborations.

CBSG as a Collaborative Network

While Seal was identified as a central force in the direction and orchestration of initiatives, of equal strategic importance for transforming the domain has been the professional network forged by CBSG, maintained and amplified in part by means of these tools and processes. If CBSG has been described as "an endangered species fire brigade which careens from crisis to crisis with state-of-the-science advice on the emergency moves best calculated to avert calamity" (Alvarez 1993, 356), it remains a catalyst and an advisor. The actual saving of the species is in the hands of local agencies in the species' range country and in the hands of the approximately five to six thousand people who have participated in the initiatives. This is the CBSG network.

In the 1993 *Futures Search Report*, drafted by a group of staff, board members, and stakeholders, the first "highly-valued characteristic" of CBSG was summarized as follows:

> One of the primary qualities identified was the priority that CBSG places on the exchange and sharing of information, with free dissemination of products and data. CBSG's communication network was seen as critical—keeping members and constituents up-to-date on technology and new programs. Its facilitation of problem-identification and problem solving, as well as being a forum for discussion of global conservation issues is highly valued. Its ability to facilitate mutual problem solving by people with diverse interests is appreciated, and was identified as the foundation upon which most, if not all, the workshop successes are based. (CBSG 1993, 34)

The CBSG office in Minnesota can be seen as the node of the international, professional network of volunteers. As such, CBSG facilitates the flow of an astounding amount of information. In 2001 alone, the central office conducted or participated in fifty-five workshops, prepared twenty-eight final workshop reports, and responded directly to more than thirty-two hundred queries from people in 120 countries. As the network grew, it demanded extensive travel on Seal's part in order to strengthen existing relationships, build new ones face-to-face, and to elicit cooperation among key stakeholders. (Up to the last year of his life, when illness slowed him down, Seal spent more than 80 percent of his time visiting far-flung points of the worldwide network.)

The CBSG network itself may be divided into peripheral and core members. Peripheral members are those who have participated in a CBSG workshop (some three thousand participated in PHVAs from 1990 to 1994), are on the CBSG quarterly newsletter mailing list (circulation: twelve hundred in ninety-eight countries), provide financial support, and/or attend CBSG annual meetings. The core of the network consists of some two to three hundred people who have been identified by Seal as primary resources who donate their time, energy, and expertise repeatedly to CBSG workshops, consultations, or new initiatives. These core representatives are self-selected and come from all levels of hierarchy and all parts of the world. Finally, within this group there are ten strategic associates who have invested considerable volunteer time in advising and assisting the organization in everything from strategy to workshop delivery.

In the last several years, individuals in this network have taken the initiative to pioneer the creation of additional, regional nodes. The first of these was set up in India. There are now CBSG "offices" in Mexico, Costa Rica, Indonesia, and South Africa, in addition to India. This expansion was not directed by the Minnesota office, rather it represented the initiative of individuals who wish to pioneer CBSG efforts in particular cultural contexts. Other such initiatives involve expanding the PHVA and CAMP process to plants (an initiative championed by a British biologist); creating comprehensive in situ and ex situ planning around a particular species (for example, a tiger program), launched by a core network member with support from Exxon, but proceeding under the CBSG umbrella; and including human demography and community-based resource management as planning tools in PHVAs.

An Overview of CBSG Strategies

In organizing the specific initiatives listed in figure 2-1, Seal and CBSG worked with a clear plan: knowledge intensification (ISIS), leading to greater collaboration between stakeholders in captive (SSP) and wild populations, allowing for better and more comprehensive species management (PHVA/CAMP). However, due to the far-flung nature of the network, the variety of groups and cultures involved, and the need to move quickly to address the crisis of rapid species extinction, the process itself has had an incremental and chaotic quality. This is partly by necessity and partly by design. Looking across the initiatives that CBSG has taken, it is evident that there are a set of consistent principles and actions by means of which Seal and CBSG have designed their tools and processes. These are worth looking at in some detail, as they shed light on CBSG's success at building the collaborations so necessary for conservation. We will look in turn at seven such principles.

The first principle is the building of informal collaborations across multiple sectors and groups in the interest of species conservation. The problem domain of endangered species touches a wide variety of stakeholders. Until recently this domain has been underorganized (Brown 1980), with little communication across sectors. There has been some success in gaining consensus over which species are endangered with the publication of the IUCN's Red Data Books (Scott, Burton, and Fuller 1987;

IUCN 2000) and the Mace-Lande classification scheme (Mace et al. 1992; Mace and Lande 1991), and protecting such species at the international level, but there has been little success in organizing stakeholders to rescue the species. More importantly, there has been little coordination of effort, particularly across stakeholder groups. It is rare, for example, that academic scientists work closely with government authorities or that information and technology is exchanged between the zoo world and the world of wildlife management.

Part of the reason for this underorganization is technical: it is difficult to clearly identify which species are endangered before it is too late to save them. The numbers of remaining individuals of a species (a head count) is far from a reliable measure, since it is the degree of genetic heterozygosity (or the number of separate genetic lines that can be found within the population) that is critical. Some species may appear to have enough individual animals to ensure continuity, but, like the cheetah, are so similar in genetic make-up that the whole population is vulnerable. Natural and anthropogenic threats as well as environmental variation also play major roles. To ensure a healthy metapopulation (a set of genetically distinct subpopulations) means maintaining sufficient habitat to support it in the wild, or, within zoos, having sufficient genetic information for individual animals in order to avoid excessive or too rapid inbreeding. Without coordination, information remains piecemeal. Without pooled and analyzed information, the crisis itself has an incipient or hidden quality, which can induce a false sense of security and blunt response.

Therefore, from a problem-domain perspective, the problem of saving an individual endangered species is extremely complex. It requires an exchange of knowledge and technology, a building of consensus, a bringing together of expertise from different professions, organizations, and sectors, and a mobilization of resources. The CBSG's PHVA process balances integrating the information required to evaluate the species with integrating, or at least connecting, the individuals from different disciplines and sectors who are centrally concerned with that species' conservation. The hope is that, at the very least, this will cause individual stakeholder groups to realign their priorities to take into account the needs, views, and initiatives of other groups.

Central to this process has been the use of VORTEX, a simulation model that provides a tangible focus for concerns about and hopes for the species and a vehicle for integrating diverse data sets. Of course, as the diversity of both information and stakeholders increases, so does the challenge of facilitating integration. However, CBSG has gone further than any other conservation organization in forging collaborations at multiple levels and in building understanding of the processes required to facilitate such collaborations. This is one of the reasons which some have described CBSG as "truly the focus of the world's conservation breeding efforts" (Tudge 1991, 110).

The second strategic principle of organizing employed by Seal and CBSG is knowledge intensification through technology. The CBSG network, as well as its products, tools, and processes, are maintained by technology and communication. The network is fueled by the technical rationality of the scientific approach; it is amplified by the integrative potential, availability, and generalizability of computer simulations and desktop publishing, allowing for instant reports; and it is maintained by rapid communications made possible by fax, e-mail, and personal computers.

This principle runs across all four of the initiatives described above. The emphasis throughout Seal's leadership has been the distribution and processing of systematic, scientific information. In the interests of accomplishing this goal, both the ISIS and VORTEX programs were designed as user-friendly software, and efforts have been consistent to distribute the software to those with the most information about the animals. ISIS was always envisioned as a data archive more than an organization. Seal and his colleagues spent considerable time and money in 1973 researching similar attempts at information systems and learned from the "systematic study of failure." From this research they learned the importance of using software developers for whom biology was first and computer skills second. The model was consistently refined to take into account data, such as death records, often ignored in traditional studbooks (records of captive animals' "family trees"). These efforts resulted in software that was easy to use and provided the information required for species conservation. The ISIS network clearly was exclusively built on the exchange of knowledge and its objective was to do this as cheaply and as effectively as possible.

This sophistication of software development has also played a role in the success of the PHVA workshops. The use of scenarios and the modeling process creates a common focus. While many may be deeply distrustful of numerical modeling features as tools, the steps of the VORTEX process unite participants in a common activity, leading to dramatic results.

CBSG staffers never travel without portable computers and generally also have portable printers, copiers, and projectors. Because of the emphasis on scientific information generation and dissemination, written reports arise from the workshop itself and are only edited afterwards at the CBSG office. This is what Seal referred to as producing "product":

> The fact is there is an enormous satisfaction on people's part in producing things and seeing their activities result in something useful.... Every working group produces reports, we go over them, recycle them several times until we get agreement on scenarios ... that has been an enormously powerful tool in terms of developing consensus. It's also been the kind of productivity that at the end everyone feels good about.... (Seal personal communication 1993)

The third strategic principle that differentiates CBSG from most conservation organizations is creative definition of resources and resource independence. While securing adequate resources has always been a challenge, Seal used a creative approach to mobilize expertise for problem solving. As Nate Flesness, director of ISIS, put it in an interview with the authors:

> We have discovered that academia is waiting to be mined. The world's top ranking population biologists, for example, like to do something altruistic once in a while. Most of them are sixties kids who are a little green anyways and they sit and work all day and publish equations and one up each other at developing the discipline.... You can call the best geneticist alive, and say, "We're going to have a meeting on the Florida panther" and he'll say "Oh yes, I have the papers, there are only 25 left." I'll say "We'll buy you a ticket, put you up, we need some expertise." He says, "sure!" You've just got two days of the world's best geneticist at no charge.

Altruistic donations of time and energy augment grants, fees, and donations from a wide variety of sources. Seal himself believed that his activities were successful in part because he never relied on a single source of funding, nor allied himself professionally with any one large organization. He used donations to the CBSG from member zoos; he used grant money; he used his own resources when necessary. In several interactions with large international funding agencies, who insist on contracts and control, Seal refused money. He avoided dependence on such organizations both because he saw their hierarchies as antithetical to the grassroots networks he was trying to create and because he sought to safeguard against control by such organized interests:

I can afford not to [accept that kind of funding] because I'm not looking for a billion dollar activity. I still think that what we are doing will
impact such large organizations in a relatively short period of time . . .
then they will have to come to us . . . as it would be hard for such
organizations to stamp us out. . . . We're so distributed. I'm not dependent on any single institution or group for my professional expertise.
There are over a hundred, I'm not dependent on any single government for activities. If we lose a whole country that isn't willing to participate it doesn't make any difference. I've got five more standing to
take their place. (Seal interview with authors 1993)

The fourth strategic principle is the rapid generation and shedding of
initiatives. As evidenced in the descriptions of specific initiatives above,
Seal and CBSG move quickly. In the past ten years, ISIS programs have
been transformed and training programs have been initiated; SSPs
and other captive conservation initiatives have been formed and put into
operation; PHVAs, CAMPs, and other integrative workshops have been
designed and implemented. The proliferation of forums that Seal and
CBSG have initiated steadily creates new clusters of interaction and
mobilizes information in slightly different contexts.

CBSG is as quick to shed initiatives as to spawn them, however. As of
October 1994, ISIS, which had been operating independently for some
years, was legally separated from CBSG, creating two independent, not-
for-profit organizations. Species Survival Plans and associated interzoo
activities are now the responsibility of regional zoo associations. Closely
linked with CBSG's self-identification as a network manager and catalyst,
CBSG refuses to define information or action as "owned" by any individual or group. CBSG is quick to hand over responsibility for any initiative to
any individual or group willing to assume it. CBSG members consciously
refuse to engage turf issues in collecting, organizing, and using the knowledge network that is ISIS, or the collective processes that are CBSG. Seal
and CBSG have argued that whoever is willing to act "owns" the problem:
"If they do it better than you then they are in charge. It's a pure capitalist
competition to do a better job and the minute you can do a better job then
it's yours, take it, we've got lots more to worry about" (Seal interview with
authors). CBSG, focusing on technological dissemination and working
within severe financial constraints, releases initiatives as quickly as possible.

The fifth CBSG principle is network amplification. The focus of Seal's
activities was always on the growth and strength of the network as a whole.

He avoided administrative structures, which he saw as a waste of resources, and concentrated on forging linkages between individuals, and, to a lesser extent, organizations. He seems to have focused on maximizing the quantity of such linkages and exchanges. If the linkages are working well, each individual collaborative initiative is likely to be a success. The success, in turn, of an individual initiative amplifies and feeds the network as a whole. For example, benefits accrue to zoos participating in ISIS depending on which other regional zoos participate and how fully. Because it is costly to move animals across oceans, and such trades are realistically limited by resources and legal barriers, a large number of participating zoos means increased access to animals at the local level. Hence, the network snowballed after a first critical mass was gathered. Zoos were initially allowed to join at the lowest level just to put them on the map; later, full-paying membership was encouraged among zoos that could afford it. The success of ISIS also has laid the groundwork for later collaborations around SSPs. Similarly, one successful PHVA workshop hopefully sets in motion new linkages, reinforces existing reciprocities, and encourages individuals to organize similar workshops in an effort to align individual activity into joint efforts.

Ideally, from CBSG's viewpoint, the network would be self-designing after CBSG's first, catalytic interventions. Evidence of activity intensification, without further direct intervention of the "node," is a CBSG goal. This appears to be occurring in India, South America, and more recently in South Africa, where champions have taken up the tools and processes that CBSG designed and are now generating their own workshops, reports, and information dissemination.

The sixth CBSG principle is a strategy of power dispersal, which characterizes most CBSG activities, at times creating conflict with other organizations. In the creation of ISIS, the SSPs, and the PHVAs, Seal and his associates worked outside of the formal hierarchy of the zoo organizations. The initiation and support for ISIS came from Seal's friendships with the zoo directors and their professional staff. While initially Seal did not go directly to the keepers to establish or expand ISIS, the keepers' role has expanded through their participation in regional studbooks and by providing data to local record keepers who handle the ISIS data. Staff who directly care for the animals have been traditionally low in zoo hierarchy, but due to the rapid professionalization of zoo personnel, they have become increasingly specialized and educated. Seal's approach directly involved this group, both because they had the necessary information and also because they readily shared his vision. "I go out and talk to the cura-tors, the young people. We go out and drink beer with them and we'll go through the exercise, the logic, and ... they can see these next steps, they're not quite as bound into the tradition [as people farther up in the hierarchy]" (Seal interview with authors).

Similarly, in the PHVAs, Seal worked to reinforce those with less for-mal status but who have critical knowledge and roles. The PHVA work-shop design ultimately creates an action focus that places the emphasis on the least powerful group, the wildlife managers. This is designed to bal-ance the perceived power of the scientists. "The very structure of the workshop is designed to set power aside as an issue ... if you're going to talk about managing an endangered species, the only people with power in one sense of the word are the [wildlife] managers. They have to do the job ultimately; they have to buy into the process ultimately. If they don't then there is no point ... in fact in most instances ... the key people will be the [wildlife] managers. The scientist and the field study people are going to be there to share their expertise and commentary" (Seal inter-view with authors 1992).

Despite this grassroots approach, Seal was sensitive to the power of hierarchies and responsive to positive initiatives of those who head them. For example, a critical factor in all CBSG activities was the support of a number of notable leaders within the zoological community (often referred to as "silverbacks" in reference to the dominant male gorilla in a gorilla troop). Despite his unorthodox approaches and his fundamental disregard for hierarchy as described above, Seal was recognized as a fel-low silverback and had the necessary support to overcome his critics. As one such silverback noted: "ISIS was opposed initially because of the cost. There was strong protest by some members including past presidents of the AZA [American Zoo and Aquarium Association], who argued that paying as much as a dollar a specimen to enter your records was an insupportable cost. One of the contributions I made [to ISIS] was to successfully defend it from AZA efforts" (zoo director, interview with authors).

CBSG's seventh and last principle is avoiding centralized control and structuring. As a network, CBSG has always considered itself without boundaries and therefore open to anyone wishing to join. This contrasts with other, more exclusive associations common in a domain. As Flesness put it in an interview with the authors: "It turns out that all other organi-zations in which zoological institutions are members define themselves by who they keep out. This is still true today. For example the American association defined itself geographically and quality-wise by exclusion . . . so do all the other zoo organizations on the planet earth."

While CBSG does have a steering committee, it does not control the work of CBSG, which remains a highly unstructured organization based on knowledge rather than policies, norms, rules, or authority. While the group produces activity reports several times a year, these are information archives that include such things as Seal's agenda, copies of his notes on meetings, correspondence between CBSG and other groups, documents that explain the rationale behind PHVAs and the other workshops he has developed, and news magazine and journal articles concerning ideas related to the work (environment, endangered species, conservation biology). There is little or no effort to orchestrate these into an overview report.

Other than a commitment to keep the steering committee as well as the network informed, Seal's institutional strategies have largely focused on avoiding centralization, external control, and boundary definitions. Rather than shaping CBSG in response to the institutional environment, Seal used his network alliances to shape the institutions. He welcomed evaluation for research purposes and avoided it for control purposes. His training of others relied heavily on those elements transferable through electronic media (software scenarios, written protocols); he had little time to articulate or educate those around him in his process strategies in any formal way, although much was absorbed by association. Indeed, Seal showed little concern for the structuring and survival of the CBSG per se. His strategy focused on the network itself, rather than on any one member organization (including CBSG).

This lack of respect for boundaries and hierarchical arrangements, the rapid spawning and shedding of initiatives, and the unpredictable, if creative, mobilization of resources have caused some concern among CBSG affiliates such as AZA, the IUCN, and the World Zoo Organization who all, at different times, expressed concern about the scope of Seal's activities. Even some members of the steering committee itself raised objections. As CBSG creates new interest groups, new workshops, and new networks, it is difficult to keep track of its activities, to say nothing of evaluating their impact. Where is the money coming from and how is it being spent? Where is the organization going in the future? What would happen when Seal withdrew or died? Consequently, at one point there was talk of setting up a central secretariat to oversee activities. Seal was resistant to the idea as he felt that such a central committee would potentially curtail his own freedom without adding much value and, more seriously, would divert slim resources to administrative ends. He was consistently able to resist such pressures to organize and structure, precisely because of his operating principles.

Why Have These Strategies Succeeded?

In sum, Seal and CBSG have employed an unusual set of strategies to build a far-flung, intercultural, interorganizational, and interdisciplinary collaboration, with the intent of consolidating and intensifying efforts to save endangered species. These strategies have largely succeeded for four reasons.

Leadership

In reviewing CBSG strategies and structures, it is impossible to avoid highlighting the role of Seal and his particular vision of science-driven collaboration, which runs through all four initiatives described at this chapter's beginning (pp. 26–27). Theorists (Cooperrider and Pasmore 1991; Brown 1991; Gray 1989) all recognize the key role of individual leaders in bringing stakeholders to the table, particularly in "underorganized" domains. Leadership is key to the coordination of loosely organized systems such as networks, which rely on the inspiration and rewarding of individuals on a highly personal level. CBSG has succeeded, in part, because the strategies described in this chapter were realized with coherence and focus. Practice was consistent with theory.

Networks are initiated by proactive individuals who create new role space around themselves. New appreciations of emerging metaproblems originate and build in such individuals as they interact with other network members, who tend to form a selectively interdependent set. Proactive individuals learn the art of walking through walls. Without carriers of this kind it is difficult to see how the processes of appreciative restructuring can either take place fast enough or go far enough to permit emergent domains to be organized in time and on a scale that will allow stakeholders to contend with the oncoming metaproblems (Trist 1983).

Globalization of Scientific Norms and Language

Cross-disciplinary and cross-cultural collaboration in CBSG workshops is clearly aided by the use of scientific language and norms. While participants in CBSG's workshops and network do not share a culture, they do embrace the notion of using population genetics to understand and to prevent extinctions. They have come to share a "thought world," if not a culture. They also accept actively, at least temporarily, the norms that CBSG sets out for participation in workshops. Participation in the workshops and the network appears to trigger the robust social compliance mechanisms described as social influence (Cialdini 1984; Vredenburg 1986; Vredenburg and Marshall 1988, 1991). Social-influence techniques utilize the psychological tendency toward behavioral and cognitive consistency. "Thought structures" tend in time to be aligned with behaviors. Agreeing to participate in a workshop or a network has been shown across numerous settings, from social movements to industrial settings, to serve as an effective inducement to comply with a subsequent larger request for involvement by the convening organization. This may partially explain the impressive diffusion rate of CBSG processes and tools across and within national cultures.

Openness to Continuous Change

CBSG's emphasis on staying lean in terms of institutional funding, and their emphasis on "shedding," means that CBSG is particularly open to an evolving definition of their processes and products, and hence represents an ideal convener of complex, multidisciplinary processes. That this flexibility is correlated to the organization's success receives some theoretical explanation in the literature on networks and strategy.

Granovettor (1985) has noted that weak internal ties (or structures) are more likely to be linked to the resiliency of a community than strong internal ties, because when internal ties are weak, individuals maintain their links to resource bases outside of the community and may draw on these to save the community. Similarly, Miller (1993) has noted that as the internal structures of organizations become more focused, tightly coupled, and "simple," the organization loses its ability to be responsive to its contextual environment. Hence, intensive internal structuring in the form of established routines, strong internal ties, and high administrative overhead ("fixed costs") seem to have an inverse relationship with adapting to the contextual environment. CBSG's openness in this regard may have allowed it to be exceptionally adaptive to the evolving needs of the conservation community, and to remain unbiased in terms of existing power and status structures among participating countries, cultures, and disciplines.

Technology as Integrating Device

To the extent that CBSG's initiatives result in increased domain organization, it is through technology that the structuring occurs. By technology we mean both the technical rationality of science and the software programs that allow for the coordination and exchange of information. It is important to note, however, that this increase in organizing remains highly decentralized, inherently unbounded, and completely informal. Here again we notice the way in which the content (scientific knowledge transmission and integration) and technology have a decisive shaping impact on structure.

The CBSG case suggests that technology can act effectively to coordinate action in interorganizational and interdisciplinary domains, particularly if it precedes formal structure. As Gouldner (1976) has suggested, technology in modern society also can substitute for cultural forms of integration. Similarly, it is interesting to note the powerful ordering and consensus-building properties of the particular technology employed. As Sandelands and Srivastan (1993) have pointed out, with truly complex and systemic problems, simulation modeling (such as that offered by VORTEX) is one of the most powerful devices to wed the theoretical to the experiential. The modeling makes the problem concrete and is a leveler of both power and cultural inequalities. From the perspective of the model, information is valuable whether it comes from a scientist or a wildlife manager. Both the specific content (a concern with the scientific evaluation of a particular species' survival potential) of this collaboration and the particular technology (computer simulations) have a powerful impact on the CBSG process and the potential for building consensus in the early stages.

Conclusions

In conclusion, it is clear that saving endangered species is an ongoing process that will never be completed. Species and their habitats are never saved "for all time," only for a particular moment. Members of the CBSG deal with this never-ending aspect of the problem by using the "watch" concept. No single human or generation of humans can be responsible for preventing a species from going extinct, only for preventing them from going extinct "on our watch" (Westley 1999).

Nonetheless it is possible to say that both in terms of CBSG's goals and in terms of problem resolution, the CBSG initiatives have been successful. Seal's own goals for the initiatives were both short- and long-term, both human and endangered-species centered. The long-term, endangered-species-centered goals are to maintain viable populations with sufficient genetic heterozygosity for long-term species health. To date, some twenty species have been temporarily "saved" by captive breeding programs in which the information from ISIS and the collaborative processes of SSPs have played an important role (Tudge 1991, 29). The PHVA and CAMP processes can be credited with helping some fifty additional species.

From the particular perspective of this book, that of interdisciplinary collaboration as a key element of conservation, the sheer (and ever-increasing) volume of these workshops qualifies as a measure of the initiative's ability to convene stakeholders. In terms of the short- and long-term mobilization of human effort, CBSG's attempts to build collaborative bridges between stakeholder organizations, in order to increase the quality and quantity of knowledge available and encourage integrated plans for managing the species, have been unique. In the remaining chapters of this book, we will continue to explore these successes and the potential of such processes for dealing with ever-increasing levels of complexity.

Chapter 3

Integrating the Human Dimension into Endangered Species Risk Assessment

PHILIP S. MILLER AND ROBERT C. LACY

Conservation biologists use population-biology models to assess the probability of extinction, the relative magnitudes of threats, and the effects of possible actions on endangered species—an approach called Population Viability Analysis, or PVA. When population data are sufficient, detailed PVA models have been found to be capable of producing unbiased and accurate representations of the dynamics of wildlife population processes (Brook et al. 2000; Lindenmayer, Lacy, and Pope 2000). Yet the conservation of biodiversity is mostly a matter of addressing processes that revolve around the interactions of wildlife and human populations. Unfortunately, experts who model human systems (such as human demography, local economics and resource use, industrial activities, social systems, and political systems) rarely interact with the biologists who model wildlife populations.

The primary theme of this book is that it is possible and necessary to integrate analyses of human systems with biological considerations in species conservation. In this chapter, we identify the linkages between human demographic, economic, and social systems and wildlife population biology. We discuss how outputs from models of each system can be used as inputs into other systems. Expansive heuristic models can help us identify pathways by which human systems exert pressure on wildlife populations via specific threatening activities. Detailed quantitative models of the relevant human systems, when available, can generate specific inputs into a PVA model of the wildlife population. Most importantly, the integration of these models can encourage broader stakeholder involvement in conservation planning and lead to more effective conservation solutions. Integrating understanding of human and natural systems will require a broader range of expertise, models, data, and perspectives than has been applied previously to risk assessments for endangered species.

Causes of Species Endangerment

Many wildlife populations are in decline, and species extinctions are occurring 1,000 to 10,000 times more frequently than the natural rate that existed prior to the domination of the world by humans (Wilson 1992; Purvis and Hector 2000). As a consequence, the diversity of the Earth's flora and fauna is plummeting, and these losses may have serious or even catastrophic impacts on the stability and functioning of ecosystems (Tilman and Downing 1994; McGrady-Steed, Harris, and Morin 1997; Naeem and Li 1997; McCann 2000). The services that humans receive from the natural world may be seriously and dangerously diminished (Chapin et al. 2000; Tilman 2000).

The primary causes for species decline are often obvious and deterministic: populations are overharvested; natural habitat is converted and no longer suitable for wildlife species; environments are polluted, with the dumping of toxins into the air, water, and soil; local and now even global climates are modified by the actions of humans; and numerous non-native competitors, predators, parasites, and diseases are introduced into communities that have never evolved defenses to the new invaders (Simberloff 1988; Caughley 1994). These primary causes of species decline will be difficult to reverse. Even if the original causes of decline are removed, a remnant, isolated population is vulnerable to additional forces, intrinsic to the dynamics of small populations, which may drive the population to extinction (Shaffer 1981; Soulé 1987).

Of particular impact on small populations are stochastic, or random or probabilistic, processes. Stochastic processes usually have little impact on long-term population dynamics, as long as the population is abundant and spread over a wide geographic range and a number of habitats. Deterministic processes, such as those listed above, predominate in widespread, still-common species, while local chance events affecting subsets of a population will average out across the broader, diverse range. When a population becomes small, isolated, and localized, however, random events can become important, even dominating the long-term dynamics and fate of a population. Extinction can be viewed as a process in which once common and widespread populations become reduced to small, isolated fragments due to extrinsic (e.g., climate, food, or habitat availability) factors; the small remnant populations then become subjected to large fluctuations due to intrinsic (e.g., age, disease) processes; and the local populations occasionally and unpredictably go extinct; and the cumulative result of local extinctions is the eventual extinction of the taxon over much or all of its original range (Gilpin and Soulé 1986; Clark, Warneke, and George 1990).

The stochastic processes affecting populations have been usefully categorized into demographic stochasticity, environmental variation, catastrophic events, and genetic drift (Shaffer 1981). Demographic stochasticity is the random fluctuation in the observed birth rate, death rate, and sex ratio of a population even if the probabilities of birth and death remain constant. Such demographic stochasticity will be most important to population viability perhaps only in populations that are smaller than a few tens to hundreds of animals (Goodman 1987; Lacy 2000a), in which cases the annual frequencies of birth and death events and the sex ratios can deviate far from the means. For example, the last five dusky seaside sparrows *(Ammodramus maritimus nigrescens)* were males, an unfortunate but not implausible occurrence that meant the end of the taxon.

Environmental variation is the fluctuation in the probabilities of birth and death that results from fluctuations in the environment. Weather, the prevalence of enzootic disease (constantly present in an animal), the abundances of prey and predators, and the availability of nest sites or other required microhabitats can all vary, randomly or cyclically, over time. Catastrophes such as epidemic disease, hurricanes, large-scale fires, and floods are outliers in the distributions of environmental variation. As a result, they have potentially dramatic impacts on wildlife populations and can precipitate the final decline to extinction (Simberloff 1986, 1988). For example, the only remaining population of the black-footed ferret *(Mustela nigripes)* was being eliminated by an outbreak of distemper when the last eighteen ferrets were captured (Clark 1989).

Genetic drift is the cumulative and nonadaptive fluctuation in allele frequencies resulting from the random sampling of genes in the production of each generation of offspring. Over time, this can impede the recovery or accelerate the decline of wildlife populations for several reasons (Lacy 1993a, 1997). Inbreeding, not strictly a component of genetic drift but correlated with it in small populations, has been documented to cause loss of fitness in a wide variety of species (Ralls, Ballou, and Templeton 1988; Lacy, Petric, and Warneke 1993). Even if the immediate loss of fitness of inbred individuals is not large, the loss of genetic variation that results from genetic drift may reduce the ability of a population to adapt to future changes in the environment (Fisher 1958; Robertson 1960).

Clearly, there are a large number of processes that can threaten species. Most of these factors interact, often in synergistic ways. For example, habitat degradation and fragmentation can lead to reduced reproductive output, easier access into the habitat by both hunters and introduced predators, greater mortality during dispersal among remnant areas of habitat, and disruptions in population age structures and social systems. Consequently, population fluctuations will be greater, and mean growth rates lower. This causes loss of genetic variation and inbreeding, which in turn reduces breeding and survival and increases susceptibility to environmental stresses. Gilpin and Soulé (1986) coined the term "extinction vortex" in their description of a heuristic model that describes the nature of these interactions that can cause rapid decline of a population to extinction.

Population Viability Analysis (PVA)

PVA is the estimation of extinction probabilities and other measures of population performance by analyses that incorporate identifiable threats to population survival into models of the extinction process (Gilpin and Soulé 1986; Lacy 1993/1994). In contrast to some earlier concepts of a minimum viable population size (MVP; Shaffer 1981), which implied to some that there were determinable population sizes above which populations would be safe from extinction, conservation biologists have more recently focused on methods for estimating the probability of extinction over defined time periods for a designated population exposed to a specific scenario of environmental conditions, threats to persistence, and future management actions and other foreseeable events (Starfield and Bleloch 1986; Soulé 1987; Simberloff 1988; Shaffer 1990; Boyce 1992; Burgman, Ferson, and Akçakaya 1993; Beissinger and McCollough 2002). Two defining characteristics of a PVA are an explicit model of the extinction process (Clark, Warneke, and George 1990) and the quantification of threats to persistence. The decision to base the World Conservation Union (IUCN) global categories of threat (e.g., threatened, vulnerable) on quantitative criteria such as probability of extinction, or on trends in such indicators as habitat area or numbers of mature individuals (Mace and Lande 1991; IUCN 2000), reflects the increased understanding of the extinction process that has accompanied the development of PVA. Simultaneously, this greater understanding demands additional progress toward developing more predictive models, gathering relevant data on population status and threats, and applying the PVA techniques to a broader suite of scenarios.

Generally, the model of extinction underlying a PVA considers both deterministic and stochastic factors. Which of the various deterministic and stochastic factors are important to consider in a PVA will depend on the species biology, the present population size and distribution, and the threats it faces. For example, orangutans *(Pongo pygmaeus)* may be threatened by forest destruction and other largely deterministic processes, but inbreeding and randomly skewed sex ratios resulting from highly stochastic processes are unlikely to be problems, at least not on a species-wide basis (Tilson et al. 1993). On the other hand, even if the remnant Atlantic coastal rain forest of Brazil is secured for the future, populations of golden lion tamarins *(Leontopithecus rosalia)* that can persist in that remnant forest are not sufficiently large to be stable in the face of stochastic threats (Ballou et al. 1998).

The identification of the primary threats facing a taxon via a comprehensive PVA is important for conservation planning. For example, tamarin populations might be stabilized by the translocations and reintroductions that are underway and planned, but an orangutan Population and Habitat Viability Assessment (PHVA) recognized that releases of confiscated "pet" orangutans are unlikely to have a conservation benefit for those populations that are facing habitat destruction. Similarly, effective management of the red wolf *(Canis rufus)* in North America must recognize and address the severe threat posed by hybridization with coyotes within the confines of a National Wildlife Refuge where habitat is not limiting (Kelly, Miller, and Seal 1999).

Shaffer (1981) suggested several ways to conduct a PVA. An elegant and general approach would be to develop analytical models of the extinction process that will allow calculation of the probability of extinction from a small number of measurable parameters. Goodman's (1987) model of demographic fluctuations, and conservation applications of the classic population genetic models describing genetic diversity lost by random drift (Franklin 1980; Soulé et al. 1986; Lande and Barrowclough 1987) are valuable efforts in this direction. Unfortunately, our understanding of population biology is not yet sufficient to provide fully adequate analytical models of the extinction process. For example, none of the existing analytical models incorporate all the sources of stochasticity (demographic, environmental, and genetic), and thus do not begin to model the full array of extinction vortices described by Gilpin and Soulé (1986).

Lacking adequate analytical solutions, most PVAs use computer simulation modeling to project the probability of possible fates of a population. Simulation models can incorporate a very large number of threatening processes and their interactions, if the processes can be described in terms of quantitative algorithms and can be adequately parameterized. Although many processes affecting small populations are intrinsically indeterminate, the average long-term fate of a population and the variance around the expectation can be studied with computer simulation models. PVAs focus on detailed and explicit modeling of the forces impinging on a given population, place, and time of interest, rather than on delineation of rules (which may not exist) that apply generally to most wildlife populations.

The VORTEX Population Viability Analysis Model

The VORTEX computer program (Miller and Lacy 1999) is one of several PVA models that have been used widely in conservation assessments and endangered species recovery planning (Lindenmayer et al. 1995; Brook et al. 1999). VORTEX has produced accurate projections of wildlife populations in complex, human-modified landscapes (Lindenmayer, Lacy, and Pope 2000). This model has been used in each of the PHVA workshops described throughout this book, as well as in many of the other conservation assessment workshops of the Conservation Breeding Specialist Group (CBSG) (see *www.cbsg.org* for descriptions and workshop reports).

While the model's core is a representation of the biological processes that are the proximate determinants of the dynamics of wildlife populations, increasingly VORTEX has been extended to allow specification of external processes, including those created or mediated by human populations and activities. The program does this by providing the capability to specify most demographic rates as functions of time, density, and other parameters. With this feature, external processes, such as trends in human populations and their impacts on the environment, can be included as inputs into the model. Below is a brief summary of the VORTEX software; more detailed descriptions are provided in Lacy (1993a), Miller and Lacy (2003), and Lacy (2000b). The program is available at *www2.netcom.com/ ~rlacy/vortex.html*.

VORTEX simulates the effects of deterministic forces as well as of demographic, environmental, and genetic stochastic events on wildlife populations. It is a tool that can be used to model many of the processes that can threaten persistence of local populations. The program simulates a population by stepping through a series of events that describe an annual cycle of a typical sexually reproducing organism: mate selection, reproduction, mortality, increment of age by one year, dispersal among populations, removals, supplementation, and population limitation by habitat availability (ecological "carrying capacity"). The simulation of the population is iterated many times to generate the distribution of fates that the population might experience.

VORTEX is an individual-based model. That is, it creates a representation of each animal in computer memory and follows its fate through each year of its lifetime. Demographic events (birth, sex determination, mating, dispersal, and death) are modeled by determining for each animal in each year of the simulation whether any of the events occur. In part because of the individual-based nature of the program, the simulation requires a lot of population-specific data. For example, the user must specify the amount of annual variation in each demographic rate caused by fluctuations in the environment. The frequency of each type of catastrophe (drought, flood, and epidemic disease) and the effects of the catastrophes on survival and reproduction must be specified. Trends in habitat availability and quality must also be specified.

Demographic stochasticity, or variation, in VORTEX is a consequence of the simulated uncertainty regarding the demographic outcomes for any given animal. To model environmental variation in birth and death rates, each demographic parameter is assigned a distribution with a mean and standard deviation. Catastrophes are modeled as random events that occur with specified probabilities. Following a catastrophic event, the chances of survival and successful breeding for that simulated year are decreased. For example, forest fires might occur once in fifty years, on average, killing 25 percent of the animals, and reducing breeding by survivors 50 percent for the year. The losses of genetic variation in small, isolated populations are modeled by simulating the transmission of alleles (genetic variants) from parents to offspring at a hypothetical gene. VORTEX also monitors the extent of inbreeding in a population, and can simulate reductions in juvenile survival or other demographic rates of inbred animals to model the effects of inbreeding depression.

VORTEX can model monogamous or polygamous mating systems, local population structure, dispersal, and other aspects of the social system of the wildlife population. When the population size exceeds the carrying capacity of the local habitat, increased mortality across all age-sex classes returns the population below this maximum level. The carrying capacity can be specified to change over time, to model losses or gains in the amount or quality of habitat. Populations can be supplemented or harvested for any number of years in each simulation. Harvest could represent managed culling (killing) or removal of animals for translocation to another population.

The model provides the following outputs as descriptions of population viability: probability of extinction at specified intervals, mean time to extinction, projections of the mean size of populations, and genetic variation within and among populations. Standard deviations across simulations are reported for these measures, providing estimates of the inherent uncertainty in population projections.

Expanding Population Viability Assessment

PVA was developed to model and understand the biological threats to population persistence. PVA models can be used to examine overharvest; habitat degradation; habitat loss and fragmentation; impacts of exotic species; increased environmental variation and even catastrophic impacts due to perturbations of the environment; demographic uncertainty, disrupted breeding systems; and the dual genetic problems of random drift and inbreeding depression. Figure 3-1 illustrates the primary categories of factors that need to be considered in a PVA model in order to assess population viability. Because of the number and complexity of the threats to viability, the predictive capabilities of PVA can be improved by recruiting expertise from many fields to understand and address these threats. These disciplines include wildlife management, population ecology, community ecology, landscape analysis, geography and Geographic Information Systems (GIS), genetics, and even statistics, modeling, decision analysis, conflict resolution, and others that are outside the natural sciences realm. In spite of some arguments about importance of one area or another (Caughley 1994; Caro and Laurenson 1994; Beissinger and Westphal 1998; Reed et al. 2002), PVA is fundamentally an analysis of multiple interacting factors (Gilpin and Soulé 1986; Hedrick et al. 1996). Therefore, PVA is necessarily synthetic and holistic, rather than reductionist. PVA models are complex and diverse, as they must be, because the essence of population viability cannot be captured with a few elegant equations or generally applicable theories.

Increasingly, endangered species risk assessment throughout the world is accomplished through PVA techniques. For example, PVA has been recommended by an Australian federal commission (Resource Assessment

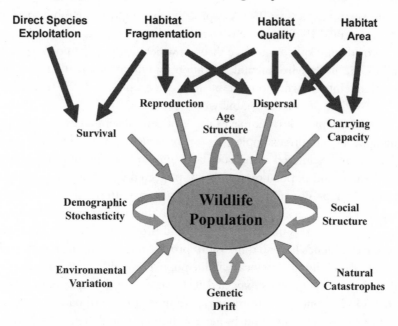

Figure 3-1. Primary factors impacting the viability of wildlife populations that need to be considered in development and application of PVA models.

Commission 1992) as one tool that should be applied to ensure that resource management decisions do not jeopardize wildlife populations. The U.S. Fish and Wildlife Service, the Kenya Wildlife Service, and the governments of Indonesia, India, Brazil, South Africa, and dozens of other nations have used PVA to address some of their most difficult species management issues. A typical PVA focuses almost solely on the biology of the target wildlife species with only a relatively vague, qualitative description of the means by which human activities affect population dynamics.

The CBSG of the IUCN's Species Survival Commission (SSC) has developed an extension of the traditional PVA approach. CBSG's non-traditional approach, known as a Population and Habitat Viability Assessment or PHVA, is a species risk assessment process involving participation by all interested parties showing a stake in the development of management plans for the species or population in question (see chapter 1). Thus, PHVA represents a broadening of the traditional PVA methodology, incorporating as much information as possible about the focal species, its habitat, and the ways in which local human populations affect this species and its surroundings (Lacy 1993/1994). There is a need to use the inclusive and expansive PHVA approach to strengthen the more specific and quantified PVA approach. To accomplish this merger requires that we provide an explicit interface through which detailed data on resource utilization and environmental alteration by human populations can be translated into information suitable for demographic and genetic modeling of wildlife population viability.

Figure 3-2 provides a simple example of the conceptual framework for this approach. In this example, the size and structure of local human populations are estimated using standard human demographic modeling techniques, and population trends are projected for some time into the future. Perhaps more importantly, the behavioral patterns of these human populations—in terms of the ways in which they interact with their local environment—are similarly estimated and projected. These analyses of the environment's human dimensions provide insight into the nature and extent of their impact on local wildlife populations either currently at risk of extinction or perhaps expected to be at risk in the near future. But what are the conceptual frameworks we must use to estimate the extent of local resource utilization by nearby human populations? And what are the precise mechanisms by which these utilization patterns have an

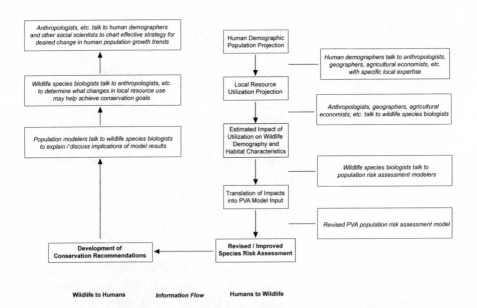

Figure 3-2. Simple flowchart illustrating the general types of people who may need to exchange and integrate information about human and wildlife population processes through a PHVA workshop.

impact on wildlife populations? Finally, what precise types of information are required to model these impacts using tools like VORTEX? Our enhanced model, discussed in some detail in this chapter, is an initial attempt at providing some answers to these questions.

Impacts of Human Activities on Population Viability

Although PVAs have been complex and multifaceted since the first models of Shaffer (1981), most PVA models still omit the primary driving force in population viability and extinctions: human populations and their interactions with the landscape. Unlike the processes in the lower part of figure 3-1, the forces listed across the top are not intrinsic to natural systems. Rather, the dominant threats to the viability of natural populations are caused directly or indirectly by human activities.

PVA models can be sophisticated representations of biological systems. However, most PVA models make simplified assumptions about what humans are doing in the system. We usually ask: what happens to the probability of population persistence (or some other measure of viability) if humans do not change in number, in distribution, or in activity patterns over time? Yet the assumption of no change in human systems is often naive and unrealistic. Even when direct human impacts on wildlife are considered in PVA models, the dominant human impacts shown in figure 3-1 are usually treated as though they were the original or ultimate causes of species vulnerability. Yet those processes are just the intermediaries, or links, between the diverse activities of the human population and the population biology of the species of concern in the PVA. For example, habitat destruction is not a spontaneous process. It cannot be fully understood if we ignore the human activities that are destroying habitat, and PVAs that consider habitat destruction only as an abstract process disconnected from the forces driving it are not likely to contribute much to ameliorating the impacts of such destruction.

To understand how PVA may be expanded to address the interactions of human populations with threats to wildlife viability, it is useful to start by examining the population biology processes that are central to most PVAs (the population biology core shown in the lower part of figure 3-1), and then to work outward to identify which human systems impact wildlife population dynamics and the mechanisms by which these impacts occur. The biological processes affecting viability fall loosely into several categories. First, there are processes that are intrinsic to population dynamics. These include the largely stochastic problems of sampling in small populations such as demographic stochasticity and genetic drift

(Lacy 2000a), and processes resulting from population-specific characteristics: age structure, breeding system, and inbreeding depression. In addition, there are largely unavoidable natural processes that are driven by forces external to the population but which can strongly impact population dynamics. This category of threats to viability includes environmental variation and natural catastrophes. Perhaps most importantly, there are the fundamental, deterministic processes driving changes in population size: reproduction, survival, carrying capacity limitations of the habitat, and dispersal (emigration and immigration). These deterministic components of population growth are the only factors considered by most of the classical models of population growth in population ecology (Deevey 1947; Birch 1948).

In some subtle ways, humans can affect almost all of the processes in figure 3-1, but the primary human entry points into the core population biology system are the impacts on the primary processes driving wildlife population growth or decline: reproduction, mortality, carrying capacity, and dispersal. Human activities affect these fundamental demographic rates through four primary modes: direct species exploitation, reductions in habitat quality, reductions in habitat area, and habitat fragmentation.

Exploitation

Direct species exploitation can take the form of harvesting for local use or external markets, incidental killing, sport collecting, or even retaliation out of spite, anger, defense, or protest. Exploitation (when it is documented and quantified) can be directly and easily incorporated into PVA models as increases in age-sex specific mortality appropriate to the particular situation being considered. For example, interviews with local villagers at the Papua New Guinea tree kangaroo (*Dendrolagus* spp.) PHVA workshop led to estimates of the rate of removal of individuals from specific populations (see chapter 9). This information could then be combined with other estimates of baseline mortality in the absence of significant human interference to gain a more complete picture of mortality in these highly threatened populations.

Discussions at the Humboldt penguin *(Spheniscus humboldti)* PHVA workshop revealed that, each year, an estimated 3–5 percent of the population dies through entanglement in fishing nets (Araya et al. 1999). Projections indicated that this additional mortality could by itself drive the taxon to extinction within the next thirty years. The analysis could be taken further by projecting the impacts of changes in the fishing industry, for example, by estimating the effects of a changeover from local fishing boats using small nets to major international factory fleets working farther offshore.

Finally, data presented at a PHVA workshop on conservation of marine turtles in Indonesia revealed that up to 99 percent of the total egg production on particular nesting beaches is harvested each year (Manansang et al. 1997). This type of activity could easily be translated within VORTEX to an effective cessation of reproduction over the short term.

Impacts on Habitat Quality and Area

Habitat quality can be reduced by human activities through impacts on food, predators, competitors, disease, local water supply, and basic geophysical resources, and on cover, nest sites, or other microhabitats necessary for species or population persistence. These elements of habitat decline are often caused by introduced exotic (non-native) plants and/or animals, the use of resources in the habitat by humans, the discharge of waste from human activities, alteration of the nature of adjacent land following landscape development or agricultural conversion, and the effects that human activities have on regional or global climate. Likewise, impacts on habitat areas are manifested through changes in the amount of habitat protected under changing socioeconomic conditions or changes in the rate of conversion or loss of unprotected habitat. Changes in both habitat quality and area can be combined into projections of carrying capacities in PVA models.

In early PHVAs on the Florida panther *(Felis concolor coryi)*, the loss of habitat was estimated as something between no loss and a total loss of the 50 percent of panther habitat that is on privately owned lands (Seal and Lacy 1989; Seal 1992). Extensive data are available for more precisely projecting changing land use by humans (e.g., Pearlstine et al. 1995), and the preferences of panthers for various landscape features have been documented (Maehr and Cox 1995). Maehr et al. (2002) describe briefly how data on changing land use could be used to inform PVAs and lead to more holistic conservation plans for the Florida panther.

In a PHVA for the golden-headed lion tamarin *(Leontopithecus chrysomelas)* in Brazil's Una Biological Reserve, researchers estimated the percent of habitat that was in "pristine" condition and the percent that was degraded due to agricultural use by twenty-four families of human settlers living within the park (Ballou et al. 1998). The density of tamarins that could be supported in each category of habitat was then used to estimate the carrying capacity of the park. In addition, fires occur in the degraded habitat, potentially causing substantial mortality of tamarins. These estimates of effects on carrying capacity and fire-caused mortality were used in comparisons of population viability under varying assumptions about the percent of the habitat that would be degraded by human occupation in the future. The analyses showed that the tamarin population would be genetically and demographically secure only if the park was fully protected and the habitat restored to pristine condition. These results were then used to argue for providing indemnities to the squatters in compensation for their removal from the area.

In a PHVA for Costa Rican squirrel monkeys *(Saimiri oerstedi citrinellus)*, increased human densities in areas adjacent to a protected park were projected to cause gradual degradation of that habitat over the next decade (Matamoros, Wong, and Seal 1995). The consequent declines of carrying capacity for the monkey populations outside of the park were then incorporated into the population models.

Habitat Fragmentation

Changes in the area, distribution, and quality of habitat can also affect the pattern and extent of species dispersal among habitat patches. Habitat fragmentation can occur either through the direct subdivision of a formerly large, contiguous habitat into multiple smaller units, or by the erection of barriers to crossing between existing habitat patches in a landscape already fragmented. The detrimental effects of habitat fragmentation might be incorporated into a PVA as increased age-sex specific dispersal mortality (which would be imposed in addition to the normal mortality of resident organisms) and/or as decreased rates of dispersal among patches. For example, Lindenmayer and Possingham (1994) explored the effects of varying patterns of forest blocks and retained corridors of habitat for Leadbeater's possum *(Gymnobelideus leadbeateri)*. Effects of logging practices on the amount, quality, and interconnectedness of habitat patches were fed into spatially explicit PVA models that simulated animal movements on the landscape.

Indirect Impacts on Stochastic Processes

Most of the stochastic processes that are intrinsic to population dynamics—demographic stochasticity, environmental variation in demographic rates, genetic drift, and inbreeding—are less likely to be the most important entry points through which human activities modulate population viability. Although the nature of these intrinsic processes is unchanged in wildlife populations perturbed by human activities, their impacts may be increased. Human activities modulate stochastic population processes and natural variation in the environment more indirectly by causing reductions in population sizes. Population instability caused by stochastic processes results from sampling effects. Large, widespread populations spread the risk of short-term threats, but small populations are not buffered from random local variation. In individual-based simulations, such processes emerge from the sampling of events in the model. Any PVA model that includes demographic stochasticity and inbreeding effects as functions of population size will capture some of the effects of human actions on increased fluctuations in wildlife population demography.

Human activities can modulate stochastic processes even without (but, usually, in addition to) changing species population size. For example, human modified landscapes may experience greater fluctuations in the biotic and physical environment because of reduced species diversity and removal of ecosystem functions that normally buffer environmental perturbations (Tilman and Downing 1994; McGrady-Steed, Harris, and P. J. Morin 1997; Naeem and Li 1997; McCann 2000). The consequent greater seasonal, interannual, and spatial variation in demographic rates can then further threaten population viability. Catastrophes, such as disease epidemics and even severe weather patterns, may increase because of human activities such as the transport of organisms, reduction in diversity of biological systems, and climate change.

Many of the links between human activities and the dynamics of wildlife populations are fairly obvious, even if they may be inadequately studied and ignored in many PVAs. However, some of the links between human activities and population processes are less direct and more subtle. For example, there are no more than a few Sumatran rhinoceroses *(Dicerorhinus sumatrensis)* within each of a number of isolated protected areas. Field surveys in Malaysia in 1995 found tracks of only one juvenile among thirty-five sets of tracks, and only one of twenty-one adult females captured in the prior decade was pregnant (AsRSG 1996). If the population were breeding as expected for a rhinoceros species, about 30 percent of adult females should be pregnant at any time, and about 15 percent of the animals should be under two years of age. It is possible that the scarcity of mates is causing a near cessation of breeding over much of the fragmented range. Thus, one impact of habitat destruction outside of parks and of poaching within parks may be a considerable reduction in reproduction. The disruption of breeding systems as an indirect effect of human activities can be incorporated into PVA models via Allee effects (e.g., Groom 1998; Groom and Pascual 1998)—the reduction of breeding that is seen in some species when population densities become low. For example, Kjos et al. (1998) concluded that the last remnant population of the winged mapleleaf mussel *(Quadrula fragosa)*, found along the Saint Croix River bordering Minnesota and Wisconsin, was reduced to a sufficiently small size to potentially restrict breeding through an Allee effect. Moreover, this process is projected to become more problematic as restrictions in water flow serve to further fragment the population. Overall, however, a review of the PVA literature suggests that PVAs rarely incorporate strong density dependence like that described above (Mills et al. 1996; Lacy 2000a).

Methods for Including Human Processes in Population Viability Analysis

Box 3-1 lists primary human systems and factors that might affect the viability of a wildlife population by directly or indirectly causing direct species exploitation, reduced habitat quality and area, and habitat fragmentation. The human systems that are relevant to species conservation include human demographics; economics; agriculture and activities of corporations involved in an area's major industries; social, ethical, and moral systems; and governance and politics. While we cannot immediately encompass all of the disciplines in box 3-1 within a PVA on a specific species, we can begin to address one or more of the factors that have the most critical effects.

The chain of effects that connects the human processes in box 3-1 to the impacts on the viability of a wildlife population in figure 3-1 can be broken down into four components: the numbers of humans; the distribution of humans; the activities of those humans; and the impacts of those activities on the fundamental inputs into the population biology of the species (direct exploitation, habitat degradation, reduction, and fragmentation). If we can marshal the expertise needed to understand each of these components (certainly no small task), then we can produce more predictive models of the viability of natural populations in a human-dominated world.

The easiest step in this process will often be to understand the changing numbers of humans. Conservation biologists often rightly bemoan the fact that we have few data with which to assess wildlife demography for any species of interest. Yet there is one species for which massive databases exist on demographic trends: *Homo sapiens*. Detailed demographic data and projections for almost all nations are available from sources such as the U.S. Census Bureau (McDevitt 1998, available online) and the United Nations Population Information Network *(www.un.org/popin/)*. Assessments relevant to specific conservation issues and species of concern will also often require estimates of the local patterns rather than the national trends (see chapter 13). Sometimes, local demographic data and trends can be obtained from published sources, the Internet, or governmental agencies. Other times, estimating local trends may require a survey of the relevant population and application of demographic models by experts in human populations (e.g., Ness 1997; Stover and Kirmeyer 1997). Perhaps all PVAs should begin by assessing the trends in the human populations that have impacts on the wildlife species of interest. This would enable the PVA to subsequently consider whether human impacts on the wildlife (e.g., poaching, road kills, habitat degradation by pollution, land conversion) are likely to change along with shifts in numbers and distributions of humans. In the grizzly bear PHVA workshop described in

Box 3-1.
Some human systems that can impact the viability of wildlife populations, principally through the four effects listed at the top of figure 3-1.

Human demographics
  Population growth rate (fertility, mortality)
  Age structure
  Distribution and movement

Economics
  Nonlocal markets and commodity prices
  Local markets
  Subsistence practices, hunting

Industry
  Extractive industries (timber, mining, fisheries)
  Agriculture

Geography, sociology, cultural anthropology
  Urbanization
  Transportation and access
  Religion and ethical beliefs
  Cultural practices

Political science
  Governance
  Land ownership
  War

chapter 10, the projection that the human population using grizzly bear habitat will likely increase by about 4 percent per year led to the conclusion that the human-induced mortality of the bears will soon become unsustainable, unless the average impact of humans on bears can be substantially decreased.

After obtaining a projection of human numbers, we need to know how the spatial distributions of those humans will be changing. For example, a PVA was completed on the Indian rhinoceros in Jaldapara National Park and Wildlife Sanctuary, India, in order to provide the local park managers and the national administrators with better guidance on conservation strategies (Walker and Molur 1994). Near the conclusion of the workshop at which the assessments were made, it was revealed by a team of human demographers who had surveyed the local villages that the number of households surrounding the park would likely double over the next twenty years, largely as a result of people fleeing the poverty of nearby Bangladesh (see chapter 14). Unlike the existing community of small-scale farmers, these immigrants would be mostly landless and therefore heavily dependent on the park as a source of firewood, grazing lands, and other resources. A PVA for the rhinoceros that ignored this pending influx of people would clearly have overlooked important threats. The wildlife biologists lacked the knowledge necessary to make a meaningful PVA, although we would have remained confident in the value of our forecasts had not the human demographers revealed the inadequacy of our PVA projections.

Projecting changes in the distribution of people requires considerations of geography, access, legal restrictions, macroeconomic forces, industry, government policy, and land ownership and tenure systems. Fortunately, conservationists do not have to start developing such fields, but can make use of the expertise within other disciplines. For example, existing models of human dispersal are more sophisticated and based on more data than are the models of wildlife dispersal that biologists are using in PVAs.

After gaining an appreciation for probable changes in numbers and distributions of humans, we need to understand what changes are likely to occur in people's activities involving the landscape. Trends in agricultural practices, the local economic base, technology available locally, nonlocal markets (which can encourage resource exploitation far beyond what would be useful locally or, alternatively, can provide opportunities to sustain a family without the harvesting of local resources), and belief systems can all alter the extent to which humans exploit a species and affect its habitat. An overview of the impacts of humans on the environment is provided in Turner et al. (1990), while discussions of some of the principles determining how humans utilize natural resources are available in Hanna, Folke, and Mäler (1996), Buck (1998), and Ostrom et al. (1999). Perrings et al. (1995a, 1995b) are examples of analyses of how governmental policies, property rights, economic policies, and value systems affect biodiversity, ecosystems, and harvested species. However, most of their treatments are still too broad and general to provide the detailed projections of changing human activities and impacts that are needed to inform endangered species risk assessments.

Broad considerations of human systems can help us identify the kinds of human processes that might be important, but details of human impacts in a specific habitat of concern have to be obtained from local knowledge or focused study. Concerning the Amazon rain forest, for example, Coomes (1995) describes how national and international markets have driven extraction of forest products, while Coomes and Barnam (1997) describe the microeconomic determinants of livelihood and resource use decisions by peoples living within the rain forest. Clayton, Keeling, and Milner-Gulland (1997) develop a detailed spatial model of harvesting babirusa *(Babyrousa babyrussa)* in Sulawesi, Indonesia, that accounts for changing road conditions (as road conditions improved, greater numbers of hunters were able to access habitat) and hunters' opportunity costs in an attempt to evaluate the viability of this endemic and endangered wild suid subjected to intense hunting pressures. At a PHVA workshop on mountain gorillas, participants projected the effects of war on fecundity, mortality, and epidemic disease, and the amount of habitat that will be available for gorilla populations in the future (e.g., see chapters 6 and 18).

Impacts of Forest Management Practices on Leadbeater's Possums in Australia

The work of Lindenmayer and colleagues on the Leadbeater's possum *(Gymnobelideus leadbeateri)* demonstrates well the value of pursuing some of the approaches we have discussed (Lindenmayer and Possingham 1994). Leadbeater's possum is a small arboreal marsupial that was thought to have been extinct since early in the 1900s until it was rediscovered in 1961. The species is confined to the wet, cool mountain forests of the Central Highlands of Victoria, Australia, and has been declared the state's faunal emblem. The Victorian eucalyptus forests contain the tallest flowering plants in the world, exceeding 100 meters in height, and Leadbeater's possums nest in hollows within these large trees in the old-growth forests. The hollows needed by the possums do not typically form until the trees exceed 190 years of age, and nest trees can be older than 400 years. The massive eucalyptus trees are very valuable timber, and economically optimal forest management practices harvest trees on a rotation of approximately 50–80 years. Fires within young forests can kill trees (and possums) before the trees are large enough to form hollows suitable for possums. Fires within mature forests (older than 200 years) can accelerate the formation of hollows suitable for possums and can lead to an increase in understory plants that are used for foraging by the possums. However, after fires, economically valuable timber can be salvaged from the forest, clearing out the old, damaged (and therefore hollow-forming) trees. Clearly, there are conflicts between management of the forest for timber and management for wildlife (Lindenmayer 1996).

Risk assessment for the Leadbeater's possum requires integrating analyses of multiple processes: the dynamics of tree growth, hollow development, and tree fall; forest fires; timber harvest regimes; nest site and other habitat requirements of the possum; breeding biology of the species; and dispersal and mortality rates. The timber harvest regime and considered options for forest management are determined by governmental policies, are tightly controlled, and are well documented. The dynamic balance between hollow formation in trees and collapse of old trees has been studied and models of the forest structure have been used to predict rates of gain or loss of tree hollows under various schemes of forest management and frequency of fire. Field studies of the species have provided information on forage requirements, territory size, dispersal patterns, and the possum's breeding system (which involves a social unit consisting typically of multiple adult males and a breeding female occupying a nest site).

By linking these disparate kinds of information in GIS analyses of landscape changes and simulation models of possum population dynamics, Lindenmayer and colleagues have been able to project the changing ability of managed forests to support the possum species. The analyses of the forest-possum system led to an understanding that numbers of the possum probably increased following a catastrophic fire in 1939 that created many tree hollows and good habitat, and this increase likely facilitated the rediscovery of the species. However, ongoing natural decay of the tree hollows created by that fire and continued harvest of younger trees prior to attainment of new hollows is projected to lead to a loss of Leadbeater's possums throughout much or all of the range.

More importantly, the analyses led to a series of recommendations regarding how to make forest management practices compatible with the legislative mandate to maintain populations of wildlife species throughout their known distribution. General recommendations included the following: old-growth and multi-aged forest stands greater than 10 hectares should continue to be reserved from logging; forest patches that serve as linkages between important old-growth stands need to be preserved as dispersal corridors; stands of old-growth timber should not be salvaged following wildfires; fire management practices should be changed; some blocks of regrowth forest should be withdrawn from production and allowed to develop into mature stands suitable for the possums. The integrated analyses allowed very detailed recommendations regarding the size, shape, slope, and age of wildlife reserves in the forests and the patterns and methods of harvest of nearby production blocks.

Conclusions

The case studies detailed in this book show that the tools of PVA can be extended to provide insights into the dynamics of interactions between human systems and wildlife viability, and can therefore become more useful for conservation. PVAs that neglect changing numbers, distribution, and activities of humans probably provide inaccurate assessments of population viability and may lead to conservation strategies that will fail to adequately safeguard biodiversity. We have just begun to explore how best to integrate models of various human systems with models of habitat and ecosystem effects and with models of wildlife populations. However, we are convinced that such exploration is essential to finding conservation solutions that will succeed in an increasingly human-dominated world (Nyhus et al. 2002).

The links between human and natural systems can be explicitly modeled within a large multicompartment model. For example, the Forest Land Oriented Resource Envisioning System (FLORES) project of the Center for International Forestry Research (CIFOR) in Indonesia is examining the interactions in that country between economics, agriculture and industry, road building, social processes, and forest biodiversity (Vanclay 1998). However, it may prove more feasible to develop separate models for the human dimensions versus those of wildlife population biology and ecology. Each of the models would then take as input information derived from the others, and in return would send its output to other systems as those systems' input. Projections from one system model might forecast the trends in a critical resource or process as a time series, and those trends would then define limiting factors influencing processes in the next model along a chain of causality (see chapter 18).

The impacts of many of the systems listed in box 3-1 on wildlife viability can be dissected and analyzed. First, we need to outline the plausible linkages between systems and identify the primary pathways of impact on the wildlife population biology at the top of figure 3-1. We then need to find and recruit the expertise needed to analyze each level of the overall system. The initial discussions among experts from the diverse fields that study these processes will be difficult. We talk different technical languages, have different mental models, ask questions differently, and get our data from very different sources. It takes time and commitment (and continuing effort) to recognize what each has to offer and how to develop collaborations to address the interaction of human populations and survival of at-risk wildlife populations. Finally, we need to synthesize the understanding from multiple models into a holistic picture of the conservation issues, threats, and options.

The extent and breadth of the list in box 3-1 is a bit frightening to reductionist scientists. We are often more comfortable when we restrict our predictions to the narrowly circumscribed areas of science in which we are experts. However, developing expanded, more realistic, and useful PVAs may not always be as difficult as it seems. In most conservation issues, only some of the systems shown in box 3-1 are important, and others can be quickly set aside. It is still a valuable process to consider briefly the broader realm, to make sure that we are not setting aside important processes. Even if important systems are identified that cannot be modeled or otherwise incorporated into a PVA, recognition of other processes, or heuristic consideration of possible effects, can be a valuable step in developing effective conservation strategies.

It is clear that the population biologists who developed and use PVAs do not have all the tools necessary for performing a fully comprehensive analysis. Fortunately, there are others who can model such things as the effect of a change in a political system on the rate of agricultural development, timber harvest, or mineral exploration. Other disciplines study the effect of access to global markets on changes in land use, as people shift from subsistence production for local use to commodity production for external markets. Some social scientists are concerned with issues such as the interactions between human demographic trends, economics, the likelihood of civil war, and mass relocation of peoples. Also, demographic and economic forecasters can project how changing national economies will change human birth and death rates, in turn affecting numbers and age structure of people and trends toward urbanization versus movements for land rights and redistribution. All of these things matter; in fact they are at the core of why we have conservation problems and why addressing those problems is important to people.

Even if we have access to a wide array of expertise, we still will not be able to develop complete, holistic models providing complete understanding of all relevant systems. Yet there is value to specifying, analyzing, and utilizing the knowledge we do have. To the extent we can expand our understanding of the human as well as natural forces affecting species viability, we will be able to more effectively conserve biodiversity.

 **Chapter 4**

Getting the Right Science and Getting the Science Right: Process Design and Facilitation in PHVA Workshops

FRANCES R. WESTLEY AND ONNIE BYERS

Collaboration is a complex social interaction that can be assisted by process design and by skilled facilitation. In this chapter we explore the ways in which process design and facilitation have contributed to the success of the Population and Habitat Viability Assessment Workshop (PHVA) workshop as a vehicle for consilience. For the purposes of this chapter, we define "design" as the intentional chaining together of interactional elements to achieve a predetermined outcome (e.g., in the case of PHVAs, the goal is to produce an integrated set of management recommendations). We define "facilitation" as the real-time intervention by individuals well versed in the nature and intention of the process design to help diverse groups of individuals to inhabit and use the design to realize their objectives.

Both process design and facilitation are skill sets that have become increasingly in demand, as complex problems require important tasks to be done outside of traditional organizational or functional groupings. People who are temporarily brought together in projects, workshops, or collaborations, without the structure of routine or hierarchy, must either self-organize (a time-consuming process that, without process skills, may result in power or resource issues dominating over task requirements) or look to those with skills to manage unstructured groups charged with solving complex problems.

Research and experience have identified generic process obstacles to successful collaboration. These include first and foremost the destructive role of power dynamics, but also the difficulties of analyzing complex problems as a group and of building consensus around and commitment to solutions to those problems. Individuals skilled in process design are aware of generic design elements (e.g., brainstorming, issue identification, timelines, paired ranking) that may be combined and recombined to secure certain group outcomes in relation to specific problems or group goals (the design). The role most analogous to a process designer's may be that of the dance choreographer. In a popular or folk form of dance such as square dancing, there are some ninety different elements that can be combined into fifteen hundred standard square dances. A skilled caller, however, can create chains of elements to match both the music, the nature of the event, and the dancers' skill. Creativity is embedded in the relation between elements. Similarly, but in a more complex way, the modern dance choreographer uses many elements to produce sequences of great originality and creativity requiring highly skilled performers. Occasionally, a sequence produces a dance number so memorable it is repeated on numerous occasions by numerous troupes. It then becomes a part of the repertoire of modern dance, and while different artistic directors may subtly change certain elements according to his or her interpretation, generally he or she works to help the dancers realize the steps and the elements according to the original design of the choreographer. In this capacity he or she acts more as a facilitator, although in a PHVA the process facilitator does not get to rehearse nor repeat the same process with the same people.

This lengthy analogy helps to identify some of the challenges around collaborations as unfolding social systems of a temporary nature. There exist certain established "whole system" process designs, including Future Search (Janoff and Weisbord 2000), Scenario Planning (Schwartz 1996), and Appreciative Inquiry (Cooperrider 1990), that have proved robust enough to be considered generic. By generic we mean that the particular pattern has taken on the nature of an artifact, a "frozen manifestation of a pattern" (Krippendorff 1998, 174). By robust we mean that the design itself is flexible and powerful enough to "survive all the successive transformations into manifestations that ultimately support themselves," irrespective of the presence of the original designer (Krippendorff 1998, 175). The PHVA workshop was such a design, created with the specific goals of bringing together scientists and practitioners, along with their information, in order to formulate scientifically based practical recommendations for saving species. Originally designed by Ulysses Seal, chairman of the Conservation Breeding Specialist Group (CBSG), the design has been used in multiple contexts and countries and has proved both generic and robust.

Original PHVA Workshop Design

In the late 1980s, when Seal designed the PHVA workshop, his intention was to develop a vehicle for building collaborations among all stakeholders around endangered species. While his World Conservation Union (IUCN) mandate at that time was to create a bridge between the captive breeding (ex situ) and wildlife management (in situ) communities, Seal understood that the "stakeholder" group was considerably broader, including government agencies, scientists, conservation groups, and involved amateurs. His original aim was to design workshops that would integrate scientific and practitioner knowledge about the species, increase the likelihood of collaboration between groups (some of whom had been actively conflictual), and align research and action agendas toward a unified effort at species preservation.

Some of Seal's ideas about the design of PHVAs came to him from an earlier request. In 1992, Seal was asked by the state of Wyoming Game and Fish and by the regional U.S. Fish and Wildlife Service (USFWS) office to chair an advisory group on actions to assist the black-footed ferret *(Mustela nigripes),* one of the animals listed as endangered by the U.S. government. The official Recovery Team for any endangered species includes representatives from the USFWS, the relevant state departments of natural resources, and the U.S. Forest Service. For the black-footed ferret, it included wildlife managers and scientists as well as government officials. In attempting to design a recovery strategy, the black-footed ferret team encountered conservation groups and dedicated amateurs with their own, strongly held values. Despite tremendous strife and conflict, the team, with help from Seal, executed a dramatic rescue of the few remaining black-footed ferrets in the wild and instigated a successful captive breeding program, which allowed for the animal's subsequent reintroduction.

Seal's experience with this program emphasized for him the critical importance of collaboration between the captive breeding and wildlife managers working in the field of endangered species preservation. He also recognized that understanding about population genetics and dynamics was relatively limited among nonscientists in the wildlife management community. Often, by the time an animal was listed as endangered, the numbers had fallen so low that inbreeding of existing animals was inevitable and the species' chances of surviving was very low. He therefore felt the need to initiate a process that would motivate action, but that would involve all parties in recognizing the crisis, in pooling existing information, in identifying information needs, and in aligning research and action orientations. In response, he designed a series of multiparty workshops to bring stakeholders to the table, which he called Population and Habitat Viability Assessment (PHVA) workshops.

By Seal's original design, each workshop had five distinct phases:

- A lengthy preplanning stage that might take months or years of building relationships, negotiating venues, and preparing a briefing book by CBSG.
- A formal opening lasting two to four hours, involving high public officials and addresses on key scientific aspects of the conservation of the species, as well as introducing CBSG processes and products.
- Small working group sessions lasting two to three days, where participants focus on the analysis of specific problems and the formulation of conservation recommendations. These working group meetings are interspersed with plenary report backs. During this phase the VORTEX program is used to evaluate extinction scenarios through computer simulation.
- A closing plenary, when recommendations are presented for consensus acceptance.
- A postworkshop period during which a report is written by CBSG, circulated to key participants for approval, and then sent to all participants. (Copies of all PHVA reports are available online at *www.cbsg.org*.)

CBSG's role in these workshops, from the beginning, was that of catalyst and facilitator. Generally, contracting or sponsoring agencies would learn of CBSG by word of mouth, an encounter with an interested party, or the success of a previous workshop. As a result, sometimes after months of informal discussions, a sponsor (who covers local costs and costs of distributing and copying briefing books and reports, sometimes in cooperation with other interested sponsors) would emerge. By preference, this would be the government agency directly responsible for the management of the species. CBSG would work with the sponsor to draw up a list of participants based on the availability of expertise, pertinent stakeholder groups, and political and jurisdictional sensitivities.

CBSG came to expect that resistance would surface early in this stage of the process, and would come from groups who feared that CBSG's relationship with the captive breeding and zoo community would result in a hidden "captive breeding" agenda. In the early days of the initiative, it was often Seal's reputation with the scientists in the larger conservation biology world that had an impact on reconciling opponents. It required sensitivity and relationship to bring all stakeholders to the table and good process design and facilitation to keep them there.

To understand the complexity of operating at a global level and across multiple national, organizational, cultural, and disciplinary barriers, an extended example of a PHVA workshop is warranted here. Each workshop is unique, but certain patterns of challenges and responses revealed by this case characterize all PHVA planning and delivery.

The Thailand Gibbon PHVA Workshop

In 1992, the Royal Thai Forest Department requested that CBSG prepare and conduct a PHVA workshop to resolve the growing crisis of too many captive gibbons (*Hylobates, H. pileatis*) in Thailand, the lack of a structured conservation plan for the species, and the desire to have an integrated national conservation program for gibbons in place. All three of the gibbon species in Thailand were considered threatened. The problem was further exacerbated by the continuing influx of captive gibbons into Thai zoos. Some Thais liked to keep baby gibbons as pets, abandoning them when they grew to adult size and became vicious. There existed, therefore, an extensive black market in these animals and a continual decline in the wild.

The time was ripe for a PHVA for many reasons. In 1991, Thailand was cited for trafficking in wild animals by CITES (Convention on International Trade in Endangered Species), the international body responsible for monitoring endangered species trade worldwide. This meant that the country lost all of its U.S. Agency for International Development (USAID) money, and an embargo was created, preventing foreign trade (U.S. foreign policy was tied to cooperation with CITES). This caused a great loss of face in Thailand, with the result that a director of the Zoological Parks Organization was appointed. The new minister was a wealthy, charismatic individual, and something of a national hero, having figured predominately in the organizing of key social programs in Thailand.

CBSG had been previously involved in helping the Thai Zoological Parks Organization do a study on the Thai Zoo, which was well received. In the process of preparing this report, CBSG became increasingly knowledgeable about the Thai political situation, as well as about the plight of the gibbons. Gibbons in Thailand are seen as a flagship species, partially because they are vulnerable and make cute pets. CBSG was able to secure the necessary international expertise in the form of veterinarians, primate specialists, and biologists. In addition, Seal had recently started working with the Social Policy Group of the IUCN and intended to bring in a human demographer to interview locals in villages bordering the gibbon habitat to try to understand the villages' economic and social structures and their potential impact on the gibbon population (see chapter 14 for further description of this data gathering process).

Other important stakeholders were nongovernmental organizations (NGOs) such as the Thai branch of the World Wildlife Fund and the Primate Protection League in Thailand (headed by two expatriate Americans women nicknamed the "witches of Bangkok," partly because of their extraordinary effectiveness and partly for their unwavering zeal). Both of these NGOs had excellent relationships with the press worldwide. It was due to their effort that the story of the illegal trading broke first in a Toronto newspaper; it was then quickly picked up by the wire services. This was embarrassing for the Kingdom of Thailand, as it potentially threatened both national image and tourism in addition to aid arrangements with the United States.

Another important group to consider was the media. The media, at least the printed media, was "fairly free" according to one informant and had tremendous power. One of the larger newspapers, the *National,* would later send a representative to the PHVA. CBSG was somewhat ambivalent about this. Because of the media's emphasis on simplicity and sensationalism, the complex social and political processes that formed the basis of a PHVA had, in the past, been poorly represented in the coverage. However, because the media in Thailand had played an important role in exposing the illegal trading of primates, they were seen as a necessary participant.

Scientists working on the gibbon, particularly those in Thailand, were a further critical presence. The leading gibbon specialist in Thailand was an American primatologist who had originally come to Thailand as part of the army's medical research corps. He had developed his expertise on gibbons in Thailand, setting up observation points in the national parks, the largest of which was in Khao Yai National Park. It was said that he knew every gibbon in the forest by sight.

As with many scientists who attend CBSG meetings, some of the field notes, which this scientist had been collecting over years, had not been put into order, to say nothing of published. It was difficult to get him to come to the PHVA, much less to release his unpublished data, his "crown jewels" as one participant called it. He had both survey and population data on the Thai gibbon, without which the PHVA could not meaningfully progress. However, he was very skeptical about CBSG's ability to gather

NGOs, zoo people, the press, and government forestry representatives in any meaningful collaboration. He felt that the output of the PHVA would only have a meaningful scientific contribution if it resulted in reintroduction plans and if these were made public. Nevertheless, his respect for CBSG was great; he agreed to produce a three-page data sheet on wild gibbon populations for use in the workshop.

The last—and in some ways the most critical—group was the Royal Thai Forest Department. There had long been animosity between the Forest Department and the Zoological Parks Organization in Thailand. The zoos had been accused of dumping animals in the forests; the Forest Department had been accused of poaching its own animals. The Forest Department's director was reputed to be opposed to processes in which the zoos were involved. However, the Forest Department ultimately had control of the habitat and the remaining populations of wild gibbons and so needed to be involved. A counter pressure surfaced: the king and queen of Thailand had come out as conservationists when uncontrolled logging in the forests had resulted in mudslides wiping out villages in the rainy seasons. The king and queen exerted pressure on the Service to at least appear to be cooperating with conservation initiatives, and Forestry's director agreed to participate in the PHVA.

The big issue remained money for the workshop. The Asian Foundation supplied nine thousand U.S. dollars, enough to cover local arrangements. However, a week before the workshop was to have begun, when the CBSG associate responsible for the workshop arrived in Thailand, no one had paid for anything yet, and no one had seen any money. By this time the representatives of the international zoo and scientific communities had started to arrive: as many as a hundred individuals. Despite prior commitments on the part of various Thai ministries, no local arrangements had been made. The workshop was to take place in Khao Yai National Park, but the hostel housing in the park, which had not been used in years, needed to be dusted out. No registration forms, bags, or badges had been prepared. The CBSG staff had to struggle, on arrival, with technical details (provision for portable computers, tables, access to electrical current) and simple things like a participants' list, which had not been prepared. Tempers flared. When Seal and other CBSG delegates arrived several days later, nothing was yet certain, but the group decided "on faith" to head up to Khao Yai regardless.

Representatives of these diverse groups arrived by bus and by car to the remote Khao Yai National Park. The park is in central Thailand and represents a large tract of rain forest with numerous wild species, including gibbons, ranging freely. Participants were housed in dormitory-like wooden cabins built on stilts, with a central outdoor meeting area surrounded by bedrooms containing sleeping platforms sleeping six to eight. Some fifty-two Thai from the Royal Thai Forest Department, universities, NGOs, and the Zoological Parks Organization participated, as did eleven American and British expatriates living and working in Thailand and fifteen foreign nationals from Australia, the United States, Great Britain, and Canada.

It was the rainy season in Thailand. After an elaborate opening ceremony, which was held in a covered area near park headquarters, the remainder of the workshop was held in a large jungle clearing, surrounded by small, open-sided, roofed meeting areas large enough for a table and chairs. One plenary presenter's feet were bleeding from the leeches he had removed minutes before the opening ceremony.

Representatives from the above groups attended the opening ceremony, bringing with them legendary animosities. The press, generally very critical of the government's conservation record, was suspected by everyone. The Forest Department and the Zoological Parks organization were barely on speaking terms. Local scientists were suspicious of the NGOs, feeling that particularly the Primate Protection people were antiresearch. Forestry was also viewed as antiscientist. CBSG was widely regarded by wildlife organizations as being too associated with zoos (until 1994 CBSG stood for Captive Breeding Specialist Group) and by other IUCN specialists groups as becoming too involved with wild populations. Seal had a reputation of moving fast and stepping on some toes as he "blew through." With a process less powerful than the PHVA it seems unlikely that any consensus could have been reached.

For the next three days, this group met from early morning until late evening. They woke early, often roused by the eerie calls of the gibbons, for whose sake they had gathered. They ate, worked, and slept together, taking refuge repeatedly from heavy rainstorms. They dealt with language barriers and repeated technological difficulties as power failed, microphones malfunctioned, and portable printers jammed. At moments of break or boredom, participants staged mock battles between the giant ticks and leeches that were plentiful on the forest floor.

As is customary, after presentations by experts, the VORTEX modeling process was explained and introduced (see chapter 3). After considerable discussion concerning input measures into the model, a modeling group was formed that, happily, included many of the Thais working for the Forestry Department. Seal worked with this group. At the same time ten self-selected working groups were formed around issues critical to the management of the gibbon population in Thailand. These included a habitat group, a human demography group, a genetics group, a captive management group, a recovery group, a gibbon disease group, a reintroduction group, and a rehabilitation group. The groups each chose two facilitators, one Thai and one foreign national. The process flowed from plenary sessions to small-group work and back to plenary.

To everyone's surprise, the modeling exercise suggested that the Thai gibbons were not as endangered as previously thought. Nevertheless, the workshop participants felt that an action plan to protect existing wild populations, their habitat, and those gibbons already in captivity was necessary, as was a plan for gibbon reintroduction. The reports and recommendations of the working groups were collated into a CBSG report and circulated to all participants.

CBSG considered this PHVA to be of average success. Thai, expatriates, and foreign nationals worked well together over the five days. Scientists and practitioners shared information. It was felt that important dialogues had resulted. Nonetheless, CBSG felt that the Thai Zoo representatives had not participated in numbers as great as would have been desirable, and all felt that the reticent Thais had hung back more than what was ideal, with some resulting overdominance of foreign nationals.

While this PHVA included a greater diversity of stakeholders than most PHVAs run in the early 90s, its dynamics were not unusual. Building relationships among multiple groups requires either a program officer or associate who is very familiar with the country involved. Without in-country contacts, important stakeholders may not even be approached, much less persuaded to come to the workshop. Despite huge efforts at planning and organization, the CBSG finds that surprise is generally the rule and improvisation to local conditions a necessity. Scientists attend the workshop due to CBSG's reputation and their personal commitment to saving the species in question but may have to be persuaded to share critical but unpublished data. It is a challenge to balance and integrate the multiple viewpoints at the workshop and to bring the group to a point where they have formulated and accepted a set of recommendations. Much work after the workshop is involved in producing a sound report that is accurate both scientifically and represents the views and recommendations of workshop participants. Remarkably, despite the wide variance in physical location, participants, and species, the PHVA has turned out to be highly robust in building consensus among disparate groups, even with the suspicions arising from multiple value orientations that representatives bring to the table. In the rest of this chapter, we will explore in greater detail the elements of process design that contribute to this robustness.

Designing for Successful Collaboration

The design of interorganizational and organizational workshop processes cannot depend on the structure of roles or the authority of position, as problem-solving processes within single organizations often do. It must depend on what Margaret Wheatley (1994) has termed the "self-organizing power of information." If the right people are in the right room with the right information and the right intentions, progress toward right decisions and actions will be made (figure 4-1). Process designs of the PHVA kind help to enhance, not inhibit, these self-organizing properties. But in general no design will succeed unless the right people and right information are in the room; the output is generally only as good as the "ingredients"—people and information.

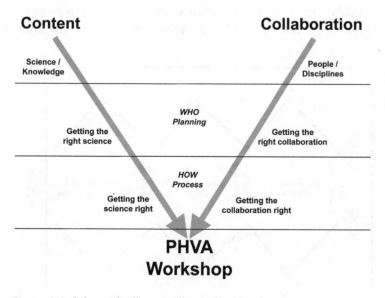

Figure 4-1. Schematic diagram illustrating the elements of successful PHVA workshop design.

A basic principle of the PHVA process and other generic designs is to provide open space and encouragement for divergent expression in the interests of creativity and then to provide the necessary tools for convergence of views in the interests of action. The thoughts and ideas of the individuals involved are the raw material of whole system processes. It is from these thoughts, ideas, and data that good analysis can be done and from which action recommendations can be built. The ideal PHVA workshop allows for approximately three cycles of divergence and convergence (figure 4-2): issue generation (divergence), working group topics (convergence); topic analysis (divergence), scenarios for sensitivity testing and modeling (convergence); brainstorming for action strategies (divergence), recommendation selection (convergence). The wider the variety of stakeholders invited to such a workshop, the greater the possible divergence and the more challenging the task of convergence. Therefore, in attempting during the Network experiment to increase the number and variety of stakeholders at the PHVA workshops documented in this book, as well as the kind of data brought to the workshop, the challenge of successfully managing both divergence and convergence increased. We will look at the new challenges of each of the cycles in turn.

Cycle 1—Encouraging Divergence: Issue Identification

Individuals come to a PHVA workshop with deep concerns, experience, and expertise. It is the job of the workshop design and of the workshop

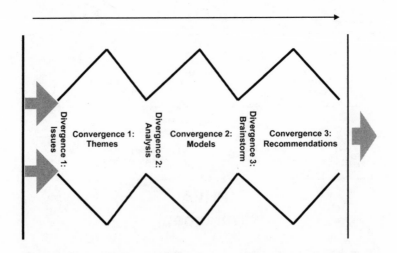

Figure 4-2. Process flow of small-group work in a PHVA workshop.

facilitator to ensure that all input is heard by everyone and given due place in the workshop process. In a number of the PHVA workshops described in this book, the workshop began with a process of listing issues as statements on flip-chart sheets. This is a time-consuming process that results in a huge amount of data being generated. Such processes allow for participants to supply the topics of the small groups and ensure that the wide variety of views present are surfaced and recorded, but the statements need to be winnowed and clustered into themes for further work (see the following discussion, Cycle 1—Facilitating Convergence: Working Groups Themes).

In situations in which there is an obvious and significant disparity in the distribution of power (professionally, academically, socially) among workshop participants, conducting issue-generation exercises in stakeholder groups can be an effective alternative. In both Papua New Guinea (PNG; chapter 9) and Ecuador (chapter 16), two cases in which Indigenous people played a major role and there was well-known animosity toward the government representatives present, this technique was used successfully. Rather than participants articulating their issues of concern in the large, diverse group setting, they were divided into stakeholder groups (i.e., government representatives, Indigenous, captive animal managers, scientists) where they were able to communicate more freely and in their own languages. This can facilitate a more open dialogue than that which might occur in a mixed-group setting, but can also risk accentuating diverse positions. It may not be advisable to use this technique in cases where there are long-standing, entrenched differences of opinions and where hostilities are high. In such instances, it may be preferable to use an exercise to identify areas of common ground early on in the process.

Language is also a challenge at the divergence stage, as lack of comfort with language will often inhibit stakeholder groups from expressing their individual perspectives. CBSG encourages PHVAs to be conducted in the language of the country in which they are held, but because participants often come from several different countries, the "official" language is more often than not English. The issue of language barriers needs to be taken into account when designing any PHVA workshop process. In only two cases, the mountain gorilla and Peary and Arctic Islands caribou, was simultaneous translation available during plenary sessions (see chapters 6 and 8, respectively). This was extremely valuable not only for the non–English speakers but for those participants who knew English but were more comfortable working in their native languages. However, this type of translation service can only be provided during plenary sessions and its value even then is limited by the accuracy of the translators. If the translators are unfamiliar with the technical, scientific vocabulary of the workshop content, the service can be a detriment rather than a benefit to the process. In addition, this service is often not an option due to financial or technical constraints.

Language problems are complicated further by the attempt to include Indigenous peoples, as is discussed in chapter 15. This was a significant issue in the PNG tree kangaroo case. We knew that the official language of PNG is English so we were not prepared for the fact that the local participants would be unable to speak or read it and had made no arrangement for translation. In fact, each of the villages represented at the workshop had its own language and the common language of the local landowners was Pidgin. Fortunately, there were three participants (two from PNG and one from the United States) who were fluent in both English and Pidgin and were willing and able to translate. This was no easy task, since Pidgin consists of a total of only thirteen hundred words, and most concepts had to be translated in the form of long, elaborate stories or metaphors (translation of the concept of stochastic events was particularly challenging!). In addition, this type of translation takes an enormous amount of time that must be built into the workshop agenda. The key to the success in this case, and in these situations in general, was and is one of attitude. A sincere desire to communicate and collaborate, along with a generous, energetic, and capable translator, makes bridging the language gap possible. Such bridging allows for the full expression of divergent views.

Related to this, but somewhat different, is the possibility of building bridges between Indigenous knowledge and Western science by the kind of preworkshop activities done in PNG (see chapters 9 and 15). Participatory Rural Appraisal (PRA), used in PNG, involves a set of techniques for charting and representing Indigenous knowledge in ways that allow for congruence with Western data-representation techniques. Such techniques also help local people to prepare to fully voice their concerns and issues once the workshop starts.

Cycle 1—Facilitating Convergence: Working Group Themes

As noted above, a successful initial surfacing of issues in a large group of stakeholders is key, in order to bring out the information that will allow analysis to proceed. However, proliferation of issues and information raises the first challenge for *convergence*. Productive problem solving works best in groups of six to eight people and expertise is best used on focused problem analysis. How to select those themes and problems for working group focus is, however, a challenge. The greater the diversity and number of issues raised, the more challenging it is to integrate the material into the five or six umbrella topics that can form the bases of working groups.

One method used at this stage is for the facilitator to theme the issues. This is efficient, but often results in some dissatisfaction from participants, who may feel the facilitator is trying to control the process or that the themes do not perfectly capture all the ideas put forward. This was the case in the mountain gorilla workshop (chapter 6). Consensus was finally reached but alternative approaches such as Mind Mapping, which allows participants to group issues as they surface (Janoff and Weisbord 2000) or using teams of participants to develop workshop themes, can maintain a sense of ownership more efficiently in large, diverse groups.

The selection of working group topics is a delicate balance of facilitator experience and participant concerns. Some topics are generic to all PHVAs (such as habitat and disease) and should be considered in all workshops to ensure a balanced outcome and adequate analysis of scientific data. Others may be particular to different groups and are reflective of "burning issues." Still others, which tend to crop up when participants are asked for topics, tend to be "red herrings" in terms of guiding productive working group sessions. The challenge is to balance the importance of participants' input and facilitators' experience derived from past workshops. However it is done, participants need to converge so as to move into the smaller working groups and begin analysis—the next phase of divergence.

Cycle 2—Encouraging Divergence: Analysis

The need to generate information to input into VORTEX can serve to focus small-group discussions in the problem-analysis stage and can keep them from jumping ahead to recommendations and actions. Analysis means to separate things into distinct parts or categories. At a workshop, this next phase of divergence is a chance to get out the details of participants' data about the species, the habitat, or the impact of human activity. It is the time to separate facts from assumptions and to surface competing data sets. This information or data can then be categorized, organized, and interpreted.

This is a time of maximum divergence that stakeholders often experience as a "groan zone" (Kaner 1996). Facts and data accumulate and interpretation lags behind. Participants, particularly those most oriented to action, can become increasingly confused and frustrated. Process design that facilitates this phase may provide specific tools, such as matrices or causal flow diagrams, to help participants order their information as it emerges. During this period, the VORTEX model may also serve as a device to surface information in a more structured way.

Without this kind of assistance, participants may seek early closure on the divergence phase, which short-circuits the quality of the analysis considerably, cutting off key information when it is most needed. VORTEX serves this stage particularly well in workshops such as the grizzly bear PHVA, which are data rich and in which the participants are comfortable with the concept of simulation modeling (see chapter 10). However, this use of modeling as a means to stimulate and integrate the small-group work produces a constant tension between the need for scientific accuracy and the need for social facilitation to drive the workshop process forward. At times, imperfect data entry or modeling with incomplete data is important in terms of the dialogue and interaction it stimulates for the working groups. At the same time, the most commonly heard complaint from biologists is that they don't have enough data to construct a model.

Cycle 2—Facilitating Convergence: Scenarios and Models

Once the full range of analysis has been surfaced it is important to interpret the data in ways that are meaningful to workshop participants. This is largely an exercise in synthesis or pattern recognition. As noted above, the VORTEX simulation itself can play a powerful role during the second divergence/convergence cycle, as it processes the data input to produce extinction scenarios and allows for sensitivity testing of various management options (see chapter 3 for a more detailed discussion of VORTEX).

Encouraging movement between working groups in the previous stage (cycle 1) also speeds convergence at this stage (cycle 2) because the cross-pollination helps meaning emerge, provides an opportunity for clarification of terminology, gives assurance that issues of concern are being addressed, provides all groups access to expert participants, and encourages ownership of workshop results by the entire group. Promoting integration of working groups also keeps a diverse group of stakeholders engaged. The use of plenary reporting sessions allows for sharing of working group results; but, in part due to the intense demands made on each group during the workshop, often the individual participants focus solely on the progress and product of their own working groups and pay little attention to deliberations of other groups.

In addition to the use of plenary sessions and modeling to integrate small-group work, synthesis groups can be convened to meld the work of the various groups into a cohesive whole. This has the downside of requiring individuals to leave their own working groups to do the synthesis work and then potentially slowing down their working groups when they rejoin. It may be worth the effort however, because the result is a richer product and increased satisfaction and support by workshop participants as a whole. The muriqui workshop may have benefited from such a synthesis group. Although various working groups in this PHVA discussed issues such as distribution and regulation across land ownership and resulting forest-cover change, which could be used as VORTEX input data, there was no clear mechanism available to facilitate the transfer of this information to the model (see chapter 7).

Cycle 3—Encouraging Divergence: Brainstorming Action Strategies

Once the working groups have completed the problem-analysis phase, participants are encouraged to open their minds to new ideas and to identify creative alternative solutions. Brainstorming techniques are used to get all potential solutions on the table. Detailed steps are required in this element of the process in order to actively control the innate desire of participants to rush to a solution before all alternatives have been considered. It is important at this stage, as well, to build ownership as the group moves toward the selection of recommendations on which they will take action.

Interestingly enough, this last cycle is the beginning of the exit phase, and it is here where issues of power will resurface. In the move toward recommendations, if those recommendations are to be meaningful, individuals and organizations need to take responsibility for taking action. If the stakeholders have neither the power nor the resources to take personal responsibility for action, the workshop may not yield a satisfactory follow-up. In the muriqui and Peary caribou workshops (chapters 7 and 8), scientists present were able to recommend research to address specific problems with some certitude that the research would be carried out. Funders present at the PNG tree kangaroo workshop (chapter 9), as well as the muriqui workshop, were able to pledge funds for establishing protected areas. Government officials at the grizzly bear workshop (chapter

10) were able to support recommendations for closing ski hills in the summer to visitors to allow for wildlife corridors. It was during the brainstorming stage that participants in these workshops were able to both reap the benefits and see the drawbacks of broad stakeholder integration.

Cycle 3—Facilitating Convergence: Selection of Final Recommendations

The workshop process allows differing perspectives to surface, but then requires that weak alternatives be weeded out, while focusing on and choosing among the strong. Prioritization techniques, such as a paired ranking, are used in this phase. In addition, participants are asked to apply SMART criteria (Specific, Measurable, Achievable, Results-oriented, and Time-fixed) to the recommendations to increase specificity and the probability of implementation. To this point in a workshop there has been movement from plenary and back into working groups, allowing for both divergence and convergence at the larger group level. At the stage of recommendations, the plenary task is particularly important, as it is at this stage that various stakeholder groups should reach consensus. Without this, the participants exit with little commitment to implementing the recommendations, and some of the collective power of the process can be lost. A good example may be found in the PNG case (chapter 9), in which the landowners were challenged by the facilitator to confirm their acceptance, a moment which was key in the successful aftermath of the case. One of the challenges of this final or exit stage of the process is ensuring that the buy-in or agreement is more than superficial and so it is reasonable to take the time to explore possible hesitations and to craft recommendations that are specific enough to be implemented but general enough to be supported.

Conclusions

PHVAs have proved robust across multiple contexts and cultures, largely because of the nature of the process design. The long period of preparation insures that the right people and the right information are in the room. The opening plenaries acknowledge the officials and high-status individuals connected to the process, while the relatively rapid move to working groups means that control is turned over quickly to those responsible for conservation action and analysis. The design of the working group process, with its movement between divergence and convergence, takes maximum advantage of the variety of views and expertise present in the room, while keeping participants moving through a process of analysis, synthesis, and recommendations. The demand for consensus acceptance of decisions in the final stage ensures that participants leave the workshop with some commitment to carrying the resolutions into the future.

While variations of this design are demanded in different settings and with different groups (the emphasis shifting, for example, between analysis and solutions), the overall flow of divergence and convergence remains true in all workshops. As we explore in the cases described in part II, the Network project did not introduce a new design but rather went about tweaking one already in progress. The PHVA process is well tested and reliable. This made it relatively easy to make minor process adjustments without risking the quality of the product. However, as part III of this book illustrates, the Network experiment did impact the process and required us to learn over the course of the six workshops.

In the initial Network experiments, we made an effort to present to participants the goals of the Network and the philosophy behind what we were testing. The Network's agenda was seen as separate from (an add-on to) the PHVA process and we felt compelled to present it as such. A slide presentation was made at the mountain gorilla workshop describing how different data are available to people from different disciplines and that they need to talk to one another, translating their information into language the other can use, before the information can be useful in species risk assessment (see the figure 3-2 flowchart in chapter 3). The presentation was clear and well executed but did not appear to have any impact on the deliberations of the workshop participants (see chapter 6).

In fact, the presentation seemed to have an effect opposite what we had intended. We wanted to make our purpose transparent and demonstrate the value of the enhanced process we were developing. It seemed instead to evoke suspicion and the sense that we were imposing this Network agenda on the participants (which in fact we were). We felt that if the Network were represented as participants in the process (not simply as observers or facilitators) then our agenda would be dealt with legitimately. However, since it was quite clear that the Network and CBSG were one and the same, our working group participation was never accepted and the group chose not to make our agenda item one of their priorities. The more we voiced our position, the stronger the resistance to it. Part of the issue was our framing of the enhanced process as an "experiment," some participants were unclear as to how our agenda dovetailed with their desire to save the species and felt that they, the participants, were under a microscope.

However, as chapters 7–11 explore, we did learn from this experience and were able to find more effective ways to integrate our project into the workshops. By the time of the PNG workshop (chapter 9), the "bubble diagram" tool (chapter 18) had been developed and was used to help identify what information and individuals ought to be in attendance at the workshop. This tool was brilliantly successful at the grizzly bear PHVA in illustrating the Network's objective of integrating various domains, data, and disciplines (chapter 10). The result was an ever-present awareness among participants of the need to translate information into VORTEX input data where possible. This seemed to strengthen the capacity of groups to diverge in analysis and then converge in synthesis.

In sum, while the content of the PHVA workshop is science and its goal is to get the science right, the success of the PHVA workshop is dependent as well on getting the process right. A widely divergent group of stakeholders are at least given the opportunity to express their views and, what is perhaps equally important, to synthesize their views with those of others into conservation recommendations that can be supported and acted upon. As we hope to have shown in this chapter, crafting a process that can carry the demand for both wide divergence and gratifying synthesis required many years of thought and experimentation. This, as much as the science-based analysis, accounts for the satisfaction of participants expressed in chapter 5. Today, the PHVA process design is robust enough to allow for variation and even some experimentation, which provided the Network with an extraordinary opportunity to learn.

 **Chapter 5**

Logic Models for Building Knowledge
and Networks: Early Evaluations
of the PHVA Approach

HARRIE VREDENBURG AND FRANCES R. WESTLEY

Interorganizational collaboration represents a challenge for organizations that work in traditional, competitive contexts. Academic scientists, nongovernmental organizations (NGOs), conservation organizations, and government agencies concerned with endangered species are no foreigners to competition. Even when the ultimate goal is shared, many attempts to save species have bogged down in ideological and political disputes (Alvarez 1993).

Despite these challenges, collaborative strategies remain critical for solving conservation issues, particularly in pluralistic and democratic societies (Clark 1989; MacNeill, Winsemius, and Yakushiji 1991; Trist 1983; WCED 1987). Biodiversity strategies demand a "cross sectorial approach" (WRI 1992) involving private- and public-sector organizations; local, state or provincial, and national governments; interest groups; and "epistemic" communities in unified action (Gunderson, Holling, and Light 1995; Wemmer 1990).

Social science theory on collaboration has focused on the microdynamics of process and on the contextual factors which facilitate that process. These include the presence of relevant stakeholders, clear problem identification, information generation, recognized interdependence, and a legitimate convener, as well as consensus building and power dispersal. Theorists (Gray 1989; Gray and Wood 1991; Kanter 1989; Trist 1983) identify three distinct stages in the collaborative process. For collaborations to succeed, participants must first successfully "identify the problem." Then participants must agree on "setting a direction." Finally, participants must find an appropriate "structure" for ongoing collaboration around the problem.

Each of these three phases of collaboration are further broken down. To successfully identify the problem, case analysis reveals that the stakeholders must all be identified and brought to the table, the problem clarified, information generated, willingness to collaborate stimulated, interdependence as well as legitimacy of stakeholders identified, and an authoritative convener recognized. To successfully set directions, a coincidence of values among stakeholders needs to be created and a triggering of joint information search effected. In addition, power imbalances must be prevented from disrupting emerging consensus. Lastly, in the structuring phase successful collaboration requires creating a perception of continued interdependence, securing a mandate for ongoing activities, building relationships with actors outside the collaboration, and mobilizing resources for continuity.

Throughout this unfolding process, both constructive conflict and commitment must be high; grassroots involvement is critical and organizations must learn to build consensus (Brown and Ashman 1995). Both implicit and explicit power struggles can derail or co-opt the collaborative agenda, resulting in unintended outcomes (Hardy and Phillips 1998; Selznick 1966). Multiple collaborations may be necessary for successful domain (endangered species conservation) transformation (Westley and Vredenburg 1997; Scheffer, Brock, and Westley 2000). Some or many of these problems can be overcome with successful design and facilitation (see chapter 3) but if ignored can result in the failure of collaborative process to secure desired outcomes.

Clearly, successful interorganizational collaboration is full of challenges for the practitioners. The problems are by definition complex and systemic, jurisdiction is unclear, multiple stakeholders must agree to find a solution, but they must do so in a nonhierarchical, problem-solving context unfamiliar to most managers (Kanter 1989). In addition, collaborations that bring together stakeholders from different cultures must deal with cross-cultural sensitivities and lack of experience with collaborative contexts among participants. These challenges are particularly present in the Population and Habitat Viability Assessment (PHVA) workshops that are the object of this book (see chapters 1–3 for the history and processes of these workshops). Fortunately, as we have noted elsewhere, the PHVA workshop design incorporates many of the qualities hypothesized in the literature as leading to successful collaboration. Therefore, the workshops offer ideal sites for exploring the potential of well-designed collaborations to result in desired impacts on problem domains.

This chapter summarizes early work done by Network members on the evaluation of PHVA workshops in general. It should be noted that the workshops on which this chapter is based are not those of the Network experiment but belong to an earlier project exploring the impact of the PHVA workshop form. However, these findings are pertinent to our current project, as they help to reveal the logic model behind the design of the workshop and early indicators of the success of these workshops that made them of interest to this project. We begin by reviewing the logic model behind the use of PHVA workshops to achieve conservation goals. Selected statistical results of an ongoing international questionnaire survey study of PHVA workshop participants are then presented as measures of how well the PHVA workshops appear to be in realizing stated principles and objectives.

The Logic of PHVA Workshops

Earlier in this book, we reviewed the history, strategy and structure of the Conservation Breeding Specialist Group (CBSG; chapter 2) and the rational and design of PHVA workshops (chapters 3 and 4). While modeling methods in general, and PHVAs in particular, have come under occasional criticism as poor approximations of natural processes, little work has been done on documenting the social or management benefits of the PHVA workshop. CBSG stresses that its goals in running a PHVA are both scientific and social. On the scientific side, the PHVA workshop provides population viability assessments for each population of a species or subspecies under question. The assessment for each species undertakes an in-depth analysis of information on the life history, population dynamics, ecology, and population history of the individual populations. Information on the demography, genetics, and environmental factors pertinent to assessing the status of each population and its risk of extinction under current management scenarios and perceived threats are assembled in preparation for the PHVA.

In addition, as noted in earlier chapters, PHVA workshop exercises are designed to assist the formulation of management scenarios for the respective species and to evaluate their possible effects on reducing the risks of extinction. It is also possible through sensitivity analysis to search for factors whose manipulation may have the greatest effect on the survival and growth of the population(s). One can, in effect, rapidly explore a wide range of values for the parameters in the model(s) to gain a picture of how the species might respond to changes in management. This approach may also be used to assist in evaluating the information contribution of proposed and ongoing research studies to the conservation management of the species.

An important feature of the workshops is the elicitation of information from the experts that is not yet readily available in published form and that may be of decisive importance in understanding the status and population dynamics of the species in the wild. This information provides the basis for constructing a simulation model of each population that evaluates the deterministic and stochastic effects and interactions of genetic, demographic, environmental, and catastrophic factors on the species population dynamics and extinction risks. The process of formulating information to put into models requires that assumptions, and the data available to support those assumptions, be made explicit. This process tends to lead to consensus building on the biology of the species, as currently known. The process also facilitates the creation of a basic simulation model for the species that can serve as a basis for continuing discussion of management alternatives and adaptive management of the species or the population as new information is obtained. The process, hence, mediates against the tendency for turf wars between the different groups that control different kinds of vital information.

Meanwhile, on the social side, CBSG recognizes that while conservation action is best built on a synthesis of available biological information, it is dependent on actions of humans living within the range of the threatened species as well as on established international interests. Therefore, CBSG offers at least four different social goals for the PHVA. First, the workshops are organized to stimulate broad-based involvement, encouraging the participation of local wildlife managers, NGOs, and concerned citizens as well as of governments and officials. Secondly, CBSG works for power equalization, designing workshops to allow all participants, from zookeepers and field workers to international scientists, to contribute on an equal footing. Thirdly, in an arena characterized by strong emotions and diversity of opinion and background, the workshops are designed to encourage consensus building and trust. Fourthly, the goal of the PHVA is to transfer skills to local participants and to encourage local solutions. Workshop reports and outcomes are the property of locals. Therefore, significant local commitment to the workshop process is essential.

In sum, the PHVAs is a tool designed to both increase scientific knowledge and understanding concerning the status of an endangered species and to build collaborative networks from a variety of international and local groups that have a stake in a particular species. The short-term scientific goals are to surface unpublished and published scientific data, and to provide opportunities for cross-disciplinary synthesis and the identification of data gaps and needs. The social or collaborative goals are to attract broad stakeholder participation, equalize power for duration of the workshop, build understanding and trust, and build consensus as to management recommendations in order, ultimately, to strengthen the capacity of the participants' organizations for action. Ideally these goals should be reflected in the recommendations of the workshops. However, the midterm objectives are to change the ways that resources and information are mobilized, to strengthen the links between stakeholders concerned with conserving the species, and to generate new approaches, programs and projects, and policies designed to achieve that end. The logic model behind the workshop design and the evaluation tools is summarized in figure 5-1.

Methodology

In 1993, at the behest of CBSG, two of the Network team designed a series of questionnaires to probe whether the design of PHVA workshops was delivering the desired outcome for both participants and designers. A questionnaire instrument was developed from theory and qualitative exploratory research in the domain in several different countries. The

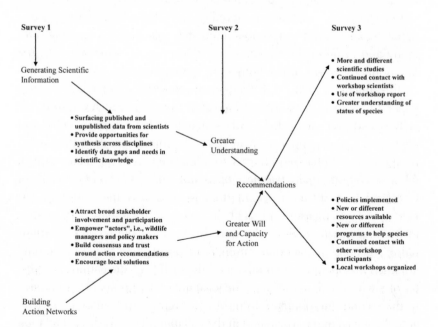

Figure 5-1. PHVA logic model.

instrument used mostly Likert-type scales and some open-ended questions. The instrument was pretested with a small international sample; questionnaires were faxed out and back, and telephone interviews were conducted with the pretest respondents to diagnostically explore question interpretation and understandability in a cross-cultural, second-language milieu. Language was kept as simple as possible so that it could be understood by respondents answering the instrument in their second language. Based on the pretest and additional exploration of the linguistic issue among experts in the field, it was decided that only English and Spanish versions of the questionnaire would be developed. It was concluded that prospective respondents in most countries would have at least a working knowledge of one of these languages as a second language. As translation of a research instrument is a resource-intensive procedure requiring translations to a language by a language's native speaker and back translation into English by an English native speaker for consistency monitoring, it was decided that only translation into these two major world languages was warranted. As most respondents from countries such as Thailand or Poland would be answering the questionnaire in their second language, data sets from these countries were carefully perused for evidence of linguistic interpretation difficulties. The small number of questions affected in this way were eliminated from analysis for the entire database.

Questionnaires were administered by the CBSG workshop facilitator at the beginning of a workshop immediately following the opening ceremony and again at the conclusion of the three-day workshop. On the questionnaire, cover instructions were written over the names and university affiliations of the independent co-researchers and authors of this paper. Questionnaires were collected in an envelope, which was immediately sealed and forwarded to the researchers for computer entry and analysis. The independence of the research project from the CBSG organization was emphasized in order to minimize bias in the data. A third, follow-up questionnaire was developed and sent by mail on University of Calgary letterhead ten to twenty-four months after the workshop was held in order to measure effects of the collaborative initiative. Responses to the first questionnaire (789) were higher than the second questionnaire, reflecting the fact that a number of people at each workshop only attended the opening ceremonies and then left (e.g., the country's minister of environment or delegate, the local university president). Responses to the second questionnaire in most cases represented all of those who actually stayed and participated in the collaborative workshop. The total number of responses to the second questionnaire was 453. In what follows, we discuss questionnaire responses from all collaborative workshops held from the spring of 1994 to 1996. This includes respondents from Thailand, Indonesia, India, Taiwan, the Philippines, the United States, Poland, Namibia, Panama, Venezuela, Costa Rica, and Brazil. The third follow-up or mail questionnaire achieved a response rate of 35 percent for a total sample size of 290.

As many of the measures discussed in this paper deal with the third data set, we worked with this last sample. Nonresponse bias was monitored by comparing the demographics of the second questionnaire (end of workshop) data set with the demographics of the follow-up questionnaire data set. There were no statistically significant differences between these two data sets, suggesting there was no demographic nonresponse bias present in the follow-up survey data set. As well, a number of open-ended telephone interviews were carried out with participants in the U.S. workshop (both respondents and nonrespondents to the follow-up questionnaire) and no biases were found between respondents and nonrespondents (the U.S. was selected for convenience, but we have no reason to believe that there would be cultural differences leading to nonresponse biases in other countries).

Results

Figures 5-2 through 5-4 give the demographic description of our sample of workshop participants. Professionally, research scientists are the group with the highest representation at these workshops, representing 34 percent of participants. Of the remainder of the participants, no group is proportionately higher in representation than others. Participants are drawn from wildlife managers, government officials of various levels, NGO representatives, zoo scientists and managers, and specialists in captive breeding of endangered species. Although we are dealing with many different stakeholders who participate in the process, a criticism might be made that in comparison to some collaborative efforts, these stakeholders are quite homogeneous. There are, for example, no representatives of major corporations with interests in the species' habitat and there are few representatives of local community populations except as to the extent that they are represented by NGOs. It is possible that if the stakeholder group was more heterogeneous that agreement on management and workshop satisfaction might be lower, as concern about the species might be less equally distributed. This question, of course, must be asked in future studies, which are already underway by the authors.

With respect to age, we find a normal distribution of working age professionals with the largest group (almost 65 percent) being between the

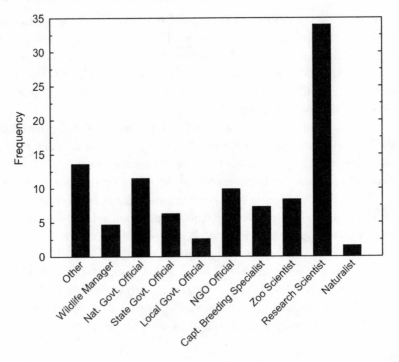

Figure 5-2. Workshop participants' professions.

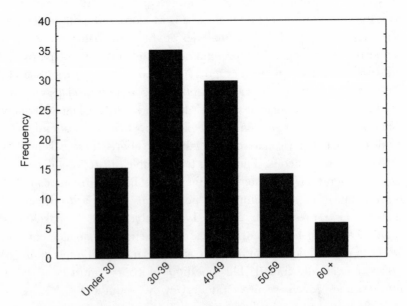

Figure 5-3. Age of workshop survey respondents.

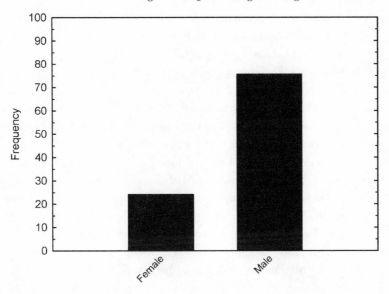

Figure 5-4. Workshop participants' genders.

ages of 30 and 49, suggesting that participants are not senior managers or scientists but rather midlevel professionals working at the tactical or implementation level in organizations rather than at the policy-making level. The gender distribution is about three to one in favor of males. This ratio should be interpreted while keeping in mind that we are dealing with participants in many non–North American, non–Western European cultures. The fact that participants are almost 25 percent female is remarkably high given that we are dealing with many Asian and Latin American cultures where women have not entered the professional world in the numbers they have in the North. We speculate, based on our observational and interview-based field work in several countries in the studies leading to the present study, that the collaborative, nonhierarchical nature of the process studied attracts a disproportionate number of female professionals, because it is a situation where merit and willingness to do the work is what counts.

Satisfaction with the Workshop: Results from the Second Questionnaire

An initial measure of a collaborative workshop's success was participant satisfaction with the workshop itself, as reported at the close of the workshop. Most of the workshops were perceived to be highly satisfactory.

Some 61 percent of respondents indicated that they were highly satisfied with the workshop, checking off ratings of either 6 or 7 on a 7-point Likert-type scale. This finding was consistent across all countries, suggesting that the CBSG workshop process appears to be successful at least at an immediate postworkshop level. An examination of the distribution of scores on this variable as well as other variables from this questionnaire suggests that there is sufficient variance to allow further multivariate analysis of workshop characteristics or attributes that seem to lead to satisfaction and other dependent variables of interest.

As one of the key desired outcomes of the PHVA workshops is agreement on how best to manage the remaining populations of the endangered species in question (figure 5-5), we ran a multiple regression analysis on this variable in order to test what the variables were that appeared to lead to such agreement. Table 5-1 shows that agreement on how best to manage populations of endangered species is a function of five variables: (1) having the key people present at the workshop; (2) having a group of participants who are equally concerned about the survival of the species in question (in other words, there are no participants who are unconcerned with the species in question and who are simply there to advance their own agenda); (3) having a workshop leader and facilitator who are well informed and have a sophisticated understanding of the problems facing the species in question; (4) the workshop participants having been able to reach agreement on what the critical issues are that will determine recovery of the species; (5) and the workshop process having not increased the participants' understanding of the views of environmental NGOs. These five variables account for 95 percent of the variance in our dependent variable (R Square) and all have statistically significant t values. The first four variables were expected, as they make theoretical sense and confirm what we found earlier in our qualitative studies (Westley and Vredenburg 1997): key stakeholders must be "at the table"; there must be a shared sense of urgency and concern about the issue; the convener/facilitator/leader of the collaboration must have credibility in the view of all stakeholders; and stakeholders must all agree on what constitutes the problem or issue. When these are in place, movement toward a solution can occur.

As the model explains 95 percent of the variance in our dependent variable, one might be suspicious that one of the independent measures (perhaps agreement on the issues) is in fact measuring the same construct as the dependent variable and thus covarying almost perfectly. An examination of the covariance matrix suggests that this does not seem to be the

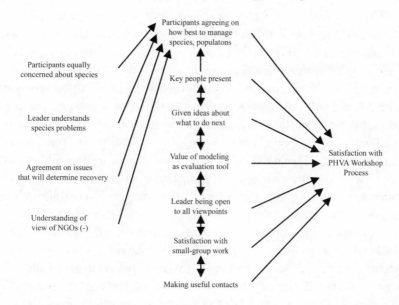

Figure 5-5. Schematic of multiple regression analyses of factors linked with agreement on how to best manage species.

Table 5-1.
Regression equation of agreement on
how to best manage species populations.

Variable	B	SE (B)	ß	t	Significance
Key people present	0.203	0.058	0.229	3.505	0.0039
Participants not equally concerned about species	-0.143	0.052	-0.228	-2.733	0.0171
Leader understands species problems	-0.340	0.056	0.374	6.016	0.0000
Agreement on issues that will determine recovery	0.742	0.095	0.563	7.795	0.0000
Better understanding of view of NGOs	-0.174	0.039	-0.289	-4.459	0.0006
Constant	-0.551	0.880		-0.627	0.5416

R Square = 0.95; F = 55.949; Significant F = 0.000

case, as none of the independent variables comes very close to covarying perfectly with the dependent variable. Theoretically, as well, agreement on the issues is conceptually different from agreement on how best to manage the remaining populations of the species.

The variable that entered the regression equation unexpectedly was the negative correlation with having a better understanding of the view of environmental NGOs. We can only speculate based on our prior field observations and interviews in several countries that this result is a consequence of environmental NGOs often participating in these processes with a strong political or lobbying position. Often these groups are unprepared to engage in the give and take required to arrive at an integrative solution, as these groups often derive their societal power from taking strong uncompromising stands in the media in order to highlight an issue. We suspect that where these groups have managed to have a strong influence on a workshop, this in fact leads to a lower level of agreement on how to best manage the remaining populations of the species in question. A strong environmental group influence may well lead to a greater polarization of participants' views. This speculation has been proposed to the scientists and facilitators from CBSG and appears to have some ecological validity. Table 5-2 shows the results of the regression analysis using satisfaction with the PHVA collaborative workshop process as the dependent variable. Although the R Square of this model is lower than the above reported one, it is respectable and the model is able to explain some 38 percent of the dependent variable variance. Again, independent variables are statistically significant (t values < .05). Satisfaction with the workshop appears to be a function of whether participants are given ideas about what to do next, whether the key people are present, whether the simulation modeling was perceived as being valuable, whether the workshop leader was open to all viewpoints, whether participants were satisfied with the small-group work, whether participants agreed on how best to manage the remaining populations of the species, and whether participants made useful contacts at the workshop. All of these results are consistent with published theory or with theory developed from the grounded theory development work carried out by the authors previously and thus far partially published (Westley and Vredenburg 1997).

Middle-Range Outcomes: The Results of the Third Survey

Figure 5-6 is the first of the figures from the longitudinal survey showing the results of our "outcome" measures. Figure 5-6 indicates that almost three quarters of respondents reported using the information documents produced by CBSG as a consequence of the workshop. When many reports are routinely reported to end up on shelves "gathering dust," this finding is quite remarkable, showing that the information put on the table

Table 5-2.
Regression equation of PHVA
collaborative workshop process satisfaction.

Variable	B	SE (B)	ß	t	Significance
Given ideas about what to do next	0.088	0.030	0.118	2.882	0.0042
Key people present	0.081	0.032	0.097	2.515	0.0123
Value of modeling as evaluation tool	0.1760	0.028	0.248	6.1762	0.0000
Workshop leader open to all viewpoints	0.099	0.039	0.101	0.541	0.0114
Satisfaction with small-group work	0.116	0.032	0.150	3.590	0.0004
Participants agree on how best					
to manage species, populations	0.170	0.043	0.207	4.903	0.0000
Made useful contacts	0.112	0.0290	0.153	3.764	0.0002
Constant	0.978	0.323		3.025	0.0026

R Square = 0.38; F = 37.997; Significant F = 0.000

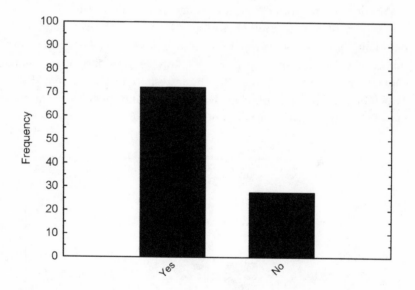

Figure 5-6. Information documents produced by CBSG used since the PHVA workshop.

and written up at a PHVA workshop is perceived to be useful and valuable to the scientists and managers dealing with the species.

Figure 5-7 shows how many other workshop participants a participant contacted since the workshop. This is a measure of what we have theoretically referred to as the network amplification effect of the collaborative workshop processes. We have asked whether the workshops and the CBSG follow-up processes serve a quasi-institutionalization function; that is, do they serve to amplify or bolster informal conservation networks that are established by the workshop process. Our results show that more than 55 percent of participants report contacting from one to three people met at workshop since the workshop was held. Another 30 percent contact from four to ten people met in the workshop. We see this as evidence that the workshop does, in fact, amplify informal conservation networks.

Table 5-3 shows the reasons reported for contact with workshop participants. More than 35 percent report contacting workshop participants in order to collaborate on species management research identified in the workshop. Another 13 percent report collaborating in order to obtain species management funding. Further, almost 30 percent report contacting participants for further species management action. Another 13 percent report organizing another workshop for a related species and more than half the participants reported contacting workshop participants to supply or ask for information. Thus, not only do most participants make contact with each other again after the workshop, they do so in order to carry out initiatives identified at the workshop and further the cause of the conservation of the species in question.

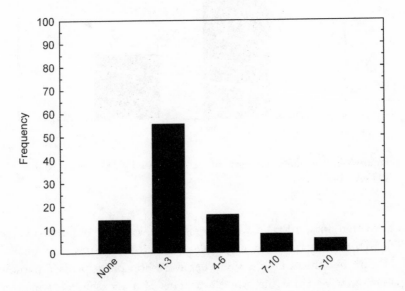

Figure 5-7. Number of contacts made among PHVA workshop participants.

Table 5-3.

Reasons for contact with other PHVA participants after workshop.

	Percentage of Respondents Contacting for This Reason
To collaborate on species management research	35.6*
To collaborate to obtain species management funding	13.1
To organize further species management action	28.3
To organize another PHVA workshop	12.6
To supply or ask for information	57.1

* Sums to >100%, as individuals checked multiple categories.

As many of the workshop participants are research scientists, a key question has been whether the workshops have any impact on the research programs carried out by these scientists. Figure 5-8 indicates that responses to this question appear to be bimodal. Almost 40 percent of respondents indicated that they agreed or strongly agreed with the statement, "As a result of this workshop, I do more or different research on the species," while almost 30 percent strongly disagreed with this statement. It may be inappropriate to interpret too much into these results, as they were drawn from the entire sample and included nonscientists as well as scientists. We are currently breaking down this analysis in order to see what the results are for the scientist subsample. It is entirely possible that the 28.4 percent reporting that they disagreed with the statement are all nonresearchers and are simply reporting that fact (or, equally possible, that the opposite is true).

One of the key objectives of these stakeholder-based collaborative workshops is to encourage greater understanding among stakeholders of the positions and perceptions of other stakeholder groups, as understanding is hypothesized to be a critical first step in affecting collaboration toward solutions. In our study, almost 60 percent reported agreeing with the statement (checking off ratings of 1 or 2 on the Likert-type scale) that they now had a better understanding of other groups and individuals (figure 5-9). We see this as evidence that the workshops seem to be succeeding in their objective of increasing understanding of the positions of other stakeholders. As alluded to above, it will be interesting to observe how well these statistics hold up when the stakeholder group becomes more heterogeneous. Again, future studies will address this question.

Figure 5-10 shows the effect of the workshop on the management of the species. We again seem to have a normal distribution anchored around

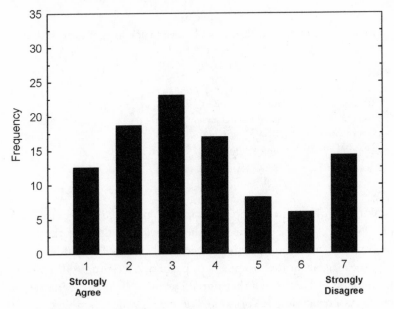

Figure 5-8. Distribution of responses to the survey question, "As a result of this workshop, I do more or different research on the species."

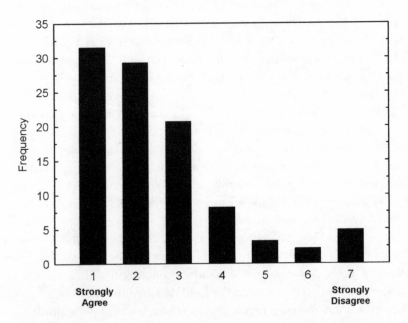

Figure 5-9. Distribution of responses to the survey question, "As a result of the PHVA workshop, I have a better understanding of other groups and individuals who attended the workshop."

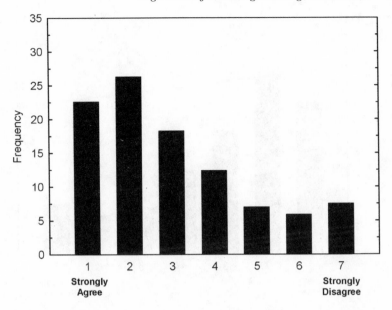

Figure 5-10. Distribution of responses to the survey question, "As a result of the PHVA workshop, I have done more/different things to help manage the species."

a midpoint of agreement with the statement and an outlier group of some 14 percent strongly disagreeing. We are at the moment doing further analysis to determine whether this finding is an artifact of having respondents who are managers and respondents who are researchers.

As a PHVA workshop is often the first collaborative stakeholder experience undertaken by participants in a country and followed by other PHVAs or PHVA-type processes, we were interested to know participants' perceptions of the process itself. Figure 5-11 shows that more than half the respondents reported that the workshop experience gave them greater confidence in the stakeholder workshop process. This finding is consistent with the finding reported above that some 13 percent had already organized more similar workshops since they attended the workshop of interest.

We were equally concerned with whether scientific and species management information was being diffused by the workshop process. Figure 5-12 shows that a large majority reported being more aware of the biology and concerns of managing species.

Finally, a major objective of collaborative stakeholder-based processes such as these workshops is to build support for action among individuals in stakeholder organizations and indirectly to build support

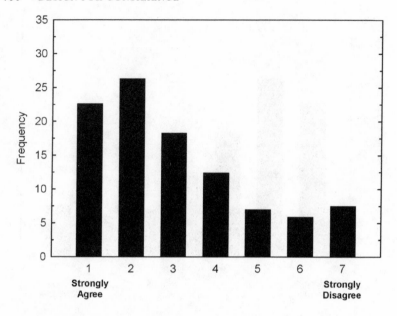

Figure 5-11. Distribution of responses to the survey question, "As a result of the PHVA workshop, I have greater confidence in species management planning workshops that involve many different organizations, professions, and viewpoints."

among stakeholder organizations themselves. Figures 5-13 and 5-14 show the results of our measures of support. At the individual level, there is a distinct skew toward being more supportive of actions to protect the species, which is less apparent at the organizational level. Nevertheless, both measures indicate that more than half the respondents reported more support at both the individual and organizational level. Analysis of responses to open-ended questions show that individuals have difficulty in moving against the inertia of their large organizations, but as more people participate in these processes this effect is mitigated.

Conclusions

This chapter completes our descriptions of CBSG and the PHVA process, grounding the discussion in early empirical findings about the success of the logic model that is built into the workshop process. Our preliminary attempts to model satisfaction with the collaborative process and then examine the longer-term effects of the collaborative process indicate that collaboration of this nature does appear to be effective in moving a domain toward action. The model identifies the primary variables that appear to lead to satisfaction with an intervention and the

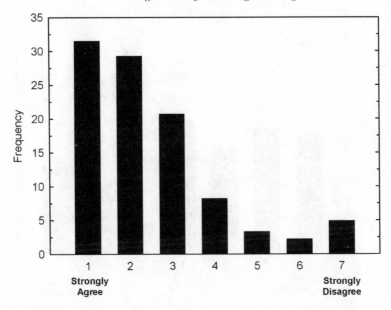

Figure 5-12. Distribution of responses to the survey question, "As a result of the PHVA workshop, I am more aware of the biology/concerns of managing species."

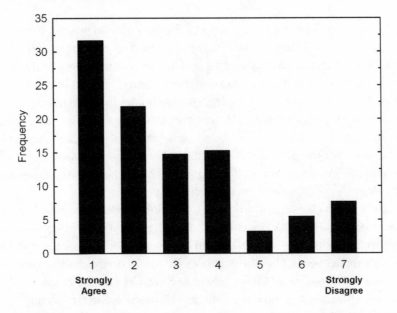

Figure 5-13. Distribution of responses to the survey question, "As a result of the PHVA workshop, I am more supportive of action to protect the species."

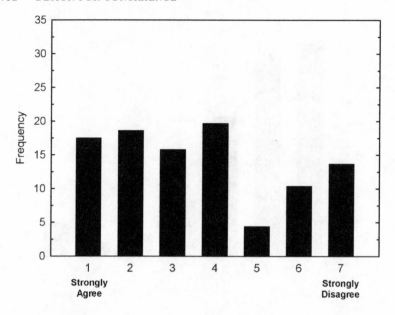

Figure 5-14. Distribution of responses to the survey question, "As a result of the PHVA workshop, my organization is more supportive of action to protect the species."

subsequent positive action outcomes of the intervention process. Primary among caveats related to this initial study is the fact that the stakeholder set involved in the process studied is relatively homogeneous. Whether these findings are sufficiently robust to hold when examined in collaborations involving a more heterogeneous stakeholder group remains to be investigated in future studies. However, results are sufficiently tantalizing to support the claims of CBSG, workshop participants, and Network members that the process design is indeed both generic and robust and worthy of further study and development. On the basis of these results, the experimental Network was launched.

In the next chapters of this book we explore in more-grounded detail six workshops that took place during the two years of the Network study. We approached these workshops with the question of how far the collaborative model holds. Can the collaborative model accommodate conflict among stakeholders and if so, how much conflict can the collaborative model withstand and how must the model described in this chapter be modified? Can the proportion of research scientists to others be shifted and still maintain the strong science focus associated with CBSG? Are there ways to render the process yet more effective?

PART THREE

The Workshops

 **Chapter 6**

Guns, Germs, and Refugees:
The Mountain Gorilla PHVA in Uganda

ONNIE BYERS, PHILIP S. MILLER,
AND FRANCES R. WESTLEY

At the first Network meeting in June 1997, we outlined the characteristics
of a workshop situation in which we could apply our expanded process
experiment. We wanted to conduct the experiments in areas of various
human demographic scenarios (high, declining, or low fertility based on
households/unit of consumption); where there is dependency of local
people on local resources (utilization/extraction); where the scale of
species distribution is clear (national versus localized); and where local
organizational contacts are available to facilitate workshop planning.
Planned in an area with high population growth rates, local dependence
on resources and localized species distribution, the Population and Habitat Viability Assessment (PHVA) for the mountain gorilla, scheduled for
December 1997 in Uganda, met these criteria but was not considered a
suitable Network case because of safety concerns related to the protracted
civil unrest in neighboring Rwanda and the Democratic Republic of the
Congo (DRC, formerly Zaire). However, when the Network met again in
November 1997, it became clear that the mountain gorilla workshop presented a valuable opportunity to test three of the primary hypotheses of
the Network: (1) increased stakeholder participation would lead to a richer
result and a greater sense of ownership of the process and the product;
(2) incorporation of human demographic information into the modeling
process would result in a more instructive picture of mountain gorilla
population viability and would lead to more useful management recommendations; and (3) a solid institutional context including government
policies, monitoring of conservation, political stability, and general well-
being of the population would be able to influence the success of conservation initiatives. Consequently, despite the late date, the group decided

to use the mountain gorilla workshop as the first Network experiment.

In collaboration with the Primate Specialist Group of the World Conservation Union (IUCN), the Conservation Breeding Specialist Group (CBSG) was invited by the director of the Uganda Wildlife Authority, the Office Rwandais de Tourisme et Parcs Nationaux, and the Institut Congolais pour la Conservation de la Nature to conduct a PHVA for the mountain gorilla in Kampala, Uganda. The recognition by the local organizers of the need to bring all available tools to bear on the problem of gorilla conservation led to their willingness to include a wide range of stakeholders and to support an attempt to incorporate a human demographic element into the workshop process.

The objectives of the workshop process were to assist local managers and policy makers to formulate priorities for a practical management program for survival and recovery of the mountain gorilla in wild habitat, develop a risk analysis and population simulation model for the mountain gorilla that can be used to guide and evaluate management and research activities, identify specific habitat areas that should be afforded strict levels of protection and management, identify and initiate useful technology transfer and training, and to identify and recruit potential collaborators from central Africa as well as the greater international community.

This chapter provides background on the mountain gorilla and the organizational context of gorilla conservation; outlines the Network's expectations and preworkshop preparation; discusses in detail the efforts of the Network during the workshop to incorporate human population information into the gorilla population modeling and conservation recommendations; and, finally, reflects back on expectations and lessons learned from this experiment.

The Context of Mountain Gorilla Conservation

Gorillas are found in two regions of the African continent: east central Africa and equatorial west Africa. One of the three gorilla subspecies, the mountain gorilla (*Gorilla gorilla beringei;* shown in figure 6-1) is restricted in its distribution to just two populations: one of about 300 individuals in the Bwindi Impenetrable National Park in southwest Uganda, and the other of about 310 animals in the Virunga volcanoes region including Mgahinga Gorilla National Park (Uganda), Parc National des Volcans (Rwanda), and Parc National des Virungas (DRC) (IUCN 1996) (see figure 6-2). Anatomical differences exist between the two populations, but there is considerable debate surrounding the proposed classification of the two populations as separate subspecies (Groves 2001; Stanford 2001). Because

Figure 6-1. Mother and infant mountain gorilla *(Gorilla gorilla beringei),* Bwindi Impenetrable National Park, Uganda. (Photo courtesy M. Robbins.)

geographic isolation has led to demographic and genetic isolation, the PHVA assessment included demographic analysis of the two populations as separate management units.

The distribution of the mountain gorilla is entirely within national parks, but despite their protected status, there are serious threats to these ecologically vital afromontane and medium-altitude forest habitats. Historically, hunting and poaching resulted in a rapid decline of the Virunga population from which it has not yet recovered. The continuing civil unrest in Rwanda and DRC—most recently demonstrated to its fullest potential in the 1994 genocide in Rwanda (Prunier 1997)—is producing thousands of refugees who are encroaching into the Parc National des Volcans and the Parc National des Virungas areas (Lanjouw, Cummings, and Miller 1996; Plumptre et al. 1997). Current rates of deforestation for firewood collection and building materials are likely to cause permanent habitat damage in the near future. Uganda's Mgahinga Gorilla National Park also has suffered from these unsustainable land-use practices. This rapid rate of habitat destruction likely will lead to a decline in the mountain gorilla population and to a long-term reduction in the viability of the subspecies as a whole. There was a recognized need for a systematic evaluation of mountain gorilla population viability and for development of a regional management plan that incorporates the needs of all relevant governmental agencies, nongovernmental agencies, and public and private stakeholders.

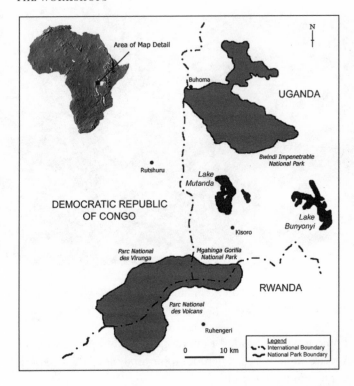

Figure 6-2. Map of east-central Africa showing the distribution of the mountain gorilla.

Since 1959, when George Schaller published his landmark studies of mountain gorilla behavior, the species has been of international interest. In 1967, Dian Fossey began her research sponsored by Louis Leakey and the National Geographic Society, keeping the plight of the mountain gorilla in popular literature through her death in 1985. Even with her death, the Karisoke research center, supported by the Digit Fund (established in memory of a gorilla in Fossey's study group killed by poachers), has remained a force in research, fund-raising, and protection of the Parc National des Virungas population of gorillas. Currently, there are two branches of the Dian Fossey Gorilla Fund: Dian Fossey Gorilla Fund–Europe based in the United Kingdom, and Dian Fossey Gorilla Fund–International based at Rutgers University, USA. Both fund the research center as well as other projects throughout the mountain gorilla's range.

Gorillas are one of the most renowned of the so-called charismatic megavertebrates—species that attract attention and human interest often because of their close relation to humans (Kellert 1985). Due to the widespread, international interest in mountain gorillas, there is no shortage of national and international nongovernmental organizations (NGOs) dedicated to gorilla conservation. The World Wildlife Fund (WWF) began support of Rwanda gorilla populations in the early 1970s with equipment donation to that country's park service. In the late 1980s, the Mountain Gorilla Project was formed in collaboration with the African Wildlife Foundation and Flora and Fauna International, and support of the Institut Zairois pour la Conservation de la Nature began to receive support at this time. By 1991, this collaboration between these three NGOs formed the International Gorilla Conservation Program (IGCP), with the goal of producing a regional management and conservation plan that included Rwanda, Uganda, and DRC.

In addition, many universities and zoos currently support research in the parks, and the Mountain Gorilla Veterinary Centre supports field studies. As a part of their commitment to conservation of the wild populations of species they exhibit, many zoological institutions also contribute to research knowledge on the gorilla as well as supporting field projects (for example, the Ohio, USA, Columbus Zoo's Partners in Conservation program works to use conservation and education as a development benefit to communities in the regions where the gorilla occurs).

Despite the collaboration of three organizations to form the most active NGO in the region, IGCP, there are multiple projects and researchers in the area competing for limited grant resources and permits, all with the same stated goal of conservation of gorillas. Differing perspectives and the instability of the governments makes it difficult for these groups to operate, and strong professional and personal commitment of individuals is necessary. Because of these and other challenges to gorilla conservation, at a planning meeting in January 1997, representatives of the major NGOs and governments met and decided to host a regional PHVA for the mountain gorilla.

Preworkshop Activities

Planning began in early 1997 for a mountain gorilla PHVA to be held in December of that same year. About a month prior to the actual workshop, the Network dedicated its efforts to developing goals for its first workshop and gathering background information to support the integration of human demographic and resource use information into VORTEX.

While the workshop was to be held in one country—Uganda—of greatest importance to the overall success of the PHVA process was the explicit regional approach taken with respect to the species' biology, habitat characteristics, and threat assessment. This approach could be successfully implemented only through continued cooperation between, and full participation by, range-country natural resource and tourism agencies and NGOs involved in the region. A university researcher worked closely with CBSG and IGCP to organize the workshop and ensure critical stakeholder and range-wide participation. The inclusion of a wide range of stakeholders was emphasized during the planning stages, although there was no special emphasis placed on ensuring social scientists' participation until the Network became involved.

Prior to the workshop, the Network members put together a package of information on the social, political, and demographic circumstances in the regions surrounding the two mountain gorilla habitats. A World Bank report (1993) provided good figures on human population size, rates of emigration and immigration, and birth and death rates in the regions that surrounded the park in Uganda. Several interesting articles on refugee data in the Virunga area gave us good insights into the habitat effects of refugee activities and the role of NGOs and management agencies in mediating that impact (Pearce 1996; UNHCR 1996; Biswas and Tortajada-Quiroz 1996). We felt that refugee impact should be considered as a catastrophic effect, noting that refugees were eradicating trees (including roots) at a significant rate. In addition, human defecation had been reported as a possible source of disease in Virunga, as increased ecotourism was in Bwindi (Pearce 1996). The challenge we identified for the workshop was to translate this valuable information (e.g., 850,000 refugees took out 410–770 tons of wood per day [Pearce 1996]) into VORTEX input data. Determining how to handle these types of data was obviously one of the primary goals of this experiment. In another attempt to build information resources at the workshop, social science and human demographers were contacted about potential participation in the workshop; the Network was unsuccessful in building stakeholder participation beyond what had already been secured by workshop organizers.

In addition, the Network developed a series of slides focused on the Network's goals that augmented the standard workshop presentation on the concepts of small-population biology and the use of modeling in Population Viability Analysis (PVA). The slide set focused on the need to center attention not only on the sheer numbers of people inhabiting a local area, but also on understanding people's behaviors concerning use of their local environment. Introductory guidelines explaining how the VORTEX population model can interface with human population demographic data were also included (see chapter 3). This presentation was designed to help workshop participants appreciate the need to understand species extinction risk in the context of quantifiable consequences of local human population growth.

The Network was keen to put our plans into action. We had identified the types of information and expertise needed, and we had national demographic data and social science information from various sources containing information on the prevalence of AIDS, local demographic trends near the parks, and a history of institutional arrangements in the parks where the gorillas lived. This appeared to be an ideal case for our first attempt to better integrate and quantify the effects of human population growth and resource use into a PHVA. We flew to Uganda with high expectations.

The Workshop Dynamics: Content and Process

Nearly eighty experts gathered at the Sheraton Kampala Hotel on December 8, 1997 for this workshop. While some individuals were unable to attend the entire five-day exercise, the majority of people were intensively engaged in the focused discussions that were to become the heart and soul of the workshop. Participants included academic researchers and park management personnel from all three range countries (Uganda, Rwanda, and DRC) as well as interested private, government, and non-government institutions from around the world.

Like all PHVA processes, the mountain gorilla workshop began with overview presentations on the biology of the species, past and current conservation efforts, and on the general workshop process. Each of the opening-session speakers identified among other things the importance of including local populations in decision making, the value of ecotourism as a viable conservation action, and the need for cooperation between NGOs.

A new workshop design element particular to the Network initiative was added: a Network member explained the intent of the project, requested assistance in generating local demographic data for the areas around the parks, and asked for ideas about how to translate human/social science behavioral data into land/resource use data, and finally into VORTEX input data. The population biologist gave a presentation on population simulation modeling in general, including some specific human population data for Uganda. Judging from comments at the break, both these presentations were well received, but as subsequent events indicated, they had relatively little impact on working-group discussions and may have been incompletely understood.

As a technique for surfacing issues around which working groups would be defined, the CBSG workshop facilitator led a problem-generation brainstorming session. More than 130 problem statements were put onto flip charts, of which 32 were directly related to community issues. Six working groups were established: revenue and economic issues, local human population issues, political governance, park and protected area management and ecology, veterinary and health issues, and population biology and simulation modeling. All groups were asked to consider their issues in terms of collaboration, research, communication, threats, and capacity building.

Increased Stakeholder Participation

The mountain gorilla workshop included a larger, more diverse group of stakeholders than CBSG's previous PHVAs. Participants included biologists, researchers, and wildlife managers from Uganda, Rwanda, and DRC and recognized experts on mountain gorilla population biology and ecology. These individuals included nationals from the range states, people from protected-area authorities and from various local and international NGOs (figure 6-3).

In spite of the stakeholder diversity, the Network experiment had in a way already failed because we were unable to garner the participation of all the necessary people, specifically social scientists. This was due in part to the absence of Network preparation time and in part to simply not knowing who the critical local experts were in advance. In seeking stakeholder involvement there also appeared to be a problem in explaining the workshop process to potential expert participants. To do this, we learned that we must speak some of each group's basic language as well as promote their interests as true stakeholders. Other than the Network members, the workshop participant most interested in the Network's approach was the representative from the Ugandan office of the community development organization, CARE, who had been working with local populations around Bwindi for some years. We had contacted this individual prior to the workshop in an effort to explain the process, the value of his in-country, social science expertise to the workshop, and the need for

Figure 6-3. Plenary session discussions, mountain gorilla PHVA workshop, Kampala, Uganda. (Photo courtesy P. Miller.)

local human demographic information. However, with inadequate lead time and an unsatisfactory explanation of the process and the role he was to play, he did not appreciate the need for the human population and human impact information we were seeking. As a result, although he participated in the workshop, this information was not available for analysis.

Incorporation of Human Demographic Data

While preworkshop research gave the Network an idea of the information required, once at the workshop our expectations for integrating this data proved hard to fulfill. The following section examines the working-group dynamics and how the available data were assembled and incorporated into the gorilla population model. The successful incorporation of human demographic information relied on the mechanisms of overall assembly and incorporation.

Data Assembly

The Network anticipated that the human population issues working group would provide the human demographic and land-use information to the simulation modeling group for translation and incorporation into the population model. Two Network members joined this working group (one as facilitator and one as a group member). Early on in this group's deliberations, it was apparent that two different foci were evolving: community attitudes, communication, and values; and determination, quantification, and assessment of human impacts/risks. Network members, rather than the working-group participants, drove the latter agenda item. In fact, there was a degree of skepticism regarding the human demographic model and the whole idea of trying to translate social information into species model inputs. Anxiety was expressed by a member of the working group that, because Network members were participating in the working group, the agenda of the Network would "hijack" the group, taking attention away from important discussions. This was a response we had not anticipated and this development significantly affected our progress.

As work proceeded, the working-group members developed a three-pronged approach to their topic: identifying the nature and impact of interactions between the human population and gorillas and their habitat, describing the types of human population pressures operating in the area, and discussing the inclusion of values and attitudes in decision and policy making. The first set of issues addressed were the actual interactions between the human population and the gorilla habitat. These interactions included resource extraction by humans, but also included interactions that are potentially beneficial for humans and the gorillas, or for the habitat itself. The second set dealt with factors that affect the makeup and attitudes of the human population, such as human population growth, economic values, cultural values associated with the gorillas and the park, movement of the local human population, and political instability; these factors can all affect the type, intensity, and frequency of interactions taking place between the local population and the forest. The third topic considered the means by which interactions between the local populations and the gorilla habitat can be influenced; the group investigated different types of community participation in the decision-making process regarding management of the park. While some progress was made, a discussion of workshop process was required to determine whether each of the model's components needed to be discussed separately or if all could be approached simultaneously. At this point it still appeared that the Network experiment's goal of exploring transdisciplinarity was to be addressed within the PHVA structure.

For the Bwindi and Virunga mountain gorilla populations, working-group participants prepared a list of interactions that exist between local human populations and the gorillas and their habitat. A high point in the working group, from the Network's perspective, was the chief park warden's presentation of such human-gorilla interaction data for Bwindi Impenetrable National Park. His profiles of those committing illegal activities were based on those arrested. Other participants, however, questioned the credibility of the warden's data because it was not written or published. For reasons unspecified, this individual left the group and did not return. A fundamental question remains concerning how to give legitimacy to different kinds of knowledge in the multistakeholder workshop setting.

By the end of this working-group session, several participants appeared quite disaffected. They felt forced into spending the majority of time on quantification of information for use in the VORTEX model even though they had not bought into the modeling process. They were much more interested in talking about community values and their impact on conservation action planning ("talking about the people"). We continued to emphasize the need to collect detailed human demographic information that could be quantified and input into VORTEX for analysis of impacts on mountain gorillas and their habitat. By this time we were concerned that the experiment was going to be derailed due to absence of specific data and the apparent lack of interest by this particular working group's participants.

As the working group appeared to disintegrate, we became increasingly concerned about the failure of the experiment. Consequently, the Network team decided that it would be legitimate (and in fact necessary if the experiment were to succeed) to split the human population issues working group into subgroups, with one working on quantification of human interactions with gorillas and habitat and the other talking about community values and action planning. We realized that we were not going to be able to surface the information needed for the human demography model within the confines of the working group without forcing our agenda on the already disaffected participants.

We were particularly aware that we did not have the necessary local people in the room, and those experts we did have were not willing to make expert judgments. This left us without any new, regional human demographic data for translation into VORTEX input data for Bwindi. The data we were able to retrieve from materials collected before the workshop were eventually used by the modeling working group to simulate a general catastrophe and war scenario as an illustration of the situation in the Virungas.

Network members made direct efforts to gather information from workshop participants outside of the working group concerning the Bwindi population and, with input from the CARE representative, they pieced together a picture of indirect, human-related threats and potential threats to these gorillas. In Bwindi, surrounding human population growth did not translate into gradual erosion of the protected area (as we had assumed before the workshop). However, increased population did seem to increase several things: pressure to change national policy due to local demands for more access to park resources (this happened when there was some sort of democratic or local empowerment process in place); human-animal conflicts; and the possibility of increased sabotage (fire, poaching) if the situation was not mediated by some sort of resource-sharing scheme (in Bwindi such a scheme was in place and serious incidence of sabotage had gone down). It also seemed possible that, when examined on a macroscopic scale, there might be some way to calculate the relationship between increased human population density and the likelihood of political instability or uprising. An apt analogy suggested by one Network member was that of a volcano, which could be predicted to erupt at a certain pressure level.

An interesting and fluid picture of human population dynamics emerged when looking at the social and ecological history of the other mountain gorilla habitat, the Virunga volcanoes region that straddles the borders of Uganda, Rwanda, and DRC. In this case, it was easier to make estimates about population pressures and the resultant impacts on the ecosystem, habitat, and species of concern. The area has been at the center of protracted warfare and the landscape has suffered accordingly. Governmental protection of the parks was completely absent and detailed estimates were available concerning the ecological impact of refugees camped on the borders of the Virunga parks. These impacts have been described in terms of the amount of wood removed from the parks, the amount of human organic waste dumped in the parks, and the increased number of animals killed. Two of these impacts could be turned into a measure of reduction of carrying capacity: rate of deforestation and rate of human depletion of gorilla food (bamboo). Two others could be turned into a measure of mortality: rate of poaching and rate of disease transmission through human contamination. These two separate equations could then be combined in a catastrophe scenario, which could be modeled as occurring at some frequency based on political upheavals in Sub-Saharan Africa (say every ten to fifteen years). The catastrophe could be simulated as having an initially severe impact, tapering off to a less severe but continuing impact.

Based on these discussions, two distinct risk scenarios began to emerge, both with potentially wider applicability than the mountain gorilla case:

Scenario 1: Ongoing armed conflict, central government(s) in disarray, and refugees massed near park boundaries. A sudden increase in local human population density, coupled with a serious breakdown in social order leads to intensive utilization of the nearby forest, estimated at 410–770 tons of wood per day (Pearce 1996). This utilization translates into loss of habitat, loss of vegetation (in this case bamboo), increased risk of human-gorilla disease transmission, and an increase in gorilla mortality through poaching or incidental killing. In such situations, habituation of mountain gorilla family groups to facilitate their study may make matters even worse, with an increased likelihood of poaching of those animals not fearful of humans. The only things that may help are if the value of gorillas for future tourism is clear to all, including incoming armies; and if park guards are provided continued professional development and support from locally based NGOs and conservation institutions, thereby inducing a more global vision of the importance of conservation (Hart and Hart 1997).

Scenario 2: Stable governmental regime with well-protected parks but an increasing human population. Land productivity begins to decline, and increasing individual empowerment (through democracy), combined with a lack of strong local support for conservation, results in local-level pressure for major policy change (e.g., degazetting a park) at the national level. Conflicts over environmental issues may increase, and general economic circumstances may improve through trusts, revenue sharing, or other means.

This scenario development was an important step and helped to identify the types of information needed for translation into VORTEX population model input data. However, it came too late in the workshop process to allow much of the necessary information to be compiled. There are probably ways of calculating increased risk to mountain gorilla population persistence that are similar to those used to predict the frequency of events such as volcanic eruptions, but there are many processes that may also ease the pressure, such as effective family planning, emigration, and reduced infant survivorship. The likelihood of uprising may also depend on the nature of national leadership.

Another unanticipated hurdle to quantifying specific human impacts on gorillas was that the overall process of species population modeling was questioned. The primary issue was one raised routinely regarding the use of models for population viability analysis: there was not enough data to properly construct a baseline model. In particular, it was argued that considering the threat to gorillas and their habitat due to refugees was purely academic since, while we had information on the impact of refugees on the forest, little information existed showing the direct impact on gorillas. When we countered this by explaining that we felt we could extrapolate to the current situation based on the excellent published data we had on the number and impact of refugees, some objected that, as the refugees are a threat to gorillas and need to be relocated immediately anyway, it was a waste of time and misleading to project refugee impact through simulation modeling. Some participants believed that following the relocation of refugees, the social situation would simply return to prewar levels (that is, Rwanda and DRC would return to the more peaceful situation found currently in Uganda).

Network members, however, thought that because there has been a history of political instability throughout the region, this needed to be reflected in mountain gorilla population risk assessments. In fact, we were looking forward to the possibility of modeling two different scenarios in the two different populations: rapid, catastrophic instability in the Virunga region, and gradual erosion of habitat protection in and around Bwindi based on a relatively long period of political stability in Uganda. These scenarios struck us as possibly fairly common in the developing world, and we felt if we could develop scenarios for each dynamic we might be able to model them to get an idea of impact on gorillas and their habitat.

Data Incorporation

The population biology and simulation modeling working group role was to develop a baseline model and incorporate impacts defined by other workshop groups into the model. This working group was composed of only four participants: two university researchers (with only one currently conducting field research on mountain gorilla behavioral ecology) and two representatives from an NGO devoted to mountain gorilla conservation. In addition, the sole range-country participant was an observer who became less involved as the workshop progressed. This lack of range-country participation in a risk assessment modeling group is by no means uncommon, but the absence of the truly local perspective in the process can only reduce its ultimate effectiveness.

Observations on the first day of the workshop led the team to conclude that the modeling and risk assessment process was not yet fully appreciated by participants. This was not wholly surprising, given that a demonstration of VORTEX and its capabilities, normally included as part of the first day's agenda, had been deferred until the second day. Consequently, the entire modeling exercise was generally viewed as nebulous. However, additional viewpoints were at work to downplay the value of the modeling process. Given that one of the central goals of this initial Network experiment was to explore integration of a broader range of data, the facilitators worked to dispel doubts about risk-based analysis in general, and they specifically worked to expand participants thinking about models and the range of possible inputs. For example, one suggestion made at the workshop was that there is no need to model the risk of mountain gorilla extinction since the two populations have been shown to be currently and historically stable or even increasing in size over time. A prospective view of the risks associated with possible future events is clearly much more difficult to facilitate than the more traditional retrospective analysis of existing data. Additionally, the simpler modeling efforts already published (Weber and Vedder 1983; Akcakaya and Ginsburg 1991; Durant and Mace 1994; Harcourt 1995) indicated little risk of species extinction (although these analyses were based primarily on internal demographic and genetic processes in the absence—with the exception of Harcourt—of considering the human element in conservation planning). There is little doubt that with explicit consideration of the consequences of increased, prolonged human population pressure in and around the Virunga volcanoes region—increased likelihood of disease transmission, potentially large-scale forest degradation and fragmentation, direct mortality (poaching or unintentional)—the risk of mountain gorilla extinction will increase.

Once the working group began developing an initial baseline mountain gorilla population model, they quickly realized that a significant amount of data were at their disposal to describe the demographics of the species. This was facilitated by the fact that one of the participants had summarized nearly thirty years of field data into a concise, quantitative description of mountain gorilla life history. Unfortunately, only mean demographic values were provided in this summary, without a statistical description of the annual variation around these mean rates. From the standpoint of stochastic population modeling, an estimate of the annual variation in a demographic rate is as important as the estimate of the mean rate itself. Moreover, the group did not have access at the workshop to the raw data used to produce the summary values; consequently, they could not calculate the annual variation in demographic rates in parallel with the development of the model. As a result, they were forced to make the same kinds of estimates of annual variation one makes in the complete absence of these kinds of data. This process element—field researchers bringing only a portion of their biological data to the table for evaluation and analysis—is a common occurrence and reflects researchers' lack of experience with using field data in a risk-based population projection. The Network is keenly aware of this limitation and must design into the workshop process a means to better inform potential participants of the various data needs in advance of the workshop.

Once the baseline model was on its way, it was decided that an appeal would be made in plenary for modeling data and increased participation in this working group. We also recognized that we needed to communicate the use of VORTEX in order to demonstrate its power in experimenting with risk and in integrating the working groups' discussions. As discussed earlier, extracting the needed data was not as straightforward as we had anticipated.

A full-scale demonstration of the applicability of VORTEX to projecting mountain gorilla population viability, using appropriate data gleaned from the population biology group members the preceding day and evening, appeared to be quite successful in helping to make the modeling process much more transparent to the workshop participants. Following this plenary session, the modeling group focused their attention on refining the baseline model and developing a preliminary demographic sensitivity analysis. Because of the working group's small size, progress was possible with relatively little internal conflict. Nevertheless, numerous disagreements over the specifics of input data arose among group members, related to particular field experiences and data collection methods. In general, these disputes were resolved in a reasonably straightforward manner, although some tensions among these members remained for reasons not completely appreciated by the group's facilitator (likely stemming from historical associations with alternative NGOs).

As the workshop progressed, it became evident that the detailed human demographic and associated natural resource use data necessary for inclusion in a refined VORTEX modeling process was not to be soon forthcoming. This was in large part due to the factors described above: (1) an inability to effectively synthesize the volumes of published information gathered by the Network prior to the workshop; and (2) problems with group dynamics within the human population issues working group, the group deemed responsible for providing this information while at the workshop. Moreover, human demographic data from regions immediately surrounding the national parks harboring mountain gorillas was not available for analysis using human demographic models such as the DEMPROJ system (Stover and Kirmeyer 1997), as was originally planned (see chapter 13).

Yet another difficulty the modeling working group encountered was reluctance on the part of the veterinarians in attendance to estimate parameters for diseases thought to affect mountain gorilla populations, either now or in the future. It was proposed by a number of workshop participants that disease epidemics could occur with greater frequency due to increased human-gorilla contact. Despite this recognition, the veterinarians were hesitant to develop quantitative descriptions of these events for use in the modeling process. One member of the veterinary group argued that this was in fact quantifiable, but was unwilling to offer a specific calculation. Instead it was suggested that sanitation engineers who work with refugee organizations or studies of interactions of wildlife and local domestic livestock might provide some answers. Only on the last full day of the workshop did the modeling group receive a list of characteristics for each of three diseases (this list was subsequently revised by the veterinary group the following day, immediately before the workshop closing ceremonies). This delay in exchange of information between groups left less time available for an iterative process of model development and analysis and, by extension, a reduced opportunity for meaningful discussion of results and recommendations emanating from them. The value of frequent and detailed communication between working groups, particularly in an expanded workshop in which additional information is being brought into the analytical and deliberative process, cannot be overemphasized.

Despite these difficulties, the modeling group was successful in testing a reasonable set of VORTEX scenarios based on the situations described earlier that included the periodic outbreak of disease as well as war in and around the Virunga region, with severe consequences for mountain gorilla demography and habitat needs. Diseases were assumed to occur infrequently and randomly and to affect both female reproductive success and age-sex specific mortality. Based on simple historical data, the working group assumed that serious civil unrest events could be expected to occur about every thirty years on average, with each event lasting on the order of a decade. During each year of war, the proportion of adult females that produced an infant was reduced by about 10 percent of the value reproducing in more stable times, and the mortality of adults and infants was similarly increased by an additional 5 percent over baseline (peacetime) values. Once a war event ended, these demographic rates would rebound to at or near their original values. In addition to these demographic effects, war resulted in a cumulative deterioration of mountain gorilla habitat that was translated into either a 25 percent or 50 percent reduction in habitat carrying capacity after each event (see chapter 18 for more information). The construction of these scenarios was made possible by a recent major addition to the VORTEX model that allows experienced users to define species demographic rates using virtually any type of algebraic or even trigonometric function (Miller and Lacy 1999).

In addition to the specific description of effects of wars on mountain gorilla demography, the models included a series of catastrophic disease events that were assumed to increase in probability of occurrence as human (refugee) population density increased during and following periods of civil unrest. These estimates on disease risk were provided after lengthy deliberations among a group of wildlife veterinarians dealing specifically with mountain gorilla health issues and their application to population management.

The modeling efforts conducted at this workshop demonstrated the significant demographic impacts that periodic war and disease can have on affected mountain gorilla populations—even when those events are episodic and, in some instances, infrequent (figure 6-4). While the risk of population extinction is relatively low over the one-hundred-year time frame of the simulation, it is important to recognize that the simulated populations are in systematic decline at an annual rate of about 2.5 percent. The low extinction risk is largely due to the long generation length (on the order of fifteen to twenty years) seen in mountain gorillas in comparison to the duration of the simulation. Despite our perceptions of the relative simplicity of these results, they nevertheless served to improve our understanding of the nature of connections between wildlife populations and the human populations with which they interact. Moreover, our ability to communicate these connections to those charged with managing the wildlife resource was likewise improved. This enhanced analysis helped to stimulate renewed efforts on the part of national agencies and international NGOs to assess the actual impacts of the recent conflicts on the local gorilla populations and surrounding habitat.

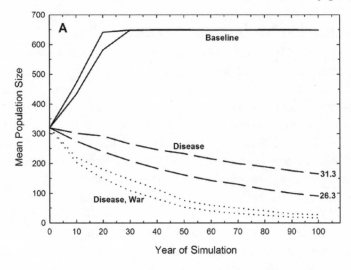

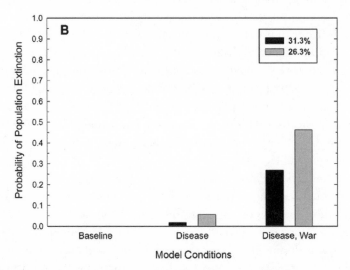

Figure 6-4. Results of population viability modeling at the mountain gorilla PHVA workshop. (A) Projections for a simulated mountain gorilla population in the Virunga volcanoes region under baseline demographic conditions (solid lines), with the inclusion of catastrophic disease events (dashed lines), and with the addition of disease and periodic war among human populations inhabiting the region (dotted lines) as derived from discussions at the PHVA workshop. Each set of models consists of a pair of projections showing alternative measures of female breeding success (31.3% or 26.3% of adults breeding in a given year) as part of a larger demographic sensitivity analysis. (B) Probabilities of extinction of the Virunga mountain gorilla population predicted from each of the scenarios. (Source: Werikhe et al. 1998.)

The Impact of Institutional Context

Members of the Network had been concerned for some time with the need to evaluate conservation efforts in terms of their broader institutional contexts (see chapter 12). The assumption is that participatory, stakeholder-based conservation processes such as PHVAs are more likely to have an impact in some institutional and societal contexts than in others. Network members went to Uganda under the assumption that a number of contextual factors, including the existence of government conservation policies and conservation monitoring programs, political stability, and the general well-being of the human population were all important indicators of the likelihood of a conservation initiative's success. In preworkshop research we also came across an interesting article by Hart and Hart (1997). The authors state that while conservation funding has tended to follow development funding (helping local governments develop capacities and then turning the projects over to local actors), in highly unstable political situations it is better to use the humanitarian model (sudden bursts of investment at times of crisis, given directly to grassroots organizations). The article also notes that the most creative conservation during civil war has come from local park guards who had worked enough with NGOs to have a global vision of the importance of their efforts and who improvised means of protection. This suggests that if a government is in risk of dissolution, NGOs active in conservation in that country may provide the only protection available.

The Network members weren't trying to model these factors as impacts, but were interested in whether participants in the workshop thought they were important. Certainly governance as an issue attracted a lot of attention (and spawned a working group). A number of the working groups, including the governance group, the local populations group, and the economics group also touched on these topics. All admitted suffering from an expertise deficit, however, as there were not enough social scientists present and only one lawyer. Discussions in these groups revealed the possibility of creating two distinctive scenarios (as further elaborated above), which may have fairly wide applicability to the developing world:

Scenario 1: The catastrophe scenario, in which active government does not exist and in which the only positive actions were stimulated by the direct intervention of NGOs or the collaboration of local park officials with the NGOs. The only management regime that seemed to offer any hope in these circumstances was one of protection, and this only worked if combatants and park officials had a global perspective (i.e., felt that the gorillas were an important resource to be conserved because of their global significance and importance to international tourism and conservation groups).

Scenario 2: The stable government regime with some empowerment of local populations, in which such schemes as trusts, revenue sharing (from ecotourism), and multiple use of the parks seemed to have fairly positive benefits.

Workshop Recommendations and Conclusions

Primary workshop recommendations included, but were by no means limited to, the following:

- Fundamental to the conservation of the mountain gorilla is the existence of sustainable national veterinary units responsible for implementation of veterinary services.
- Work should be done with humanitarian agencies to ensure their emergency plans fully address conservation concerns. In addition, conservation agencies (governmental and nongovernmental) must prepare their own emergency plans that address identified critical interactions of humans with gorillas and their habitat.
- When human-gorilla population conflicts are at a relatively minimal intensity, recognition should be made of the potential for resilient gorilla population growth. However, human population pressures resulting in severe loss of gorilla habitat and reduction in gorilla population growth potential require an even greater recognition of the acute risks facing this subspecies in order to minimize extinction risk.
- Lead conservation agencies should encourage ministers of range countries (Uganda, Rwanda, DRC) to meet and discuss legal issues pertaining to mountain gorilla conservation.

Throughout the workshop, we were aware of the need for additional experts (local people with Indigenous knowledge, resource economists, agricultural economists, human demographers), that surfacing the data we had hoped to use in the human demographic model was more difficult than anticipated (the national human demographic data we did have proved impossible to translate into local habitat/species impact), and we realized that it may be inappropriate for the human demographic element to be subsumed within a human population issues working group. The major challenges to the Network came out of these limitations. While the results of this first Network experiment were mixed, the experience was worthwhile and a tremendous amount was learned, which strengthened subsequent activities. Of particular value are the findings regarding the inclusion of human demographic data, the impact of including expanded stakeholder groups, the impact of institutional context, and concerning the experiments with process.

Network Learnings

The Challenge of Integrating Human Demographic Data

The Network went to Uganda with the hopes of utilizing human demographic modeling software (DEMPROJ) simultaneously with VORTEX. It quickly became apparent, however, that the technical sophistication of such translations, coupled with the difficulty in obtaining detailed data highly specific to the geographic region of interest, made it impossible to conduct such an effort at this workshop. For example, the macrolevel data that was available through public sources was not fine grained enough to interpret the dynamics of human population at the borders of a park.

Although this first experiment resulted in a failure to use detailed human demographic data in interaction with VORTEX (at least in the context of the workshop itself), it stimulated much thought about how to do this constructively in the future. It was suggested that more-qualitative data might be gathered in advance, by sending a team of researchers into critical locations (see chapters 9 and chapters 14 for more reflection on this approach). The failure at this workshop also suggested, however, the importance of modeling resource use as an important variable in species survival. We concluded that Network members and a human demographer and modeler will need to determine what specific information is needed for the species model and gather as much of that information as possible in advance of a workshop. We also need to ask social science experts in advance specifically to provide information in the form needed for the model.

The Challenge of Increasing Stakeholder Participation

The Network failed to bring the full set of stakeholders to the workshop, but our efforts stimulated the creation of a protocol for improving the chances for success in future workshops. It is obvious that this workshop would have been strengthened by additional expertise in various fields, as many of the working groups complained about lack of knowledge or information. The complexities of economic and conservation needs loomed large in a number of working groups; while much information seemed to be available, discussions would have benefited from more expertise in data translation and extrapolation. In the case of this species, the biology was fairly well established, and people were eager to deal with some of the social science topics.

The particular expertise lacking at the mountain gorilla workshop fell into the following categories: resource economists (to help introduce the impact of natural resource utilization on species survival), natural resource management (e.g., forestry) experts, social anthropologists (with an interest in conservation), demographers, and human epidemiologists.

To avoid similar problems in future, the Network hypothesized that people with this expertise should be contacted early in the workshop planning process. The Network experiment needs should also be built into the original workshop proposal so that the organizers are comfortable with it and can assist with gathering of contacts and data. Local academics, researchers, and social science experts could possibly be tracked by working through the IUCN Social Policy Program local office. Bringing in several "outsiders," particularly those having global experience with similar ecosystems and social systems and expertise in comanagement or natural resource management, might introduce fresh ideas. It is hypothesized that several development workers with experience working in these communities could contribute vital information. Community representation remains a challenge; despite efforts at inclusion, Network members were told that including representation below the district level would be difficult, as people would not have the sophistication to deal with this kind of workshop. (See chapters 8 and 9 for a different perspective on including community representatives in the PHVA process.)

From interviews with participants and workshop organizers, it appears that in addition to literature searches conducted before the workshops, the way to get the needed local data to the table in a developing country may be to hire an in-country consultant to dig for epidemiological, resource use, regulatory, or demographic information.

The Possibility of Researching the Institutional Context
Information concerning the institutional context seems fairly easy to obtain through journals and Internet sources. The literature search conducted at McGill University prior to the mountain gorilla workshop

turned up some interesting data of good quality. This particular process, however, suggested that in addition to information on the amount and kind of government regulation and monitoring, details on the general well-being of the human population, the political stability of the regime, and the presence and active NGO involvement are critical contextual factors, particularly in unstable scenarios. In the stable scenarios, the presence of some kind of resource and revenue-sharing regime, in which the local community benefits, is also critical if conservation is to work. In particular, this case study suggests that developing some ideal/typical scenarios about institutional context could be used to enhance workshop discussions and analysis of conservation alternatives (see chapter 12 for a fuller discussion of this possibility).

Process Lessons

This workshop provided significant process lessons as well. Since the necessary expertise was not present, Network members experimented with trying to get the relevant issues onto the agenda by participation in the various working groups. But it became clear that other working-group members did not see the logic for including these issues on the agenda, and so they were dropped or relegated to a less important status. We also learned that Network interests could not be forced, either by a facilitator or working-group member. We lost credibility with the human population issues working group early on in the workshop. They did not see Network members as participants with shared goals, but as outsiders forcing our own agenda. This unfortunate situation may be avoided if workshop participants include the appropriate people (resource economist, demographer, social scientists, local people), if their participation is built into the workshop process from the beginning, and if they make up a separate working group (e.g., a human demography and modeling working group). It also was decided by the end of the workshop that it would be a good idea to have a "human process modeler" working side by side with the species modeler in the modeling group. This person could then try running demographic models or economic models and appeal to the working groups for input.

Another interesting process issue arose during one of the final plenary sessions that may provide some insight into risk assessment workshop process design. The war and disease scenarios discussed previously were the products of numerous discussions within and between many of the working groups and the models' results were to be an important starting point for formulating conservation recommendations. In order to foster a greater sense of workshop product ownership among the working-group participants, the modeling group's facilitator suggested that one or two of the other group members should present these important results in plenary. Unfortunately, due primarily to the presenters' inexperience in discussing results of relatively complicated simulation models with an audience having little background in the topic, much of the essence of the intended message was not conveyed. In fact, the presentation placed undue emphasis on the ability of a mountain gorilla population to increase in size in the absence of catastrophic civil unrest. Additional statements made during the presentation diminished the impact of the modeling group's results and compromised the credibility of the modeling process as a whole. This example points out the difficult conflicts that can arise when attempting to maximize group ownership through plenary presentations at the potential cost of scientific clarity and/or accuracy. Nevertheless, the body of workshop participants appeared to recognize the potential severity of these threats and to focus some of their subsequent discussions on how to deal with them in the context of gorilla conservation.

Aftermath of the Workshop

The mountain gorilla PHVA workshop resulted in a series of specific recommendations:

- Improving veterinary care
- Improving regional and institutional stakeholder cooperation
- Generation of tourism and revenue to support gorilla conservation
- Promotion of community participation in gorilla conservation
- Implementation of research on poaching of forest products, crop raiding by animals from the park, impacts of tourism and habituation of gorillas and impacts of resource sharing, and on the amount and quality of habitat in the two gorilla populations

The NGOs participating in the workshop, while extremely active and with a long history of international collaboration, historically were uncooperative with each other at a management level. Mountain gorilla conservation programs have been so dominated by NGOs that the need for improved cooperation was one of the primary issues addressed during the workshop. The workshop resulted in the formation of the Mountain Gorilla Foundation (MOGOF), whose first annual meeting was held in

January 1999 in Rwanda. MOGOF will bring together top management representatives annually to encourage collaboration in developing mechanisms for implementing the numerous action steps detailed in the workshop report. This is considered by some to be the most significant accomplishment of the workshop.

Since the workshop took place, two and a half of the three countries addressed in the workshop have become involved in civil war, so implementation of recommendations has been slower than would be otherwise. Nevertheless, the workshop restated goals for gorilla conservation and developed new ideas (a Peace Park, for example) and recommendations are now being used as guidelines for those working in gorilla conservation. Tourist activities have been restructured to ensure that guides adhere to the allowed number of visitors and that visitors remain five meters from the gorillas. Research foci in the area are being determined by the workshop recommendations. In addition, a ranger-based monitoring program was recommended and has now been implemented. Finally, the workshop resulted in the formation of an organization dedicated to cooperation among NGOs working for gorilla conservation.

This is a story of an experiment. It centers on the problem of conserving the planet's endangered species, but it also tells the story of a new form of organizing for effective risk assessment, recommendation, and action. It focuses on the challenges of cross-disciplinary analysis as well as cross-functional, cross-disciplinary, and cross-sectoral action. Most centrally, it is the story of a sustained project in action research and the learning that resulted.

 **Chapter 7**

Linking Monkeys, Biologists, and Palmito: The Muriqui PHVA in Brazil

JENNA S. BOROVANSKY AND EMMANUEL RAUFFLET

The Population and Habitat Viabilty Assessment (PHVA) for the muriqui was the Network's second experiment. The workshop was held in Belo Horizonte, Minas Gerais, Brazil, May 23 through 27, 1998. The primary Network goal in this PHVA was to introduce a human demographer to the PHVA process. We hoped that including this expertise would help supplement the discussion of human population growth and resource use, an important issue in conservation that, until the Network experiments, had not been meaningfully addressed in PHVAs. We also wanted to continue the effort begun in Uganda of having workshop participants generate location-specific data to be used in the modeling exercise.

In this chapter, we reflect on preparations for and events during this workshop and the lessons we learned. This case is organized into three main sections. In the first section we briefly provide the context of muriqui conservation in the Atlantic rain forest. In the second, narrative section, we highlight Network expectations and pre-workshop preparation efforts; we present what actually happened, with a focus on the dynamics of the workshop process and on the participants' interactions. Finally, we discuss the consequences of the workshop dynamics for the Network experiment. In the third section, we reflect on why many of the Network's expectations were not fulfilled and briefly present the lessons we learned from this workshop.

The Context of Muriqui Conservation

The muriqui *(Brachyteles arachnoides)*—also known as the woolly spider monkey or *mono carvoeiro*—is endemic to the Brazilian Atlantic rain forest and is the largest primate in Brazil (figure 7-1). The species is thought to exist in two geographically distinct subpopulations, with an estimated total population of 1,000–1,200 individuals distributed in twenty-two Atlantic rain forest locations in southeast Brazil (Strier and Fonseca 1996; see figure 7-2). The northern population occurs mainly in the states of Minas Gerais and Espírito Santo (about 32 percent of the total population) and the southern population is in the states of Rio de Janeiro and São Paulo (about 68 percent of the total). The largest northern population (90 individuals) is located at the Biological Station of Caratinga, Minas Gerais. Of the southern populations, the largest (400–500 individuals) is in the Carlos Botelho Park, São Paulo. In general, the population densities are lower in the southern forests than in the north. Because of its relatively small, dispersed populations and threats to its habitat, in 1997, the muriqui was listed as critically endangered in the World Conservation Union (IUCN) red list.

The Mata Atlântica

As a forest-dependent species, the fate of the muriqui is closely linked with the future of its only habitat—the Atlantic rain forest of Brazil. It can inhabit both primary and secondary growth forests. The Atlantic rain for-

Figure 7-1.
The muriqui
(Brachyteles arachnoides).
(Photo courtesy L. G. Diaz.)

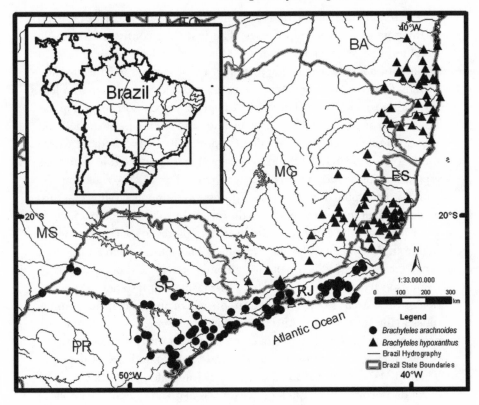

Figure 7-2. Map of southeastern Brazil showing known locations for the muriqui and a related taxon. (Courtesy Andre Hirsch, from Hirsch et al., forthcoming.)

est, or Mata Atlântica in Portuguese, covered 1.5 million square kilometers when the Portuguese arrived in Brazil in 1500. Five centuries later, only 3 percent of the original forested area is left, mainly in high altitude or steep and unconnected areas.

Several historical and present-day settlement patterns led to the degradation of the Mata Atlântica. First, the Portuguese conquest of the native Tupi Guarani Indians often meant the destruction of their Atlantic rain forest territory. Second, building roads into the immense rain forests led to cutting of the forest to open new agricultural and colonization fronts. Third, a long tradition of extensive cultivation techniques—as opposed to intensive ones—combined with the abundance of forests resulted in few efforts to renew the forest and widespread harvest was followed by migration into formerly forested areas.

The remaining Atlantic rain forest and muriqui habitat can be grouped in three general ownership classes. The first class consists of relatively small forested areas of less than 1,000 hectares, where muriqui exist in enclaves of a larger fazenda (farm estate). These sections include Fazenda Montes Claros, Caratinga, Minas Gerais; and São Sebastião in São José dos Campos, São Paulo. These enclaves generally support small populations (10 to 50 individuals) on private lands and are the most studied and best documented populations. The Caratinga Biological Station, where most of the research on the muriqui has been conducted to date, was established inside the Fazenda Montes Claros. The second ownership situation is exemplified by the Mata do Sossego, Simoneia, Minas Gerais. Multiple small coffee growers share the ownership of 800 hectares, including 150 owned by Fundação Biodiversitas (Biodiversity Foundation), a nonprofit Brazilian conservation organization. In these steep and remote areas, coffee growing is the main economic activity and other extensive cultivation techniques have required increasing deforestation. Preservation of these areas results largely from the will of the small farm communities living nearby. The third ownership situation is state parks. These areas are large, around 30,000 hectares or at least thirty times larger than private forests, and are owned by the São Paulo State Forestal Institute. These parks are Carlos Botelho, Fazenda Intervales, Alto Ribeiro, and Serra da Bocaina. In chapter 12, we discuss in greater detail the implications of different governance models for conservation practices.

Muriqui Conservation and Research

To date, muriqui research sites have been largely determined by physical accessibility. Most of the research on the muriqui has been conducted in small, privately held forests, and in Caratinga in particular. Larger forest areas have received little research attention to date, as large parts of these areas are impenetrable and conducting research on muriquis in these areas is no easy task. Research done in these large park areas is limited to small muriqui groups in limited areas of the park system.

Because the muriqui is endemic to the Atlantic rain forest and its dramatic story of decline is linked to decline and fragmentation of the forest, leading international conservation organizations have seen the muriqui as a symbol for conservation of its forest habitat. Though not as world renowned as its tamarin cousins, the muriqui's significance as the largest primate in the diminishing Brazilian Mata Atlântica has not been missed entirely by the international conservation world. For instance,

issues of *National Geographic* and an introduction to a recent primate narrative (Peterson and Goodall 2000) have featured the muriqui. This
significance eventually led to the workshop held in Belo Horizonte, Brazil.

Preworkshop Activities

The Network entered this workshop with a focus on investigating two
issues: integration of a human demographer and human population
model into the workshop process, and incorporation of social science
and "human" information into conversations through broader stakeholder participation. This workshop was chosen because of the Conservation Breeding Specialist Group's (CBSG's) history of work in the
country and the perceived potential for involvement from an expanded
range of professionals.

Network Expectations

We came to Brazil with several expectations as a result of our work in the
mountain gorilla workshop (chapter 6). In Uganda, participants (mostly
local officials and international conservationists) did not have the information necessary to provide specific details on how resources were used
in the small, localized areas where mountain gorillas live. In addition,
discussion of the effects of humans in the context of a human population
issues working group had been somewhat polarized between international
conservation representatives and biologists. This permanent tension
had to a large extent impeded the inclusion of any nonbiological information into the conversations. Data on the social dynamics of the locale
that affected the conservation chances of the gorilla had thus largely been
ignored.

Therefore, the Network hypothesized that we needed to actively recruit
into the workshop process local stakeholders and social scientists familiar
with local resource uses. Specifically, we wanted to translate human population and resource use trends into a format compatible with the animal
population model. Statistical data are helpful to map human population
trends and resource use at the municipal, state, or federal levels. However,
we also realized that what is often needed is to go beyond statistical data
and to translate these data into specific relationships between statistics on
resource use, the muriqui, and its habitat.

As a result, the Network tried to expand the workshop process in Brazil
by aiming at including two specific groups. The first group was local
resource users, such as local landowners and ranchers. The Network had

the following reasons for trying to include these local resource users. First, we expected these various stakeholders to be able to provide fine-grained and local information about the agricultural and industrial practices that affect the animal and its habitat. For instance, a local coffee grower might explain how, how much, and why the forest is cut in areas key to muriqui populations, while even the most specific statistical data will only describe how much forest has been cut at the municipal or state scale without further details. Identification of relationships between resource use and risk (or effects) on the animal species are needed to translate the numerical or statistical data into meaningful inputs for the animal population model. Second, as we explore in greater depth in chapter 16, we thought that including the resource users in the decision-making process would generate more "solid" (i.e., practical) and implementable recommendations from the workshop.

The second group we were trying to include was local social scientists. We thought that if this Network goal was shared with the workshop organizers, it would be relatively easy to expand the expertise base of workshop participants and open the debates on the social dimensions of the problem. We were trying to include in the process local Brazilian social scientists interested and committed to conservation, knowledgeable about local practices related to resource use, and, ideally, acquainted with the muriqui. By including these social scientists, we hoped to link the evolution of political, economic, and social systems with the status of the animal and its habitat. We hoped to reflect on local practices and representations related to the animal and its habitat. The potential role we perceived for social scientists was that of translator; they would know which questions to ask in order to extrapolate local information into the broader context of the status of the muriqui and its habitat. We expected our own Network human demographer to assist in this "translation" during the muriqui workshop.

We expected CBSG's existing, successful working relationship with the Fundaçao Biodiversitas, the local workshop organizers in Belo Horizonte, to create favorable preconditions for this social science experiment. Fundaçao Biodiversitas and CBSG had had a successful collaboration on the lion tamarin since the beginning of the 1990s and two workshops had already been hosted in Brazil prior to the muriqui PHVA.

Workshop Planning
While the muriqui workshop was in the planning stage since 1993, final preparations occurred about one year prior to the actual workshop, with

the submission of a grant to fund the workshop. The original grant for the workshop, and its stated purpose, was heavily focused on biological science as a basis for management recommendations, which influenced the selection of the limited number of "key" participants who would be funded to attend.

Five months before the actual workshop the Network decided that the muriqui workshop would be used as a case study and three months prior to the workshop we began our efforts to bring social scientists and local stakeholders to the table. Because of the short time frame, Network efforts were concentrated on broadening participation, gathering human demographic information across the muriqui's range, and securing the participation of a human demographer to help translate the data for the model at the workshop.

Efforts were made to prepare for the workshop according to our expectations. First, a letter was sent to Fundaçao Biodiversitas encouraging them to open the workshop process and invite participants from various academic disciplines as well as from various stakeholder groups. Gayl Ness, Network human demographer from the University of Michigan, also contacted Biodiversitas to find out about the local demographic dynamics in the areas close to the muriqui habitats and requested this data before the workshop. Second, one month prior to the workshop, a personal meeting with the key researcher on the muriqui was held to introduce her to the workshop process and the Network goals for this workshop, and to inquire about contacts with Brazilian social scientists. While she knew some of the local landowners from her research area, she was not familiar with local social scientists who might be available. She was sympathetic to the need for broader participation and made an attempt to contact the organizers in Brazil and have some local landowners added to the invitation list. Nonetheless, the invitation list remained substantially unchanged.

Finally, in choosing Network members to participate in the process, we included a human demographer. We expected him to contribute to linking the dynamics of human demographics with the evolution of the muriqui population in general (see chapter 13). More particularly, we also expected him to help "tease out" information relevant for the species that was not explicitly formulated in the working groups.

Four other Network members were present at the workshop and acted as a working group facilitator (1), interviewers (2), and observer (1). The Network interviewers and observer gathered most of the interview and observation data used to write this chapter. In addition, the chairman of

CBSG was actively involved in the Network and facilitated the meeting and acted as the species modeler. While we presented our goals explicitly in the workshop opening session, there was ambiguity among the Network members as to what role to play in the process, with an introduction being given as social experts, but only the demographer and facilitators explicitly participating in the workshop.

The Workshop Dynamics: Content and Process

As we said, our overall intention was to expand the workshop process by having social scientists and local resource users around the table. In this section we review what actually happened at the workshop, as compared to our expectations. First, we examine the demographics of the group participation. Second, we briefly highlight some features of the process, including the working-group process and modeling integration efforts. Third, we review the outcomes of the workshop.

The Group Composition

Most participants at the workshop were currently engaged in or had previously done field research with the muriqui and were familiar with the scientific data on the muriqui. The typical participant's expertise was focused on one or two particular populations in the states of São Paulo or Minas Gerais. These conservationists and biologists were generally in two different phases of their careers: senior conservationists, or graduate students in biology and conservation. The expectations of these two groups and their contributions to the workshop reflected their different perspectives and responsibilities. On the one hand, the major players (faculty and senior conservationists from North America and Belo Horizonte) had contributed funding or social capital to convene the workshop. On the other hand, the young graduate students in conservation biology generally came to the workshop with expectations that they could (1) learn about the workshop process and share their current work; and (2) cultivate international contacts and funding for their academic or personal projects related to the muriqui or its habitat. For instance, one participant was looking for funding for his doctorate abroad, another for some funding for a nongovernmental organization (NGO) he was establishing to protect the muriqui. This difference in perspective and participation was perhaps most noteworthy because most of the Brazilian graduate students' research projects had been or were currently funded and directed by the senior conservationists and there was a noticeable power differential to Network observers. Beyond this homogeneous group of senior conservationists and their Brazilian graduate students, few nonconservationists were active participants in the three and a half days of the workshop.

The representative from the Brazilian Federal Agency for Protection of the Natural Environment (IBAMA) inaugurated the workshop and left after introductory remarks on the first morning for another city. Also, the only social activist, who worked for Fundaçao Biodiversitas in the town of Intervales, left after one and a half days of discussions. In interviews, she expressed the feeling that the workshop discussions were very biology and conservation focused, and that she could not contribute to such a technical discussion.

Unfortunately, no resource users, such as landowners, coffee growers, fazenda owners, or industry representatives were present. Neither was there anyone present from the Minas Gerais state and São Paulo forest and park administrations, although a forest administrator had been expected—in fact, he spoke with the senior conservationists over the phone to negotiate a land purchase for a possible muriqui reserve during the workshop.

Process

The homogeneity of the group, as described in the former section, had implications for the process of the workshop. The process was generally characterized by a relative absence of conflict and adherence to a problem definition centered on biological conservation issues and, in some cases, predetermined solutions. The need to preserve the muriqui was not a hotly debated issue but rather a postulate all participants shared, since all had strong emotional links with the animal and its plight and were moved by a passion for conservation. The only disagreements appeared to be in determining the degree of risk to each population. This narrow problem definition was evident in the opening issue-generation session at the workshop, with only eight of the forty-three issues identified being social or economic in nature and these were very broad problem statements (e.g., the role of education, the role of resource use in communities).

Working Groups

Essentially, three working groups dealt with the biological information on the species and one small group, lacking local social science expertise, attempted to address the human interface with muriqui issues. The participants self-selected into these four working groups and expertise and agendas were brought to the table to address selected issues from the plenary. A brief discussion of each group follows, with a focus on highlighting human use issues that arose.

A distribution and status working group worked to map and inventory data for each of the current and historical populations of muriqui throughout the country. In addition, a population and habitat management working group identified twenty-four scenarios to describe management of lands where the muriqui occurs. The scenarios were based on forest size, conservation status, and muriqui density. For each case, management recommendations were made based on the identified threats to the population and conservation potential. Ecotourism potential and environmental education and additional research were recommended in every case, with much thought put into identifying priority sites for action.

While the management group addressed general conservation situations for the muriqui populations throughout its range, the modeling working group worked with extensive data sets on the Caratinga population of muriqui, as well as data on the captive population. This group served a dual function of building a base population simulation model and teaching the population modeling software. The workshop facilitator served as modeler for this group and therefore time for integrating additional data from other working groups into the model was limited. A few scenarios were run at a cursory level to examine the effects of timber harvest or forest conversion in the habitat of the Caratinga population.

The majority of participants were involved in these three working groups and they focused mainly on consolidating and integrating biological information and building scenarios for the future. It became clear from very early in the workshop process that there was not enough information on human-related issues in the room. The problem had been framed mainly as a biological issue and participants came ready to contribute this type of information. This can be illustrated by three different dynamics that occurred in the human impact working group.

The first dynamic concerned the mismatch between biological and social data. This human impact group, as the only nonbiologically centered working group, was facilitated by two Network members, including the human demographer, and was composed of three conservationists and the only social activist in the room. The social activist had an extensive working experience in local rural development and had specific knowledge on agricultural practices in the area of Intervales, where the local situation is driven by coffee-growing activities initiated by small landowners. Yet the working group did not specifically discuss this particular region. Most of the research available on muriqui to date had been conducted in Caratinga, a part of a private fazenda and the biologist participants of the working group had conducted research in the Carlos Botelho State Park, São Paulo. There was a mismatch between the biological information available (on Caratinga mainly), the personal research experience of the biologists in the working group (in Carlos Botelho), and the social-economic information available (on Intervales). As a result, the social activist from Intervales left the room after the first day and a half. In an interview, she said that she could not see her contribution in a group mainly composed of biologists and conservationists and in a debate centered on these issues.

The second and third dynamics in the Human Impact group occurred around human-related data on the local social-economic dynamics. The human demographer attempted working with existing statistical data while a Network member worked backward to generate data from the implicit knowledge of the participants. The human demographer worked with United Nations population data for Brazil, which showed an overall reduction in fertility and population growth rates and an absolute decline in Brazil's rural population since 1970. These suggested that overall, mere population growth would not be a likely cause of pressure on the muriqui. A CD-ROM made available at the workshop provided population data on all states and counties of Brazil. This enabled us to determine the overall pattern for Brazil found in each of the specific muriqui habitats. Discussions with local conservationists confirmed that in all of the habitats there was a real out migration of people to urban areas. Thus it was clear that threats to the species and habitat came from specific human activities, and not from sheer numerical pressure. It was further determined in discussion with local conservationists that there were two very different kinds of human activities that threatened the species. One of these could be quantitatively estimated, the other could not, but could easily suggest a policy that would protect the muriqui.

The third dynamic involved the generating of fine-grained data on human impact on the muriqui habitat. A Network working-group facilitator tried to generate data backward. For instance, researchers mentioned that the harvesting of palmito (palm tree heart) was detrimental to both the quality of habitat and the muriqui in the Carlos Botelho State Park. Both muriquis and humans compete in consuming palmito.

Harvesters collecting palmito usually degrade the habitat of the muriqui and they even sometimes kill muriquis for human consumption.

Similarly, participants identified that the central problem in Intervales was the spreading of coffee cultivation, which affected the muriqui habitat. The facilitator then tried to estimate the habitat destruction caused by the increase of coffee-growing activities. He asked participants to estimate roughly how many hectares of forests were transformed every year into cultivation areas, and how this affected the muriqui habitat. In one habitat it was clear that land clearing for pasturage was reducing habitat and threatening the species. Using past data on land clearing, it was possible to make a projection of expected habitat area for the near future. This quantitative estimate could be fed into the PHVA modeling.

In protected habitat of Carlos Botelho, it was not habitat destruction but direct harvesting of the muriqui for meat that was identified as threatening. Here the problem was illegal harvesting of palm hearts, as the harvesters shot the animals for meat for consumption during their stay in the park. However, if obtaining this somehow generic information was straightforward, documenting it further turned out to be a challenging task for both the facilitator and the participants. The Network facilitator teased out specific information from the participants, for instance asking about the monthly frequency of harvester incursions into the forests and of muriqui casualties, in order to obtain data usable in the modeling exercise. To answer these specific questions, participants made rough estimates based on their personal experience in the park and made phone calls to local informants to complete this deficient information.

Unfortunately, it was not possible to make a quantitative estimate of the extent of either palm heart or monkey harvesting, but the discussion produced one important policy alternative. The state could legalize and promote palm heart plantations, which would provide a larger and far cheaper source of palm hearts, thus diverting activity from the illegal harvesting that incidentally led to monkey harvesting. Perhaps an agricultural economist or rural sociologist would have moved the discussion further toward quantitative estimates, although the absence of resource users at the workshop most likely would have hampered this discussion as well.

Altogether, these efforts to shift from general to specific information and to generate data for the modeling exercise did not bear the expected fruit. It turned out that the data generated by the Human Impact group was partial and was finally not integrated into the modeling process. This absence of social data in the modeling further confirmed the biology-dominated debate.

Overall Dynamics

At the more general level, the workshop process was characterized by a dual level of participation and involvement by younger and more senior conservationists. The (generally) younger graduate students were industrious and actively involved in both working groups and plenaries. They were actively involved especially in the human impacts working group. On the other hand, the more senior conservationists seemed to be less directly involved in discussions and plenaries. Some seemed to play other roles, since they had the final say and were in control of what was going to be implemented. In general, the power dynamic between the two groups was a dependent one. The case of the forest administrator who was not present to share information at the workshop, but communicated via phone with the senior conservationists to negotiate a land purchase is illustrative of the dual system of operation. The majority of participants were working on the issues of individual working groups, while a few senior members were floating in and out of the working groups—discussing the workshop, research funding for ongoing projects, and a land purchase with other decision makers like this forest administrator. While the workshop lacked the forest administrator's direct input, the power of a partnership with him was present in the senior conservationists side operations. It is very likely that this dual operating system had consequences for the actual dynamics and recommendations of the workshop.

Workshop Recommendations and Conclusions

The workshop may have been a success from the biological conservation science point of view. However, its success was more limited in terms of social science and strength of the recommendations that were generated. Many participants agreed that important and quality biological information was synthesized at this workshop. For instance, by systematically reviewing the status and distribution of the species throughout its range, the data suggest that the status of the species could be revised downward from critically endangered to endangered. In fact, Dr. Seal observed that the Caratinga population had the most complete population data set for the modeling exercise that he has seen. Unfortunately, the working groups were not able to go much beyond the biological recommendations and integrate human use information into this model. As a consequence, recommendations were also research centered and not related to conservation problems as a whole.

Network Learnings

In Montreal a month following the workshop, we reflected on our expectations for the workshop and highlighted the lessons we learned. The
Network had expected that integration of a human demographer and
model into the workshop process and incorporation of social science
and "human" information into conversations through broader stakeholder participation would enhance the workshop recommendations and
information available for the model. The following lessons were learned
from this workshop: the Network and CBSG need to refine preparation
efforts and provide a broader context when defining workshop goals to
interest other stakeholders; we refined our expectations for the role of
expert knowledge and information needs; and we identified the need for
strong facilitation to tease out needed information and encourage a broad
problem definition, especially when the experience of participants is
homogeneous.

Lessons on Preparation and Participation

We had expected that expanded participation would advance Network
goals. We were not able to test this hypothesis because we were unable to
affect the participant makeup. While the goal of the Network was to have
local resource users and social science professionals at the table, the only
additional participants were: the Network members themselves. The
desired expansion in expert participation was limited due to the lack of
local professional counterparts and resource users.

This led the Network to address the preparation phase of the workshop in general and of the muriqui workshop specifically. The two questions we raised were: (1) what was done during the specific preparation
phase of this workshop that resulted in predominantly biologist participation? and (2) what can we do in future preparations to broaden the participant base?

In-country preparation for this workshop was done with little contact
or direction from CBSG. While the Network initially perceived the organizers' familiarity with the workshop process as a benefit to expansion,
it may have actually hindered efforts to expand the process. We hypothesized that established personal relationships would create a more responsive environment for Network suggestions. Instead, organizers had a
preconceived idea of the workshop process as mainly a species conservation–centered meeting. In the lion tamarin workshop conducted in the

early 1990s with the same organizers, two workshops were conducted, with the first workshop focusing on the biology of the species. Nearly six years later the second workshop occurred and it addressed in more detail local human population interactions and included a broader range of stakeholders than the first tamarin workshop. In the case of the muriqui, the Network expectations were different, attempting to include a broad range of stakeholders from the beginning of the process.

This workshop led the Network to hypothesize new ways to improve the preparation phase and more specifically to sensitize local organizers to human issues that impact the conservation of a species. The first solution proposed was to increase lead time in order to work with the local organizers to deliberately explain the expanded workshop process. The second was to improve the message sent to the organizers. This includes rephrasing the letter sent to the organizers and invitees, making explicit the reasons for expanding the process to social scientists and local stakeholders, and also proposing a list of questions for the local organizers, a heuristic model to help the local organizers frame the problem statement a broader way. (See chapter 18 for discussion of this model.)

In addition to raising questions about why a broader range of professionals were not present at the workshop, the Network noted that the biological basis of this process was unable to engage professionals from a more human-centered domain; this was noted by both the Network human demographer and the Brazilian social activist who left the process early. This inability to attract social scientists led to discussions of the focus of workshops.

Within the Network, questions were raised about the overall workshop process—should the issues addressed be broad to incorporate new stakeholders, or should we maintain the biological species focus of workshops? Up to this point, Network efforts had revolved around the "right" people providing the "right" questions. After the muriqui workshop, discussions moved toward locating or developing a tool to use in the workshop process that would extract or organize the broader information on species-specific impacts inherently available in any given workshop.

Lesson on the Role of "Experts"

Our second lesson concerned the role of experts. As in Uganda's mountain gorilla workshop, we attempted to bring human demographic data into the workshop discussions. It was the Network's hope that having this data would facilitate more-detailed discussion and quantification of resource use and other threats by humans to the species and its habitat. The challenge in gathering site-specific relevant information proved especially difficult.

First, the Network revisited the relevance of our efforts to include social scientists in the workshops. The first question we explored was related to the contribution social scientists may give to the workshop process. Was it possible to find the "right" social scientist or community activist? What kind of knowledge and what kind of specialty do we think will contribute the "right" information? Were there local social scientists or activists interested in or working on the relationships between human institutions, the use of local natural resources, and the preservation of biodiversity?

The case of the social activist from the Intervales region is illustrative on this point. While the animal population data from Caratinga was modeled and could possibly be used to extrapolate to other regions, there was little correlation between the discussions of human threats to the species around Caratinga and Intervales. That is, one could not say that because small landowners banded together to protect the muriqui in a community reserve, with land area dependent on the price of coffee in Intervales, that habitat availability at Caratinga was dependent on coffee prices. The site-specific nature of the biological data in Caratinga could not be linked to the site-specific nature of the community data in Intervales. The Caratinga muriqui population survives almost exclusively on one family's forest reserve area and is dependent on the family's stewardship of its habitat. Even if we are able to find social scientists whose work overlaps with animal habitats, will the site-specific nature of social research be able to significantly contribute to PHVA workshops and how will social scientists reluctance to extrapolate data to other regions affect that contribution?

Secondly, we reflected on and identified two factors that may have influenced the lack of social scientist interest: the workshop format and the scale of the species problem definition and information needs. Both in Uganda and Brazil, the Network discovered a lack of interest from the traditional social science disciplines in the applied, short-term workshop format. It seems that this cross-disciplinary research is generally discouraged by the way academic work is structured. Furthermore, the scale of information needed for the biological model differs from that of social research. There is a need for data that is site specific, yet generalizable across the species range. To provide both the scale of data necessary and the translation and modeling ability, we would need to find social scientists who are interested in conservation issues and who have conducted studies in the specific landscape we are dealing with or in a similar, comparable situation.

These many conditions are unlikely to be met, making the "right" person with the "right" local information a rarity. It seems that while biological scientists are running up against the human influences on "their" discipline, the same interactions are not yet forcing social scientists to move beyond their focus and seek answers in partnership with other disciplines. In addition, it seems the species scale of the problem definition may be creating some of the lack of interest among social scientists.

It also became clear then that in addition to developing new tools for integrating social information into the workshop process, identifying local stakeholders beforehand and including them in the process was a task more likely to bear fruits than attracting professional experts.

Lessons on Facilitation during the Workshop

We observed that participants, who broadly agreed on the need to conserve the muriqui and came from a homogeneous milieu (all were active in conservation or biology), tended to shortcut the phase of problem analysis and propose recommendations at the start of the workshop. To find reasons for this, we further investigated the relationship between facilitator and modeler. The first facilitation note is that in Belo Horizonte, the primary workshop facilitator was also doing the modeling. This has been a unique situation in Network workshops. This may have focused the discussion quickly into the modeling phase and away from broader problem analysis and brainstorming.

Second, there was quite a lot of information on the biological issues of the muriqui available in the room. We also found that even though there was very little apparent social information available at the workshop, and we had to "tease" informants to get information available in the public arenas or based on their local experience, the necessary information does often exist. What is needed is a "broker," a challenger, to elicit the information relevant for conservation—a facilitator with the ability to translate human activities into biologically relevant data. For instance, participants in the management working group identified palmito harvesting as a threat to the muriqui. The role of the "broker" then consisted in asking for specific numbers related to specific places and occurrences so as to construct data that could be used to get an improved and detailed picture of the situation.

This case highlights the need for a much more intensive facilitation and for persons to translate all of this local information into data that can be used to understand the connections between humans, the muriqui, and its habitat.

Aftermath of the Workshop

The main outcomes of the workshop from the Network perspective may have been unexpected ones. We were trying to include human demographic data and test whether human in-migration mattered and it turned out that the areas close to the muriqui habitats had out-migration trends instead. There was a human demographer present to help with modeling, but without local resource-use information, it was difficult to translate the results of the demographic model and data into something meaningful in the species population model (see chapters 13 and 18). We tried to encourage social scientists and resource users such as landowners to participate and instead, a homogeneous population of conservation biology specialists attended. However, it is likely that contacts between senior conservationists who have access to resources and junior conservationists were made in Belo Horizonte. These contacts may have helped conservation initiatives around the muriqui. A Brazilian NGO centered on the muriqui was to be established at the end of the meeting. This case highlights the potential role of conservation entrepreneurs, or boundary spanners, that move across places and mobilize resources. The workshop provided a situation that made this mobilization of resources possible.

From the point of view of the biologists and conservationists involved in the muriqui workshop, it was felt that the analysis had provided a good opportunity to pull together data on the muriqui as well as to explore threats. There was discussion about setting up a biosphere reserve as a means of habitat protection.

From the Network perspective, this workshop illustrated the need not only for social data but for brokers who can connect human dynamics with the dynamics of the habitat and of the species. One potential way to address this problem is to visit the area of interest before organizing a workshop, as a means to introduce more diverse participation and information at the workshop. This is how and why we decided to conduct an abbreviated Participatory Rural Appraisal in order to prepare for the workshop on the Papua New Guinea tree kangaroos that took place in September 1998 (see chapter 9).

 **Chapter 8**

Building the Back Loop: Community Decision Making and the Peary and Arctic Islands Caribou PHVA in Northern Canada

GEORGE FRANCIS

This Population and Habitat Viability Assessment (PHVA) case study was of interest to the Network because it addressed a cross-cultural situation set within a transition of governance in northern Canada. The transition arose from the Inuvialuit Final Agreement (DIAND 1983) and the Nunavut Settlement Area agreement (DIAND 1993), both of which require comanagement arrangements with the Inuit (Inuvialuit), and provide for a degree of self-government. The region of interest for the Peary and Arctic Islands caribou was defined by the geographic distribution of five metapopulations (populations linked by dispersal) in the high-latitude west-central Arctic Islands; these caribou populations come under a divided jurisdiction based on the new governance arrangements (figure 8-1). These populations are considered to be either endangered or threatened, and a "recovery team" first formed in 1990 had been developing recovery plans for the caribou. The PHVA was to help expedite this work toward developing and implementing plans in the context of comanagement.

At the Network meeting in Banff, Alberta, in November 1997, members discussed the need to incorporate more societal considerations into the PHVA design, including those that are carried out in a cross-cultural context. The Peary and Arctic islands caribou opportunity arose rather suddenly after this meeting when a Government of the Northwest Territories (GNWT) representative in Canada contacted the Conservation Breeding Specialist Group (CBSG) directly about convening a PHVA that might help expedite the completion of a recovery program for these caribou. The Network was interested in this situation, as mentioned above, because such a PHVA would have to address governance transition and

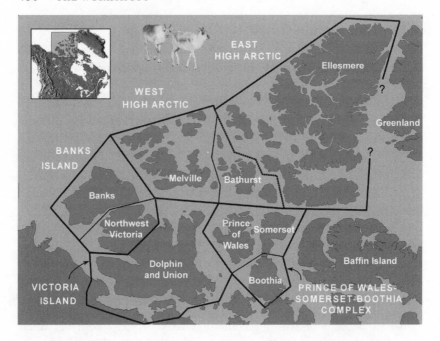

Figure 8-1. Distribution of the Peary and Arctic Islands caribou *(Rangifer tarandus pearyi)* in the northern islands of the Northwest Territories and Nunavut, Canada. Boundaries identify the individual metapopulation units that became the focus of working-group discussion and detailed VORTEX-based analysis. (Courtesy of Department of Resources, Wildlife and Economic Development, Northwest Territories.)

cross-cultural issues. Governance was moving from the centralized policy-making and sector-based administration of the GNWT and the federal government toward comanagement arrangements with the Inuit Indigenous peoples.

Comanagement places greater emphasis on the use of Inuit traditional ecological knowledge systems and on community-level decision making (see chapters 12 and 15 for further discussion of this). Community decisions are especially important for matters relating to wildlife, and particularly for caribou, which are highly valued animals in the Inuit cultural tradition. Governance, therefore, requires a more balanced approach between "top-down" and "bottom-up" initiatives, set within a cross-cultural context. This presented a challenge to both the Inuit and the Euro-Canadian participants.

This chapter summarizes a Network member's observations of this PHVA, with particular focus on the influence of governance systems on the conservation management plans discussed at the workshop. The chapter discusses the background of the species and its habitat, a summary of events at the workshop, and finally, the lessons and observations on governance and conservation as they relate to the larger Network goal of broad participation.

The Context of Caribou Conservation

The profound institutional changes in governance in the Canadian Arctic region began in 1973 when Inuit organizations started negotiating a greater degree of independence from both the federal and the territorial governments. The Inuit realized that they constituted a relatively homogeneous cultural majority in the region, but they had no treaties recognizing their aboriginal rights and they were geographically remote from government administrative centers in Ottawa and Yellowknife.

The Inuvialuit Final Agreement (IFA) for the 1.3-million-square-kilometer western Arctic region of the Northwest Territories (NWT) was officially approved in 1984. Besides the extensive marine component (Beaufort Sea), it covers some 435,000 square kilometers of lands traditionally used and occupied by a combined total of 2,500 Inuit in six settlements. The IFA created two major policy and administrative bodies. The Inuvialuit Game Council, along with community hunters' and trappers' organizations (HTOs), deal with wildlife and renewable resource issues. The Inuvialuit Regional Corporation deals with development and business interests associated with six nonprofit community corporations. The Game Council oversees five comanagement bodies, including the Wildlife Management Advisory Council for the NWT (DIAND 1987; WMAC 1988).

In 1993, an agreement was reached for The Nunavut Settlement Area (DIAND 1993), a geographic region in the central and eastern Arctic of some 2.1 million square kilometers, with an Inuit population of about 20,500 grouped into twenty-seven widely dispersed communities (with non-Inuit residents, total population for the area is about 25,000 people). This agreement came into force in April 1999, creating a new Canadian territory. It provided for a Nunavut Legislative Assembly, ten administrative departments, and eight agencies. These were decentralized into three subregions, including Kitikmeot, which covers the part of Nunavut associated with the region of interest for the caribou. The departments, especially the Department of Sustainable Development, were expected to work closely with five comanagement boards created by the agreement (Legare 1997). One of these was the Nunavut Wildlife Management

Board, which had identified studies on wildlife harvests at the community level as a top priority, working closely with community HTOs.

The Peary and Arctic Islands caribou included five distinct metapopulations distributed over some 758,000 square kilometers of the high-latitude west-central Arctic islands. The human population in this region was almost entirely Inuit. The 1996 census recorded just under 5,000 people living in eight highly dispersed settlements ranging in size from 135 to 1,351 people (622 on average). Although relatively small, this population was growing quite rapidly. The 1986 census of the same settlements reported 3,800 people, and the settlements ranged in size from 114 to 1,002 people (425 on average).

The new governing structures arising from the two major agreements provided for a greater degree of self-governance, within the constraints of a near total dependency upon the federal government for funding and for comanagement forms of administration. Issues of formal self-government, as distinct from comanagement, remained ambiguous. One result of these agreements is that two of the five metapopulations of Peary and Arctic Islands caribou fall under divided jurisdiction of the Inuvialuit region and Nunavut and thus, under two sets of comanagement arrangements. Two settlements and one metapopulation of caribou occur entirely within the Inuvialuit region, while six settlements and two metapopulations of caribou occur entirely within the territory of Nunavut.

The conservation status of wildlife in Canada is determined by the Committee on the Status of Endangered Wildlife in Canada (COSEWIC). Created in 1979, COSEWIC includes representatives from the twelve Canadian provinces and territories, the Canadian Wildlife Service (part of the federal Department of the Environment) and three national nongovernmental organizations (NGOs), the Canadian Wildlife Federation, the Canadian Nature Federation, and the World Wildlife Fund (Canada). COSEWIC commissions recognized specialists to conduct status reports for taxa that may be at risk in order to provide the documentation needed for recommendations on conservation status. One of its first decisions was to list Peary caribou as "threatened," meaning that the species were likely to become "endangered" (or "extirpated") if limiting factors on them were not reversed. In 1991, the two most northerly populations of Peary caribou in the High Arctic and on Banks Island were reassessed as endangered (faced with imminent extirpation), while the remaining populations in the Low Arctic were still classified as threatened (COSEWIC 2000).

Whenever there is sufficient information to enable consideration of conservation measures, a "recovery team" of specialists is convened by COSEWIC under the REcovery of Nationally Endangered Wildlife program (RENEW), which was created in 1988 in order to respond to the growing COSEWIC list of taxa (usually species) at risk. In 1990, a five-person recovery team was formed for Peary and Arctic Islands caribou. In September 1992, the team issued a discussion paper in the form of a draft recovery plan, for federal and territorial government officials and Inuit organizations and communities to consult. The recovery team also continued to conduct seasonal field work in order to get better information on the population size and movements of discrete populations of Peary and Arctic Islands caribou over their immense geographic range in the Canadian Arctic. A revised draft recovery plan was prepared in 1994 and circulated for comments. Caribou numbers remained difficult to pin down. Severe winter conditions in different parts of the Arctic during 1994, 1995, and 1996 decimated some of the known populations, especially on Prince of Wales Island, northwest Victoria Island, and the Boothia Peninsula.

In 1993, a biologist with the GNWT had suggested that a captive breeding option also should be considered. Arrangements were approved in 1996 to capture twenty-five animals from Bathurst Island for transport to the Calgary Zoo, but the operation was cancelled at the last moment because of bad weather. Objections from some Inuit communities lead to cancellation of a plan to repeat the attempt in 1997; this may have stimulated interest in a PHVA, which then led to the invitation received by CBSG in late 1997.

The Department of Resources, Wildlife, and Economic Development (RWED) of the GNWT viewed the workshop as an opportunity to identify management options for long-term caribou recovery plans that would then be taken to communities and wildlife management boards for their consideration and input. Some staff within the host department apparently had apprehensions about the workshop. These partly concerned potential Inuit sensitivity to importing "outsiders" who might not know or respect the cross-cultural context in which the workshop consultations and follow-up would have to be done. Other underlying concerns came from doubts of at least one field biologist (and perhaps others) about the scientific value of population-modeling exercises; because modeling focuses on the risks faced by populations once they became small, the ecological processes that caused the population decline in the first place might be overlooked. The RENEW process had been underway for caribou for the previous seven years, and some disagreements were emerging among the Inuit over the process being followed. This conflict was compounded by frustration inherent due to difficulties in obtaining good data on the populations of some caribou herds and the legacy of disagreement over the captive breeding option.

Pre-Workshop Activities

There was less lead time for the caribou workshop than is typical for a CBSG PHVA. The Network had not planned to include this workshop as one of our cases but because of its multi-stakeholder and comanagement character, it struck us as particularly interesting. We had hoped to send two Network members, in addition to the author and the two CBSG members, but were unable to do so due to the short lead time. We were also unable to do the same kind of background research that we did for the mountain gorilla and muriqui workshops. Nonetheless, the caribou workshop represented a unique opportunity to observe a process in which Indigenous stakeholders played a central part.

The Workshop Dynamics: Content and Process

Organized by the RWED staff, the workshop was held in Yellowknife, NWT, from February 27 to March 1, 1998. Besides three people from CBSG, there were thirty-two participants, about one-third of whom were Inuit from seven communities across the region of interest representing regional or community HTOs. Other participants were affiliated with the RENEW team, GNWT departments, and the three main comanagement bodies for wildlife (the Inuvialuit Game Council and its Wildlife Management Advisory Council, and the Nunavut Wildlife Management Board). Simultaneous interpretation was provided for the plenary sessions. Logistically, there were a few complaints about insufficient consultation with the comanagement bodies about the workshop, and about the fact that the briefing books were not available to most participants beforehand.

The opening plenary presentations and subsequent rounds of discussions to identify issues and concerns took up much of the first day. Five of the participants were asked, or volunteered, to be facilitators of working groups and reviewed the "flip chart" listings to identify major themes for the working-group sessions. These were population biology and modeling, factors affecting the caribou, information needs, and comanagement. The participants self-selected into the different working groups, and some groups reformulated their provisional titles and themes as they got underway. Except for the population modeling group, the Inuit participants opted for involvement in each of the other three groups, especially the one discussing factors affecting caribou populations. This group may have been perceived as the most conducive for drawing upon the traditional ecological knowledge that people from the HTOs brought to the workshop.

There was good participation in both plenary and working-group sessions throughout the workshop (figure 8-2). The process of engaging everyone in discussions and seeking consensus or common ground conclusions seemed to resonate well with the two cultures present. CSBG occasionally helped facilitators move discussions along so that the sequences of problem analyses, needs analyses, and suggested actions were duly discussed. As one working group reported "everyone did take part and there was sometimes frustration but no personal animosity."

Workshop Recommendations and Conclusions

Several themes of particular interest emerged from the workshop and underlie the recommendations that came from it. They may be summarized as follows.

Figure 8-2. Field researchers discuss demographic data to be used in the VORTEX analyses, Peary and Arctic Islands caribou PHVA workshop, Yellowknife, NWT, Canada.
(Photo courtesy P. Miller.)

Comanagement issues emerged in several ways. The working definition of comanagement adopted was similar to definitions found in the
literature. Operationally, understandings of comanagement came down
to different views about the process for involving communities in the
design of research, monitoring, and management programs in which traditional ecological knowledge would be prominent. There was an implicit
questioning of the legitimacy of centralized initiatives, especially those
from the federal government, as exemplified by questions raised about the
RENEW recovery process and the then pending federal endangered
species legislation. (This legislation was subsequently delayed for a number of reasons.)

The importance of community decision making is reflected in the
report from the working group that discussed factors affecting caribou
populations. For those factors that may be susceptible to a management
response, the group added a proviso: "Appropriate actions at this stage
consist of providing the communities with information on certain items
so that more informed decision-making can take place." Items for which
information was especially needed were highlighted in the group's report.

The captive breeding question was a special case in point, and served
as a lightning rod for issues about how consultation and decision making
should be carried out. Both the factors and the comanagement working
groups discussed captive breeding, noting that opposition to translocating
caribou to the Calgary Zoo arose because the process for deciding to
undertake the plan was disrespectful both to the caribou and to the knowledge of Inuit elders. An appropriate approach was suggested along the
lines used to translocate five muskoxen calves for a research herd at the
University of Saskatchewan in May 1993. In that case, elders were con-
sulted, some local residents were taken to see the facilities for the animals
at Saskatoon, and periodic reports on the well-being of the animals were
made.

Of the many factors affecting caribou populations, few could be dealt
with at the community level. Possible climate change effects are an
example. There are still critical needs for more information on population dynamics, harvest rates, and the impact of weather events on the
productivity and survival of animals. Questions were raised about the
appropriate scale for conservation planning (i.e., metapopulations or
individual herds) and about whether the purpose of drawing up management plans was to meet the sustained need of communities, or the "long
term ecology of the caribou."

Biological problems were addressed in terms of possible biological solutions, as noted in the report from the factors working group. The Inuit, however, have developed economic solutions to the problem of low caribou numbers for many years at a time. There is an extensive kinship-based trade in "country food" over large areas extending from the high-latitude Arctic Islands to mainland areas in both the Inuvialuit region and Nunavut territory. The implications of such strategies for managing caribou or possible ecological effects of this practice were not addressed.

Network Learnings

While in other workshops the Network added new steps to the PHVA process, the Network role in this case was more as an observer. This workshop occurred early in our learning process and it is valuable for reflecting on governance and representing native interests in the PHVA. In terms of the Network's interest in the workshop, the following conclusions might be drawn:

- The workshop processes were compatible with the two cultural traditions present, but the VORTEX dependence on numerical data was not easily reconciled with the Inuit tradition of narrative storytelling to communicate their traditional ecological knowledge.
- The workshop succeeded in identifying many of the issues that needed some attention, especially ones that required local community understanding and involvement.
- The hope that a PHVA might expedite completion and acceptance of a recovery plan for the caribou was not realized. Biological studies and discussions with stakeholders continue, as they have since 1992.
- It is not clear whether or how the larger contextual issues, including different perceptions and interpretations of the caribou situation, are being addressed. The continuing work is focused on issues of population biology.

Aftermath of the Workshop

The Peary caribou recovery team (which had been increased to nine people) has continued field studies to estimate the size of different populations, track seasonal movements, and obtain mortality estimates (especially during seasonal ice-crossings); it has also continued genetic analyses of population structures and relationships among caribou on the different islands. Community-based monitoring of winter conditions and the diet of wolves has been initiated. The team reports meetings with stake-holders to draft regional implementation plans; it was scheduled to final-ize and approve action plans for the Inuvialuit Settlement Region during 2001 and to draft plans for the Nunavut caribou populations (RENEW 2000). The aerial surveys to assess calf production and overwinter sur-vival of caribou are a continuing priority (www.gov.nt.ca/RWED).

Is there a management solution for caribou recovery? Based on the biological model of population dynamics, recovery might be enhanced by controls over human-caused mortality. Other than hunting restrictions, the feasibility or effectiveness of other population protection measures seems limited. The proximate causes of caribou die-offs are associated with weather events, notably deep snow or icing conditions that prevent the animals from grazing at critical periods in the spring and fall. The VORTEX simulations generated a number of extirpation outcomes under different combinations of caribou reproduction and survival rates, har-vesting practices, and weather-induced catastrophes (Gunn, Seal, and Miller 1998).

The environmental and ecological effects of atmospheric and climate change are expected to be particularly severe for Arctic ecosystems. Deeper snows, along with more frequent thaw/freeze events and freezing rain may increase episodes of starvation among caribou. Warmer sum-mers may enhance the production of forage plants and provide a longer growing season, but may also increase the harassment of caribou by biting insects (e.g., Wahl 1995; Brotton and Wall 1997). There is also some spec-ulation about disruptions of migratory patterns. In reference to the pos-sible impacts of climate change in the Canadian Arctic, Maxwell (1997, 51) notes in passing that "North of the mainland, High Arctic Peary Cari-bou and muskoxen may become extinct"; changes in sea ice conditions could disrupt migration of caribou between some of the islands.

While some protection of caribou calving areas may be appropriate, the extensive movements of the animal make parks and protected areas relatively ineffective as a conservation tool. Indeed, this was not a topic considered at the workshop. Inuit communities have introduced voluntary hunting bans during periods of scarcity of caribou. The country food trade may partly compensate for this. Collings, Wenzel, and Condon (1998) review the history of food-sharing networks among Inuit in the central Canadian Arctic as well as current practices in Holman, a ham-let within the Inuvialuit agreement region where one of the metapopula-tions of caribou also occurs. They note that the practice of kinship-based obligatory sharing of seal meat had declined somewhat, but informal exchanges of country foods (including caribou when available) were still widely practiced among people living in Holman. If population cycles are comparable to the four-phase cycle of ecosystems described by Holling and Meffe (1996), then the meat trade can be seen as a management strategy for adapting to the "back loops" leading from population crashes to population build-ups, a process that may take two decades or more for particular caribou herds.

Rodon (1998) notes that comanagement can be interpreted as (a) integration or co-optation into a state model of management, (b) an intercultural transaction process that blends different values and knowledge, or (c) empowerment leading toward greater self-determination. From this perspective, the Nunavut Wildlife Management Board was deemed by Rodon to exemplify the integration model. The most extensive experience of comanagement for caribou comes from the Beverly-Kaminuriak Caribou Management Board, established in 1982. Usher (1993) reviews the experience of this board, which presides over a large migratory herd of barren-ground caribou on the mainland southeast of the Peary and Arctic Islands caribou. The board had been successful in directly linking user groups to management, in assuring that subsistence hunting of caribou remained a priority, and in coordinating research and monitoring among the jurisdictions through which the animals seasonally migrate.

However, from an aboriginal perspective, Usher notes that the board did not fulfill their aspirations for self-government or self-management, and reliance upon a "management biology model" to incorporate field data on population numbers and productivity relegated traditional ecological knowledge to little more than anecdotes; "so long as the management biological model is hegemonic, traditional knowledge cannot be an effective guide for action" (1993, 117). The modeling exercises at the workshop may also have been perceived this way, since the Inuit participants avoided the modeling group in favor of other groups addressing other themes. In chapter 15, the challenge of working with traditional ecological knowledge will be discussed at greater length.

Stevens (1997) summarizes issues of conservation measures on Indigenous peoples' lands in "frontier" regions from a global perspective. He notes that the key concerns of Indigenous peoples are about ". . . sovereignty, self-determination, land tenure, resource use rights, fulfillment of agreements, culturally sensitive enforcement of protected areas regulations, development, coordination with national NGOs and governments, and tourism management" (280). This is quite similar to the context in which issues concerning the Peary and Arctic Islands caribou now have to be addressed.

In the end, given the environmental conditions under which these animals survive and the possible effects of climate change on them, it is not immediately apparent that a management solution is feasible.

Chapter 9

Incorporating Local Knowledge: Landowners and Tree Kangaroos in Papua New Guinea

PHILIP J. NYHUS, JOHN S. WILLIAMS, JENNA S. BOROVANSKY, ONNIE BYERS, AND PHILIP S. MILLER

New Guinea is one of the largest, most rugged, and most biologically diverse islands on the planet. Its mountainous topography, limited infrastructure, and remoteness have restricted the study of the island's flora and fauna. Among the least studied of the island's marsupials are the elusive tree kangaroos (figure 9-1) and little is known of the abundance, distribution, or the threats facing these animals. One, the Scott's tree kangaroo is thought to inhabit no more than 40 square kilometers; another, the Matschie's tree kangaroo, is found only on the Huon Peninsula of northern Papua New Guinea (PNG), where much of its potential habitat has already been degraded (figure 9-2). Given the limited information available about wild tree kangaroos and the very real risk of extinction facing several species, the need to summarize existing information and to develop a concerted research and conservation plan is urgent (Kennedy 1992; Flannery, Martin, and Szalay 1996).

It was in this context that we began one of the most logistically ambitious experiments of the Network's efforts to incorporate greater stakeholder involvement in the Population and Habitat Viability Assessment (PHVA) workshops. To assess the current state of knowledge about the abundance, distribution, and threats to the eight species of tree kangaroos currently thought to occur in PNG, a workshop was held August 31–September 4, 1998, in Lae, PNG, with an emphasis on assessing the status of the Matschie's tree kangaroo (Bonaccorso et al. 1999).

The Conservation Breeding Specialist Group (CBSG) was invited to hold the workshop by the PNG National Museum, Rainforest Habitat, and the PNG Department of Environment and Conservation (DEC). The workshop was intended to be a forum to enable collaboration among

Figure 9-1.
The tenkile, or Scott's tree kangaroo
(Dendrolagus scottae).
(Photo courtesy M. Vincent.)

local agencies, stakeholders, and international scientists. Several international and local groups and individuals suggested the workshop process as essential to ensuring the survival of tree kangaroos in PNG. Dr. Timothy Flannery of the Australian Museum, researcher of tree kangaroos and co-author of *Tree Kangaroos: A Curious Natural History,* was the first to call attention to the need for such a workshop and the strategic planning that would follow (Flannery, Martin, and Szalay 1996). The Marsupial and Monotreme Taxon Advisory Group of the Australian Zoo Association and the PNG National Museum endorsed the need for a workshop to be organized and held in PNG.

The Network chose the tree kangaroo workshop in PNG as an ideal case study; because relatively little published biological information was available on tree kangaroos, information from local people would take on added significance. In addition, PNG represented an example of a developing country with low human population densities. The decision to include an interdisciplinary Network field team in this workshop was made approximately three months before the workshop. This evolved out of concerns about how to incorporate local community hunters and landowners in a meeting that was organized and designed by scientists and resource managers. Many of the world's most endangered species are located in areas that are remote and where scientists know little about their basic biology or threats. We hypothesized that Indigenous knowledge

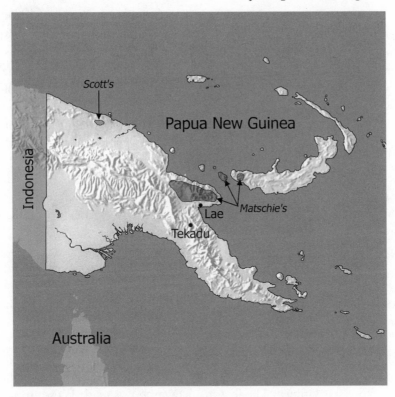

Figure 9-2. Map of Papua New Guinea showing (major provinces) the location of the workshop in Lae, the location of the preworkshop field research in Tekadu, and distributions of Matschie's and Scott's tree kangaroos.

could augment and guide the collection and use of data in the workshop, and broader stakeholder involvement could help to drive recommendations and actions that would be locally relevant and politically and culturally feasible.

This chapter explores the context and process of the tree kangaroo workshop, describes the Network's involvement before and during the workshop, and explores the lessons we learned about incorporating local knowledge and broader stakeholder participation into the wildlife assessment process.

The Context of Tree Kangaroo Conservation

Papua New Guinea, covering the eastern half of the island of New Guinea (the western half is Indonesia's Irian Jaya Province), encompasses almost half a million square kilometers (figure 9-2). The main island and its satellite islands are home to more than 200 species of mammals, 700 species of birds, and 25,000 species of plants, many found nowhere else on the planet (Alcorn 1993; Paijmans 1976). Tropical evergreen forest still covers nearly 70 percent of the island's area. A number of mammal species, including one species of tree kangaroo, have been scientifically identified only during the last two decades (Flannery 1995).

PNG also is one of the most culturally diverse countries on the planet and is home to an estimated one-third of the living languages on earth. The country is unique in that an estimated 97 percent of its land area is still held under customary tenure (Lynch and Marat 1993). As a result, the central government exerts relatively little control over how resources are used and relatively little land (3–4 percent) is designated as formally controlled protected area. Shifting cultivation still predominates, creating a complex matrix of forest and agricultural lands (Gagné and Gressitt 1982).

Grassroots participation and approval of methods and plans by landowners is considered essential to the success of wildlife management and conservation efforts in PNG because wildlife is generally considered the property of the landowner, not the government.

Biology and Status of Tree Kangaroos

Tree kangaroos (*Dendrolagus* spp.), are found only in Australia and the island of New Guinea. The first tree kangaroo was made known to science in 1836 (Groves 1982); the most recent was made known in the 1990s. Tree kangaroos are members of the family Macropodidae, which is represented in New Guinea by eighteen living species. Tree kangaroos are the only member of this family adapted to an arboreal existence. Six species and ten taxa (species and subspecies) of tree kangaroo are known to inhabit PNG (Flannery, Martin, and Szalay 1996). All six species are endemic to the island—that is, they are found there and nowhere else. Relatively little is known about the ecology of these elusive animals because of their arboreal lifestyles and the inaccessibility of their rugged habitat (Hutchins et al. 1991). They range in size from 6.5 to 17 kilograms, have long, coarse fur; short, broad feet; a tail densely covered with long hairs that often form a brush at the tip; and are found in both lowland and mountainous habitats (Flannery 1995). Three species occur in very small geographic ranges totally within PNG: Matschie's *(D. matschie)*, lowlands *(D. spadix)*, and Scott's *(D. scottae)*. Three other species, Doria's *(D. dorianus dorianus)*, grizzled *(D. inustus inustus)*, and Goodfellow's *(D. goodfellowi goodfellowi)*, have somewhat larger distributions that include

Table 9-1.
Status of tree kangaroos (after Bonaccorso et al. 1999).

Taxon	IUCN Status
Doria's *(D. dorianus dorianus)*	Lower Risk
Finsch's *(D. inustus finschi)*	Vulnerable
Golden-mantled *(D. goodfellowi pulcherrimus)*	Critically Endangered
Goodfellow's *(D. goodfellowi goodfellowi)*	Vulnerable
Grizzled *(D. inustus inustus)*	Vulnerable
Ifola *(D. dorianus notatus)*	Vulnerable
Lowlands *(D. spadix)*	Lower Risk: Near Threatened
Matschie's *(D. matschie)*	Vulnerable
Scott's *(D. scottae)*	Critically Endangered
Seri's *(D. dorianus stellarum)*	Lower Risk
Timboyok *(D. goodfellowi buergersi)*	Vulnerable

Irian Jaya Province, Indonesia. Table 9-1 outlines the conservation status of tree kangaroo populations.

In the last few decades, logging, mining, agriculture, and the development of infrastructure to meet the needs of the country's growing and economically developing population have put tremendous stresses on the country's ecosystems. International logging companies in particular have managed to obtain government sanctioned concessions for large areas, even when the concessions violated existing forestry legislative provisions and landowner rights. The threats of large-scale deforestation are particularly significant for tree kangaroos. Because of their slow reproduction and vulnerability to hunting, even moderate habitat disturbance and hunting pressures can have significant impacts on their populations (Flannery, Martin, and Szalay 1996).

Preworkshop Activities
The Network team members arrived in PNG one week before the workshop. The team was organized to use rapid participatory appraisal tools to carry out a village-level assessment of community knowledge about the tree kangaroo and to evaluate how local communities affect tree kangaroos and their habitat. The team had prior experience with human demography, wildlife management, community conservation, and rapid Participatory Rural Appraisal (PRA). The group was assisted by a representative of a local nongovernmental organization (NGO), the Foundation for People and Community. This person had worked in the

community since 1994 in a conservation and development project and served as a guide and interpreter. The team hoped to develop a better understanding of local-level issues affecting tree kangaroos and to gather village-level data on human population growth, resource use, and habitat change and hunting of tree kangaroos that could be incorporated into the modeling process at the workshop. Plans were also made to work with the village representatives that were to attend the workshop to enable more effective participation in the workshop process.

The team's research objectives were to gather qualitative and quantitative data on human impacts affecting tree kangaroos and tree kangaroo habitat, including habitat quality, such as food, predators, water, and pollution; size of protected tree kangaroo habitat; habitat fragmentation (e.g., the number of habitat areas and barriers that might prevent tree kangaroo dispersal); and species exploitation, such as subsistence hunting and trade for profit.

The field site was the remote village of Tekadu and the surrounding settlements in the Lakekamu Basin of Morobe Province (figure 9-2). This area was selected because it was logistically accessible on short notice, we had access to a guide and interpreter, and two villagers from this area were scheduled to take part in the workshop. This area was inhabited by at least two species of tree kangaroos, but not by Matschie's tree kangaroos.

The team members first met with their guide the night before the trip to the village. There was an extended discussion of the purpose of the visit and the logistics and methods to be used. There was spirited discussion among the team about the workshop methods to be used and the role of the team members as researchers versus a more activist role (e.g., making suggestions regarding ecotourism or family planning). The member of the team with the greatest experience with PRA techniques supported a more activist approach, while other team members supported a more passive, participant-observer approach. By the time the team reached the village, it was agreed that the team would use participatory methods but focus on the biology of the tree kangaroo, its habitat, and human threats. Box 9-1 outlines the tools used for rapid participatory appraisal while in the village.

Arriving at the village's small grass landing field by small aircraft, the team's plane was immediately surrounded by a large group of people; these people were in the village to meet judges who were making their first visit to the community in order to try a number of court cases in Tekadu. The judges' visit in some ways made the arrival of the research team less

Box 9-1.

Tools for rapid participatory research used in preparation for the workshop.

The Network team used the following tools during the pre-workshop visit to Tekadu and the Lakekamu Basin. These methods were used to gather information about tree kangaroos, basic village demography and land- and resource-use patterns, and other information affecting tree kangaroos and their habitat.

Village mapping: For each of four small communities in the area, knowledgeable individuals were asked to identify each household; the number, age, and sex of individuals in each household; and the major livelihood strategies associated with each household (fishing, hunting, and gold panning).

Interviews: Interviews were held with representatives of the local health clinic, the local school, the regional court, and with the community "councilor." Additional interviews were carried out with village elders and a number of "average" villagers.

Village transect: Several villagers walked the team from one end of the village to another, and pointed out major geographic and cultural features (rivers, gardens, hills, forests, houses, government buildings) in and around the village.

History: A village history was completed with the help of several village elders from approximately World War II to the present. Questions were asked about major natural and cultural events that have helped shape the village.

Landscape-level map: A map of the surrounding villages, landscape features (rivers, mountains, valleys), and major cultural features (airplane runway, trails) was made with the assistance of several local community members.

Wildlife resource use: A group of hunters was asked to identify all the different species of mammals commonly hunted in the area. These animals were identified in the local language, and a description of each animal was written as well. These hunters were then asked to rank these animals by their relative abundance. The list was narrowed by asking them to first rank each animal as 3 (very abundant), 2 (abundant), or 1 (not abundant). The animals identified as 2 or 3 were then ranked using a "paired ranking" technique by comparing each animal to every other animal. The most common (or at least most commonly hunted) animals then received the largest total score.

Tree kangaroo knowledge and hunting: To examine current knowledge about tree kangaroos and hunting practices, interviews were held with hunters and other villagers in which they were shown pictures of tree kangaroos and asked to identify the common species. Landowners were also asked to recall how many tree kangaroos they had killed, where they had killed them, and the type of animals they killed.

Seasonal calendar: A one-year calendar was made that identified major rainfall patterns, agricultural patterns, and specific hunting practices, generally and with regard to tree kangaroos.

Miscellaneous activities: Interviews were held wherever possible, such as during the local market and with individuals who came to talk with us. Several group exercises, such as a map showing the altitudinal distribution of tree kangaroos, were also carried out.

intrusive. The court proceedings taking place during the next few days proved an unexpected backdrop to our work, making it more difficult to spend time with some villagers, but enabling the team to more easily pursue its activities unobtrusively and to gain access to people from surrounding villages.

The team began with a village transect walk to better understand the underlying economy of the valley and its geography. This served as a base for later mapping, interviews, and group interviews with hunters and other community members (figure 9-3). The team was hampered by not having any accurate maps of the area. In hindsight, the team could have included additional areas in the transect walk.

In order to obtain information on the distribution, movement, and size of the local human population, the team interviewed a number of individuals, including the school master and the health worker. In addition, a local artist was introduced to us and we were able to persuade him to draw maps of both the communities and the surrounding landscapes. These maps became an important tool to encourage the active participation of other villagers, who provided information to place on the maps.

Figure 9-3. Villagers in the Tekadu area of eastern Papua New Guinea work on activities during the PRA conducted by Network members before the PHVA workshop. (Photo courtesy Jenna Borovansky.)

With the aid of these maps, we were able to complete a detailed census of the local population of about 80 households by sex and age. This information became the starting point for the analysis of hunting, population pressure, and human population change. The maps were helpful in defining the areas of hunting; the maps also provided a backdrop to the stories that were later told about hunting in general and the characteristics of the tree kangaroo population in particular. However, the maps were less helpful for determining the size of the area being hunted. If these locally drawn maps had been linked to aerial photographs or other existing maps, we may have been able to determine more accurately the nature of hunting pressure, the abundance of animals, the presence of mining or logging in adjacent areas, and to better estimate the population of tree kangaroos.

The discussions around hunting started with the maps. A "paired ranking" technique was used to determined the names and relative abundance of the mammals in the hunting area. After the completion of this task, the group relaxed and entered a storytelling mode. But it was mostly in the storytelling mode that we found out about how the hunting took place and that only a very few hunters in the village, those with skilled hunting dogs, ever saw or killed tree kangaroos. These discussions proved useful at the workshop in helping guide our estimates and analysis of hunting pressure on tree kangaroos.

We came out of the village with some general information about the population of tree kangaroos in the Lakekamu Basin area. We had no evidence that there had been a significant decline in the tree kangaroo population during the previous two decades. We learned that the village's population had increased slowly over the decades, but guessed that hunting pressure on the tree kangaroos may not have risen because a smaller proportion of men in the village were now hunting. Fifty years ago, all the men of the village reportedly hunted, now fewer than half did. Firearms were not widely used and it seemed likely that the traditional hunting (with bows and arrows) still in use in Tekadu did not yet pose an immediate threat to the future of the tree kangaroos in that area. On the other hand, it was clear from discussions with some villagers that if roads opened up and large-scale logging occurred, or if human population growth continued, a situation common in areas closer to the coast, the impact of these activities could have a disastrous effect on local populations of tree kangaroos.

Workshop Dynamics: Content and Process

The workshop itself was held at the University of Technology (Unitech), Lae. Forty-five people attended the workshop. Participants from PNG, Australia, and the United States, including specialists in population biology, captive management, reproduction, veterinary medicine, and human demographics, compiled and analyzed both published and unpublished information on all six tree kangaroo species of PNG. Thirteen local landowners, representing several regions of PNG where tree kangaroos are found, participated in the workshop. While the PHVA focused largely on the Matschie's tree kangaroo, this workshop was unique among the Network cases in that a general conservation assessment was done concurrently for all of PNG's tree kangaroo species.

The workshop began with formal introductions and background presentations about the modeling process and tree kangaroo biology. In the afternoon, the floor was given to a hunter from the Huon Peninsula, who brought out a plastic sack containing samples of food plants that provided the preferred diet for the tree kangaroo. This presentation immediately sparked a discussion among the landowners about the similarities and differences between plants in their areas, and also sparked a barrage of photo-taking by the foreign participants (figure 9-4). This watershed event appeared to "break the ice" and changed the tone to a more accessible discussion of issues surrounding the tree kangaroo. It appeared that having "one of their own" take center stage helped to empower the landowners and to encourage them to participate more actively in the workshop process.

Figure 9-4.
A landowner (left) and American field researcher (right) present information on tree kangaroo ecology to workshop participants, tree kangaroo PHVA workshop, PNG.
(Photo courtesy P. Miller.)

Following the introductory presentations came an issue-generation session that set the stage for collaboration on all issues raised by stakeholder groups. This straightforward, grounding exercise identified the interests of each stakeholder group. The groups were self-identified as landowners, captive managers, and biological and social scientists. The issues highlighted during this exercise served as a reference point to ensure consideration of all stakeholder groups' concerns throughout the workshop. After this initial issue-generation session, participants self-selected into mixed-stakeholder, topic-based working groups, including life history and modeling, government and legislation, and socioeconomic issues. Plenary consensus sessions each day served two critical purposes: to help keep all issues at an equal level and to assure that responsibility for management and policy recommendations was taken by all participants.

Landowner Participation

Thirteen landowners came from communities located across the country and spoke different languages, so they communicated with each other and with other workshop participants using Pidgin. Three workshop participants who could speak Pidgin and English served as translators.

The landowners provided a significant amount of information that became a valuable part of the workshop process. For example, considerable information about the distribution and life history of tree kangaroos was obtained from those with hunting experience. Without the landowners present, much of the information available during the workshop would have been limited to a relative handful of publications and one researcher's preliminary field information. The landowners also expressed a number of unique concerns and issues that may have not otherwise arisen (see box 9-2). For example, they expressed concern that many of their fellow villagers were not following existing wildlife regulations.

Throughout the meeting, landowners also helped to suggest specific methods for carrying out recommendations at the village level. By actively taking part in the workshop, the landowners could return to their villages and share information and recommendations with a wider audience, expanding the benefits of their participation during the actual workshop. Formally, the landowners only represented themselves as individuals, but indirectly they represented the concerns and issues of their communities.

On the final day, the landowner stakeholder group reconvened to evaluate the progress of each working group in relation to the needs their

Box 9-2.

Landowner stakeholder group issues generated at the workshop.
(Source: Bonaccorso et al. 1999.)

The goal of this issue-generation exercise was to define the concerns and needs of each group member. All group recording and discussion was done in Pidgin; what follows is an English translation.

- Many habitats of the tree kangaroo are vanishing (due to mining, logging, oil palm).
- People go against the Wildlife Management Area regulations for the tree kangaroo and other animals; they want to hunt as much as possible; landowners don't have enforcement power.
- There is not enough information and experience for experts to teach locals and those responsible for looking after tree kangaroos in the wild and in zoos.
- The population of tree kangaroos is going down.
- There is additional hunting pressure from immigrants and increasing human populations (e.g., from Wau—people coming in to mine).
- New technology is being used to hunt (e.g., guns).
- (As a landowner who hunts tree kangaroos, he has hunted with dogs now and in the past,) he doesn't know how many tree kangaroos are in his bush.
- Landowners have information about the tree kangaroo (e.g., what tree kangaroos eat, where they are, when they breed) to share and haven't been given the opportunity to share.
- There is little time to think about conservation because of other social needs (health, transportation, education).

---

stakeholder group had expressed on Day 1. The group met for nearly an hour and then a plenary session was held during which all groups presented their conclusions. It was a tense moment. The landowners were asked if they were pleased with the progress of the workshop. There was silence. They were asked again and still silence. It was made clear to us that we could not end the workshop unless the landowners expressed their opinion to us all. Finally, one landowner stood up and spoke passionately for quite a long time. The non-Pidgin speakers held our breath wondering what was being said. In the end, it was translated that the landowners were very pleased with the workshop results and by the attention being paid to their wildlife. All the landowners began clapping and the Network team almost passed out from the relief!

There was one other tense moment in this plenary session when one of the landowners said that they were wasting their time talking to DEC because the government would not help them. The DEC representative responded saying that he knew that everyone was angry but that landowners do not understand that DEC wants to help but cannot due to lack of funds. Participants were reminded that on the first day of this workshop they had all agreed to a set of guidelines including that they assume good intent on the part of all participants. Everyone confirmed that they had made that agreement and then each stakeholder group was asked directly, beginning with DEC, if they were willing to work in good faith to realize the implementation of the workshop recommendations. They publicly agreed to do so and, from that moment on, DEC representatives were much more engaged and animated. It was a shame that this conflict was not aired on the first rather than the last day, but it was a significant incident and another turning point that was extremely rewarding.

The workshop highlight was the input and level of participation of the local landowners. We were told repeatedly that the local landowners had never before had the opportunity to voice their concerns and have them addressed in this type of forum. (In fact, the PNG Conservation Needs Assessment conducted in 1993 was reportedly picketed by landowners who were angry at being excluded from the process.) We learned an incredible amount from the landowners and they told us that the level of commitment and assistance shown by the international conservation community during the workshop pleased them.

Workshop Recommendations and Conclusions

The inclusion of both the landowners and a Network team specifically contributing data on the human dimension of species conservation had several impacts on the workshop process. The field research team had gained a basic understanding of how a village worked, the institutions present, general household livelihood strategies, and how resources were used in one area. In particular, the team was able to determine the proportion of households that hunt, when they hunt, and what they hunt, estimating how many tree kangaroos households hunt and the relative importance of these animals to the hunters.

Several tools were used during the workshop to better incorporate information from the landowners, the team's field activities, and additional threat information that surfaced during the workshop. Box 9-3 outlines these tools, which included several mapping exercises, the population projection program DEMPROJ (Stover and Kirmeyer 1997), and discussions with Geographic Information Systems (GIS) specialists at Unitech in Lae to gather habitat and other spatial data about Matschie's tree kangaroo habitat.

Box 9-3.
Interdisciplinary tools used at the workshop to incorporate
local knowledge into the population model.

Human population projections: Data from the 1980 PNG census were collected from the Unitech library to augment data collected during our village trip. These data were analyzed with the computer program DEMPROJ (Stover and Kirmeyer 1997) to develop a thirty-year projection of the population in that area. A second set of projections was made using data from the Huon Peninsula, home of the Matschie's tree kangaroo. These data were obtained indirectly from a GIS database at Unitech.

Tree kangaroo extraction rate estimates: Based on interviews with two landowners and a field biologist from the Huon Peninsula, and on additional information from our field activities, crude estimates were made of tree kangaroo hunting rates for the Huon Peninsula.

Habitat quality and area estimates: Information about land use and hunting pressure from interviews, published data, and Unitech's GIS database were used to estimate total habitat area for the Matschie's tree kangaroo.

Catastrophe estimates: Interviews with landowners and government officials at the workshop were conducted to estimate potential natural and human-made catastrophic events that could significantly affect the Matschie's tree kangaroo. These included the prevalence of major drought and flood events and the likelihood of major deforestation by logging.

VORTEX modeling: Relevant data gathered before and during the workshop were shared with the population biology modeling group and integrated into the VORTEX modeling activities.

Several constraints limited the team's ability to incorporate some of this information into the larger workshop process generally and into the VORTEX simulation models specifically. The field data were not from the Huon Peninsula, the home of the Matschie's tree kangaroo. The team had some qualitative information but limited numerical data due to a lack of access to maps and quality population data; and we were concerned about applying information from Tekadu to the Huon Peninsula. Finally, the ability to share information among groups during the workshop did not work as well as was expected because there was limited communication among working groups.

Nevertheless, several specific benefits came out of the Network team's participation and suggest how the pre-workshop activities were able to provide unique data for the actual workshop. The following examples describe how we were able to ask qualitative questions in the field and translate them into quantitative applications within the workshop. These in turn resulted in specific and significant conservation recommendations.

How Important Is Hunting to Livelihood Strategies?

To answer this question, a map of the villages was drawn and knowledge-able villagers were asked to identify houses and the primary livelihood strategy of people living in those houses. We found that most villagers in this area have a garden where they plant a range of annual and perennial crops. Fishing is the most important activity, but hunting is a principal household strategy for almost one-third of the houses in four villages (table 9-2).

Table 9-2.
Principal household livelihood strategies in PNG by village.

Village	Number of Households				
	Hunt	Fish	Goldpan	Other	Total
One Mile	4	6	6	0	16
Yenuwe	4	9	8	1	22
Korama	8	11	4	3	26
Piarogamango	3	4	3	0	10
Total	19	30	21	4	74

What Is the Estimated Human Population Size and Expected Population Growth Rate?

To answer this question, we summarized the number of people per household, their approximate age, sex, and marital status. Population projections were made based on past census data, estimates and projections of mortality and fertility, and anecdotal evidence with regard to patterns of migration. Using the demographic software package DEMPROJ, the population of the Tekadu area was projected over the next two decades. The projection showed that the population was expected to increase approximately 100 percent, from 320 to 614 people. Over recent years, the proportion of villagers who actively hunt has declined.

The population projections for the Huon Peninsula showed that the population was growing faster, that the area was less sparsely populated, and that there could be significant increases in human pressure on the Matschie's tree kangaroo over a twenty-year time horizon.

What Do Villagers Hunt?

During the field research, village elders and other recognized "expert hunters" were asked to summarize the principal mammal species they hunted. This list of animals was then ordered using a paired ranking technique whereby each species was compared to every other species and the more commonly hunted was given a mark. In total, villagers identified more than forty different mammal species that they commonly hunted, from those living in the treetops to ground-dwelling animals. Information regarding the relative number hunted and locations of species by altitude was described. (Had the team been more familiar with PNG species, the descriptions of where tree kangaroos were found and their numbers relative to other mammals could have been used to determine population information and habitat ranges.) Little information was available to the team to cross-reference the villagers' knowledge. Additional information was collected to summarize the total number and species of tree kangaroos that selected hunters had killed over the years.

How Much Do Villagers Hunt and What Is the Impact on Tree Kangaroos?

To derive the estimated number of tree kangaroos killed by hunters throughout the Huon Peninsula, the team began by talking to one of the landowners from that area over several hours one evening at the PHVA to determine how many tree kangaroos he had killed or captured, the size of the area from which these were obtained, and the time it took him to kill them.

First, we found that he had killed a total of seventeen tree kangaroos (ten males and seven females) in a six- to seven-year period. Two were juveniles (one male, one female). Nine of these were taken from his land and the remaining eight were taken from other people's land in the immediate area. These data are summarized in box 9-4.

Second, to derive an estimate of the number of households in his area that hunt, the team asked him to describe all the households in his village (a total of forty-one) and then the number that hunt (eleven). He described four hunters as "serious hunters" and seven as not serious hunters. He described himself as a serious hunter and, in fact, probably one of the best. This information was used to estimate that approximately 25 percent of the village population hunts. This was a similar estimate derived from detailed household data from four different villages in the Tekadu area, gathered by the same team. Moreover, based on this study of four villages

Box 9-4.

Example of translation of local information from
landowner interviews into VORTEX input data for the hunting
of Matschie's tree kangaroos around villages in the Huon Peninsula.

Expert Hunter's Tree Kangaroo Harvest			
Extraction Type	Male	Female	TOTAL
Killed on his land	5	1	6
Killed on other's land	4	2	6
Captured on his land	1	2	3
Captured on other's land	0	2	2
TOTAL	10	7	17

Village Hunting Information	
Total households	41
Serious hunters (estimated 2 tree kangaroos/year)	4
Less-serious hunters (estimated 0.5 tree kangaroos/year)	7
Proportion of hunters	11

Data Summary	
N	= Total population size living at > 1,800 m
Villages near suitable habitat	= 0.5 (estimated that ½ villages suitable)
Households per person	= 0.2 (estimated 5 people/household)
Hunters per household	= 0.25 (proportion that are hunters)
Hunting rate high	= 1 tree kangaroo/hunter/year
Hunting rate low	= 0.5 tree kangaroo/hunter/year

Formula for VORTEX Input	
High extraction rate	= [N] x [Suitable Villages] x [Households] x [Hunters] x [Rate]
	= [N] x [0.5] x [0.2] x [0.25] x [1]
	= [N] x 0.025
Low extraction rate	= [N] x [0.5] x [0.2] x [0.25] x [0.5]
	= [N] x 0.0125

in the Tekadu area, the team concluded that the average household size
was about five people. This estimate was applied to the Huon area.

Third, assuming that the four serious hunters mentioned above killed
tree kangaroos at a similar rate as the landowner being interviewed
(2 per year), and the less-serious hunters killed only a quarter as many (0.5
per year), we estimated an average of 1 tree kangaroo killed per hunter
per year (box 9-4).

Fourth, the team obtained from both landowners and a researcher with
considerable field experience an estimate of how many villages in their
area had forest cover that was suitable for tree kangaroos. Out of the seven
villages listed, three probably had no tree kangaroos in the forests surrounding them, three had adequate forest, and one had excellent habitat.

When similar questions were posed to a second landowner, he estimated that eight of twenty-one villages in his area were "close to bush." The team used these figures to estimate that roughly half of the villages in the peninsula could be expected to have tree kangaroos in the surrounding forest habitat.

With the above information and estimates in hand, the team began deriving a way to estimate the rate of extraction of tree kangaroos from a particular habitat segment. For the Matschie's tree kangaroo, we assumed that only those villages situated above about 1,800 meters would be in or near suitable hunting habitat. Therefore, if the total human population in such an area is known, the total number of hunters within that population can be calculated and, assuming a particular annual rate of tree kangaroo extraction per hunter, the total number of animals removed can be estimated. For example, if the total human population in the region of interest is 10,000, the annual number of tree kangaroos extracted from a particular population would be between 125 and 250 per year.

Obviously, a critical piece of information required for this analysis is the total human population size in the region of interest (N). Unfortunately, we were unable to gather this information during the workshop. As a result, we were forced to make a more simple set of assumptions concerning the total rate of extraction of tree kangaroos from a given area. However, the discussion presented above provides a set of guidelines that can serve as a starting point for extended analyses on the severity of hunting pressures on wild tree kangaroo populations throughout New Guinea.

To evaluate the effect of hunting on tree kangaroo populations, we assumed that adult females were preferentially removed from the population, primarily because they are slower when they are with joeys (offspring) and therefore easier to catch. In addition, when a female is caught the joey that is with her is also removed from the population (i.e., either eaten or kept or sold as a pet). We assumed that the joeys were male or female with equal probability. Finally, we assumed that hunting resulted in an additional 2 percent mortality among adult females. Therefore, if the baseline adult female mortality rate is 8 percent in a nonhunting scenario, this rate becomes 10 percent when hunting is added to the model. Once again, this estimate of additional hunting-based mortality is largely based on educated guesswork; the primary piece of data required for a more precise estimate of hunting mortality—total tree kangaroo population size within a given forest patch—was not available. Given the total land area available within individual patches and very crude preliminary estimates of tree kangaroo population densities, we surmised that a hunting rate of 2 percent among local villagers was a reasonable estimate. We realized, however, that this estimate could easily be in error by a factor of two to five times.

The models developed from this information demonstrated that current hunting rates could indeed be a major force influencing the future viability of tree kangaroo populations in PNG, particularly when those populations are quite small (figure 9-5). In addition, workshop participants were able to observe the serious demographic consequences of preferential hunting of adult females over males in this polygynous species, an unexpected result for those engaged in hunting throughout the country. Despite our interest in developing increasingly insightful models of wildlife population viability, it is important to remember that neither sophisticated demographic models nor comprehensive field data are always necessary to generate insights vital to the conservation decision-making process.

Other Conservation Recommendations

Additional focus was directed towards the country's most critically endangered population: the tenkile, or Scott's tree kangaroo. This species is reduced to as few as 50–100 individuals in a single locale within the Torricelli Mountains in the Lumi District of northwestern PNG. The precipitous rate of recent population decline and, more importantly, the serious risk of imminent extinction, as revealed by the data collection and population modeling efforts, were instrumental in the immediate formation of a multidisciplinary "rapid-response team" at the workshop. This new team (Team Tenkile) was composed of representatives from zoo and wildlife management organizations in PNG and Australia and was tasked with traveling to the Torricellis in order to update the conservation status of this rare taxon (see Aftermath of the Workshop, below).

Network Learnings

In our postworkshop assessment, several key process elements were apparent. A significant challenge faced by the Network team was the relatively limited time available to prepare for the trip, the limited amount of data available before and during the trip, and the limited knowledge the team had of PNG and tree kangaroo biology. Team members cumulatively had experiences carrying out similar activities in several countries, but were hampered somewhat by the lack of experience in this very unique country. On the positive side, the different backgrounds and

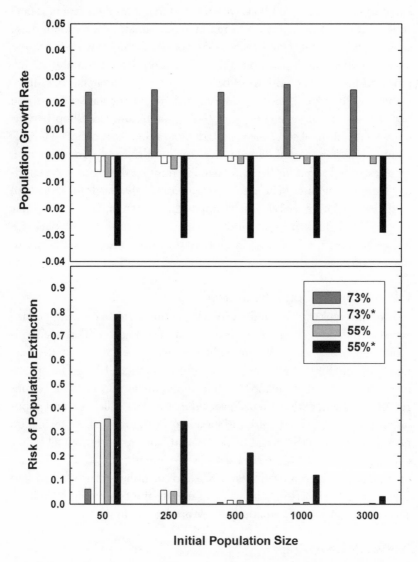

Figure 9-5. Results from simulation modeling conducted at the tree kangaroo PHVA workshop. Stochastic population growth rate (top) and probability of extinction (bottom) for simulated PNG tree kangaroo populations under alternative levels of female breeding success (73% or 55% adults breeding in a given year) and additional mortality of adult females and juveniles of both sexes resulting from hunting by local human populations (designated by * in the legend).

(Source: Bonaccorso et al. 1999.)

perspectives of the team members strengthened the overall effectiveness of the team.

The most significant constraints were the difficulty of coordinating and planning our activities with counterparts in PNG, our limited access to data and maps, and our reliance on external factors (e.g., availability of a guide and transportation) that drove our choice of location.

In comparing the PNG experience with the earlier mountain gorilla and muriqui workshop experiments (chapters 6 and 7), several common comments and suggestions were identified (table 9-3): a need for more lead time and preparation; a need for more nonbiologists with a stake in the animal to be present at the meeting; discussions about the need for and utility of having separate human impact groups, modelers, or facilitators; and the difficulty of translating human impact information into VORTEX.

In any endeavor such as this, two overriding concerns are whether the cost and time necessary to carry out field research prior to a workshop are balanced by the value added to the workshop process. On one hand, this exercise provides the opportunity to better understand the situation at the village level and to gather information and insight that might otherwise be unavailable, especially if villagers are not present at the workshop. On the other hand, the activity takes time and money and there is a risk in placing significant weight on data gathered rapidly by people with limited experience in the host country.

The inclusion of social issues into the workshops requires a considerable degree of planning and preparation, which requires time and commitment from both CBSG and from the workshop organizers. This planning should begin at the outset and take into consideration the following questions:

1. Which are the communities that have the greatest impact on the species?
2. How may stakeholders living in these communities should be represented at the workshop?
3. Should field work be done in these villages prior to the workshop, and how should that be supported?
4. Who will gather analysis of human population, hunting and other local practices, GIS information, and bring that to the workshop? (These data could include census data, health and mortality data, GIS maps, and information on extraction activities such as mining, timber, and hunting.)

Table 9-3.

Comparison of the workshop process experience with
Network involvement in Uganda, Brazil, and PNG.

| | Workshop Location and Species | | |
Network Experience	Uganda (Mountain Gorilla)	Brazil (Muriqui)	PNG (Tree Kangaroo)
Human demographic data	Available, difficult to translate into impacts or to integrate with VORTEX	Available, but numbers alone not a major impact on muriqui	Country-level and some local data available, projections made
Expanded stake-holder groups	Few nonbiologist experts, lack of understanding about role of nonbiologists	Lack of experienced nonbiologists and few social scientists	Excellent local landowner participation, little additional nonbiological participation
Role of Network	Difficulty in facilitating due to lack of expertise, need for dedicated human process modeler	Limited value of Network participation due to language differences and small groups, need for dedicated project leader?	Little information shared with population biology group, some value to landowner participation, need for dedicated project leader
Preparation	Lack of lead time to prepare and include human dimension focus	Lack of lead time, need for planning checklist	Lack of lead time, better development and incorporation of planning checklist
Problems	Lack of exchange of information with biology group, disaffection of human impact group members	People and data available, but need longer lead time and broader stakeholder involvement	Field visit to "wrong" location, lack of maps or good biological information
Human dimension information for VORTEX	Limited quantitative human dimension information	Some habitat and species-specific quantitative data	Limited quantitative human dimension data, some hunting data

5. What social scientists might be brought to the workshop: anthropologists, demographers, economists, community development specialists?

While PNG is unique in having most of its land area owned by the people (family groups and clans) and not the government, the successful inclusion of local stakeholders at this workshop seems to lend support for efforts to include similar people at future workshops, especially when little research has been previously completed on the target species. Further expanding the number and experience of participants would probably have enabled us to gather more threat information. For example, we had little expertise regarding logging and mining, two serious threats to tree kangaroo habitat in PNG.

Aftermath of the Workshop

One criticism of the PHVA workshop process, and others like it, is that these meetings often have little tangible, lasting, and measurable impact. In PNG, however, several important products came directly from the workshop. Several recommendations were implemented during the workshop itself, including the development and translation into Pidgin of educational materials specifically requested by the landowners. In addition, a rapid-response team was formed at the request of landowners to address the urgent need for conservation action directed toward the critically endangered Scott's tree kangaroo. This team committed to visit Lumi District, Sandaun Province within thirty days following the workshop.

As a result, Team Tenkile was formed (an interdisciplinary, multinational recovery team initiated at the workshop that has since evolved into an incorporated NGO called the Tenkile Conservation Alliance) and traveled to the Torricelli Mountains in Lumi. This initial visit resulted in a two-year moratorium on hunting tenkiles that was signed by thirteen villages. Plans were also developed to establish a Tenkile Field Station and project based at Lumi Station; the captive breeding and rerelease of tenkile, provided specimens are not removed from the Lumi area; and a community-formed management committee. The Tenkile Conservation Alliance logo will be used on letterhead, T-shirts, and promotional materials, which will help communities identify with the conservation work they are doing in partnership with the Alliance.

In addition, a draft species recovery plan was published in March 2000 (Vincent, Slater, and Clark 2000), and efforts are underway to accomplish some of the tasks identified in the recovery plan. Work is beginning on the production of a community education booklet, funds are being generated, and Taronga Zoo in Australia is working with a major supplier of rice to PNG on a set of collectable tree kangaroo information cards that will go in the bags of rice.

Conclusions

The tree kangaroo workshop in PNG was the third workshop to explicitly address the Network's efforts to incorporate the human dimension into the PHVA workshop process. This workshop was unique because a team of three Network representatives traveled to a village area before the workshop and a significant number of workshop participants were local landowners.

Meetings were held in Minneapolis and Montreal before the trip to discuss goals, objectives, and integration of human dimension data with VORTEX. In PNG, the team gathered qualitative and quantitative information about village demography, institutions, healthcare, agriculture, tenure, household strategies, specific information about the hunting of tree kangaroos, and other biological and nonbiological information from the Tekadu area in southern Morobe Province.

At the workshop, some information from the field was shared with the population biology modeling working group and other information was used in the social issues and education working group. Constraints faced by the Network team included limited information from the Huon Peninsula, home of the Matschie's tree kangaroo; limited quantitative nonbiological data that could be translated into VORTEX; and limited communication among different working groups. Landowner participation was strong at the meeting and the inclusion of local knowledge and concerns were an integral part of the workshop.

The value of involving local landowners included but was not limited to their significant knowledge about tree kangaroos and their ability to raise unique concerns, questions, and recommendations. This experience contradicts the assumption by workshop participants at the Uganda mountain gorilla workshop that the "sophistication" does not exist at the local level to participate in a science-based workshop (see chapter 6). The tree kangaroo PHVA workshop highlighted the robustness of the PHVA process in its ability to absorb and incorporate local knowledge. This exercise illustrated the value of including local representatives who are knowledgeable, understand the goals and objectives of the exercise, and can communicate their perceptions and those of their constituents.

 **Chapter 10**

Uneasy Guests: The Grizzly Bear
PHVA in the Central Canadian Rockies[1]

EMMANUEL RAUFFLET, HARRIE VREDENBURG,
AND PHILIP S. MILLER

The Network viewed the grizzly bear Population and Habitat Viability Assessment (PHVA) workshop as an excellent possibility to examine the conditions for involving larger stakeholder groups, particularly industry, and for integrating human-related and biological data. Favorable conditions for larger stakeholder participation included the involvement of two Network members and existing links with a local organization already working on grizzly bear conservation in the Canadian Rocky Mountains: the Eastern Slopes Grizzly Bear Project (ESGBP). Headed by bear biologist Steve Herrero of the University of Calgary, the ESGBP had already created an arena around grizzly bear conservation in the Central Rockies Ecosystem. In addition, scientific studies indicated a strong correlation between human access to range territory and grizzly bear mortality. The intent of this workshop was therefore to focus on the VORTEX model in order to integrate habitat and bear biology data with regional human socioeconomic activities.

In this chapter, we reflect on preparations for and events during the workshop, the lessons we learned, and the aftermath of grizzly conservation eighteen months after the workshop. The chapter is organized into six main sections: the regional context and main challenges for grizzly conservation; the preparation for the workshop, including the Network's expectations; what actually happened at the workshop; workshop recommendations; our learnings; and the aftermath of the workshop.

Context of Grizzly Bear Conservation

Alberta, Canada, is thought to have hosted about 6,000 grizzly bears *(Ursus arctos)* nearly three centuries ago; the current census estimates suggest the grizzly population is just 800 individuals (figure 10-1). Both human settlements and activities have increasingly affected grizzly habitat and population over the last decades. The demographic growth of the Calgary area and patterns of spatial use—including extractive and outdoor- and tourism-related activities—have had a strong impact on grizzly populations. Increased human population due to the booming oil industry and an expanding tourism sector created rapid growth in and around Calgary, including in Banff National Park. A younger population attracted by industry and a Rocky Mountain lifestyle is augmented by an aging population attracted to the area for retirement and recreational pursuits. In the area of the Banff National Park (6,641 km²), questions about grizzly bear management arose in the early 1990s in light of the enlargement of the Trans-Canada Highway and increasing recreation developments in the Bow River Valley.

The individual and cumulative effects of these factors threaten the ecological integrity of the grizzly bear's home range. Over more than twenty-five years of radio data, 31 out of the 32 (97 percent) recorded grizzly bear deaths have been directly related to human activities. Of the 25 mortalities where location could be determined, 24 (96 percent) were within

Figure 10-1. Eastern Slopes grizzly bear *(Ursus arctos horribilis).*
(Photo courtesy J. McCormick.)

500 meters of a road or 200 meters of a trail. This demonstrates the very strong relationship between human access and grizzly bear mortality.

The ESGBP was founded in 1994 as a concerted effort to respond to the increasing scientific and societal concerns for grizzly bear survival. In addition, the ESGBP was established in relation to three societal elements: (1) federal and provincial legislative changes that broadened the scope of environmental assessment and protection by including cumulative effects at the landscape scale; (2) new research findings that evidenced the declining grizzly bear population and the need for interagency management; and (3) the growing awareness of the discipline of conservation biology. The ESGBP consists of researchers and various regional stakeholders who underwrite the research. Its focus is on cumulative effects assessment regarding current and proposed developments on bear populations and habitat, mainly in the Central Rockies Ecosystem (CRE), which encompasses Kananaskis Country and the Banff, Yoho, and Kootenay national parks (figure 10-2). The "greater ecosystem" perspective of the Bow River Valley began when the ESGBP focused on 11,400 square kilometers of the watershed region within Alberta. Within this region, the grizzly bears' home range crosses many different land management jurisdictions, including the Banff, Yoho, and Kootenay national parks; Peter Loughheed, Bow Valley, Assiniboine, and Elk Lakes provincial parks; Kananaskis Country in Alberta; provincial Crown lands; the Stoney First Nation; and private land holdings. This region is roughly bisected into east and west by the Continental Divide, and into north and south by the Trans-Canada Highway. Grizzlies move across these sections and exchange individuals within a variable dispersal frequency. Grizzlies have been trapped and fitted with radio collars and their movements are monitored over the entire Bow River watershed.

Preworkshop Activities

In the Network meeting immediately preceding the grizzly bear workshop, there was considerable excitement about the potential for wider stakeholder inclusion. Here was a far-ranging indicator species of western Canada that spanned two Canadian provinces, three American states, national parks, provincial parks, forest reserves, and ranch grazing lands. In this instance, the stakeholders were already collaboratively organized through the ESGBP. We anticipated, therefore, that this workshop would bring together a richer set of expertise than was typical of PHVAs, as the ESGBP included graduate students, park managers, representatives from

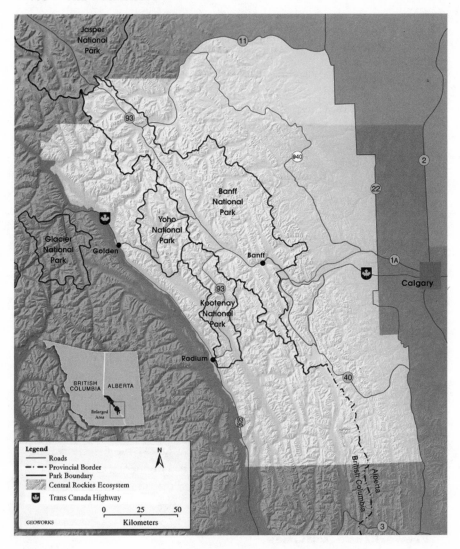

Figure 10-2. Map of the Central Rockies Ecosystem, western Canada.
(Courtesy S. Jevons, Geoworks.)

the Ministry of Natural Resources, oil and gas company environmental managers, local forestry saw mills, tourist industry representatives, and nongovernmental organizations (NGOs) concerned with grizzly conservation, including the World Wildlife Fund, Canadian Parks, and the Wilderness Society. We saw that our challenge would be to include the ranchers and aboriginal groups in the region who were not involved in the ESGBP. We also hoped to encourage industry representatives to bring detailed information on tourist activities and numbers, road and pipeline construction, and human demographic data. Since Alberta has a relatively low population, we did not anticipate any problems getting this kind of information.

We expected that the institutional and stakeholder complexity would play a significant role in the workshop and felt that an understanding of the institutional overlap between territorial jurisdictions and mandates might prove useful. A separate initiative, spearheaded by two Network members, gathered detailed information about the institutional overlaps in the CRE. Information was gathered about what kinds of plans and policies were in place, what kinds of jurisdictions were in place, who was responsible for the implementation of what programs, and which types of activities were allowed in what regions. The idea was to overlay these institutional arrangements on grizzly "hot spots" (areas most important for grizzly survival) to determine where pressure on institutional arrangements might have the most beneficial effects in terms of enhancing survival of grizzly bear habitat and populations.

As usual, the Network members worried about how the various disciplines could best integrate their knowledge. Our idea was to focus on modeling to encourage knowledge integration. For example, future recreational activities and future commercial developments would be articulated in terms of impact on habitat. These scientific concerns, however, led to process concerns. Our past efforts to get working groups to test sensitivities using the VORTEX model and to quantify impacts for input into VORTEX had been frustrated. We experimented with how to organize small groups to reflect the data requirements. Would it be better to organize groups around human activities or impacts? Should working groups focus on expertise (knowledge sets) or activities (e.g., industrial and research)? Other questions concerned working-group composition, whether more structure of the group process would make the goals more clear, and whether facilitator expertise would be useful in making the groups function better.

A Network human demographer was not able to attend this workshop. However, we did not anticipate a need for this expertise or for human demographic data, since we assumed that population pressures would be relatively unimportant (see chapter 13). Because of the ESGBP's rich connections in the Calgary area, we also anticipated that the stakeholder mix would ensure data availability on population growth, tourism, development, and cultural values. In general, however, it was difficult to make this decision without knowing exactly who was going to attend.

It was argued, then, that the focus of modeling should be on resource use and its impact. Reference was made to the heuristic "bubble" diagram (for examples, see figure 3-1 in chapter 3 and figure 18-3 in chapter 18), which shows details of the quantitative links between human population processes and wildlife population dynamics. We felt we had an outstanding opportunity with this group to make significant progress in understanding how these links can be affected. Organizing participants into knowledge groups based on the different bubbles (i.e., human population, economics, and industry) was a discussed possibility. We also considered using the "bubble diagram" (detailed in chapter 18, figure 18.3) as a heuristic device to better set up the modeling part of the workshop. We agreed that maps would be useful for integrating knowledge and that the VORTEX model should be used as a central focus. To these ends, Network members would actively facilitate small groups.

The Workshop Dynamics: Content and Process

The workshop extended over three and a half days, January 28–31, 1999. It was held in a rustic camp setting outside of Calgary, in the foothills of the Rockies. The seventy participants included scientists, conservationists, modeling and technical experts, as well as researchers, field biologists, environmental lobbyists, and wildlife managers from Alberta, British Columbia, and the United States. In addition, seven Network members attended, amounting to the largest Network involvement in a workshop. Our initial intention was to have five members act as facilitators (including the three Conservation Breeding Specialist Group staff members), one as a modeler, and to have three members "float" between working groups, with the two non-CBSG staff Network facilitators conducting interviews with participants.

The opening plenary began with a brief history of the ESGBP, specifying that its goal is to develop a more complete understanding of the cumulative effects of human development on grizzly bears throughout the Project's study range. This was followed by presentations on mapping efforts (designed to identify suitable habitat and the impact of human activities in that habitat) and presentations on mortality rates (documenting a strong relationship between human access and grizzly bear mortality). The Network experiment and the bubble diagram were introduced and explained, and two modeling presentations were made. Finally,

CBSG led a vision exercise, based on identifying desired outcomes for the workshop. Despite a history of considerable conflict among the stakeholders, the goals articulately reflected a closely shared vision of the need for careful grizzly bear management.

Small-Group Work

Working groups had been pre-decided in consultation with the workshop organizers. They included two modeling groups (life history and habitat), a habitat and distribution group, a group on human activities in moderate-use protected areas, and a group on human activities in core protected areas (secure areas). Network members facilitated the small groups, each using tools appropriate to the needs of the group (figure 10-3). These working groups began work late in the afternoon of the first day. What follows is a discussion of the highlights of each group process.

Landscape, Mortality, and Risk Modeling Working Group

In the case of the modeling group, after much debate, we decided to separate into two subgroups: a VORTEX and life history group and a "big picture" habitat group. This latter group seemed rather disaffected with the prospects of jumping into a VORTEX modeling effort and wanted to talk

Figure 10-3. Working group session, Eastern Slopes grizzly bear PHVA workshop, Canmore, Alberta, Canada. (Photo courtesy P. Miller).

about larger issues, centering on how to make the most explicit link between human-mediated changes in landscape habitat and grizzly bear demography using more sophisticated directed models. The Network concern was that these folks—comprising much of the habitat mapping and spatial analysis expertise in the workshop—were going to work in nearly direct conflict with the VORTEX effort.

Despite misgivings, we began the general effort by agreeing on an appropriate geographic scale for the modeling effort—centered first around the smaller Kananaskis Country/Banff National Park region and then broadening out to the CRE. The next step was to work through the preliminary model input data set to develop firm baseline demographic parameters and the uncertainties surrounding some of them. It then became clear that adult female mortality was a driving force influencing grizzly bear population dynamics; this led the later demographic sensitivity analysis to be centered on this parameter. Given our best parameter estimates (based on female mortality in Gibeau and Herrero 1999), the CRE grizzly population is expected to grow at an annual rate of about 2 percent. However, as Gibeau and Herrero's (1999) estimate of adult female mortality was judged by participants to possibly be an underestimate and is the lowest among other studies of grizzly demographics in the Rockies, models using higher estimates of mortality quickly led to very little population growth or even population decline.

Once the more general demographic sensitivity testing was completed, we were ready to develop specific models relevant to current and future scenarios in the CRE. The facilitator suggested that the two subgroups reconvene to update each other on progress: by this time some participants from the big picture subgroup were trickling back over to the VORTEX and life history table. The big picture subgroup was working hard to develop a mathematical formulation for the relationship between grizzly mortality across the CRE landscape and the likelihood that a given individual would be in a particular type of habitat at a given point in time. By extension, the subgroup was working out an equation relating human-mediated changes to the landscape and changes in grizzly mortality within that landscape. To do this, intensive Geographic Information System (GIS)–based spatial analysis was necessary, and a local private consultant with extensive GIS expertise and an ESGBP GIS expert were brought in to complete the data manipulation and analysis. Three leading scientists worked closely with the GIS experts; it was quite exciting to see the level of intensity of this group's analysis.

Of greatest interest was the fact that they were developing this equation for explicit incorporation into VORTEX. Network concerns about the intent of this subgroup were no longer grounded. In fact, the participants were the driving force behind coming up with the types of information that formed the heart of our Network's enhanced risk assessment process.

The modeling working groups got excellent information from the habitat and distribution working group that helped us set parameters for a metapopulation model in which the CRE was divided east-west by the Continental Divide and north-south by the Trans-Canada Highway. Once again, accurate estimates of adult female mortality were shown to be critical in assessing the long-term viability of this four-patch metapopulation. Additionally, a working subgroup on human access issues in core protected areas developed a set of scenarios designed to look at the impact of rapid human population growth in Calgary's environs: despite Canada's low overall population growth, the area around Calgary is projected to grow at a rate of 4 percent per year for the next 10–15 years. This can have major impacts on the projected demography of local grizzly populations (figure 10-4).

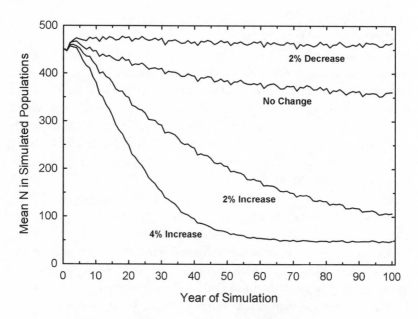

Figure 10-4. Impacts of human activity on grizzly bears in the Central Rockies Ecosystem. Projected grizzly population size in four scenarios, each with a different level of human impact on bear mortality.
(Source: Herrero, Miller, and Seal 2000.)

Habitat and Distribution Working Group

The objective of the habitat and distribution working group was to determine how best to incorporate habitat quality, effectiveness, supply, and distribution numerical data into a habitat-based population viability model in VORTEX. The group defined terms, identified the scale of assessment, and analyzed existing information on the spatial distributions and broke them down into measures of population density. Both the human dimension and habitat management issues were considered.

Discussions culminated in an identification of high-quality or potential quality habitat overlaps with secure or insecure areas. A recommendation was made to validate and refine the VORTEX model to input values based on existing and new data before implementing the group's concluding recommendations. The revised PHVA model needs to include habitat quality polygons (ground-truth Landsat greenness polygons); linkage zones; and population density, home range size, and other values by habitat class. These tools could be used to determine where conservation efforts should be focused when implementing specific strategies.

Human Activities—Moderate Use
Protected Areas Working Group

This group was focused on semiprotected areas where some human activities, including recreation, ranching, and resource exploitation, were allowed. The group contained representatives from forestry, oil and gas, cattle industries, and some provincial managers, all with different vested interests in the question of grizzly bear habitat conservation. As the working group was large and the concerns diverse, it was decided to subdivide further into one group concerned with access impacts (roads) and a second group looking at other habitat uses.

Human Access Impact on Habitat Subgroup: This subgroup began by trying to define what the problem of access actually was. From the grizzly perspective, critical variables in access character seemed to be motorized versus nonmotorized, concentrated versus dispersed, frequency, and density.

At this point, the management-oriented participants argued that the group should just go to the most critical aspect: grizzly mortality. In particular, they insisted that increased mortality was being caused by a combination of factors: increasing access (more roads), increasing frequency of use, increasing availability of firearms, and ignorance about grizzly behavior. This intervention led to some disarray in the group.

A breakthrough occurred in the late afternoon when a participant presented his data on sources of mortality. We then began to talk about the frequency formula and determined that (a) we were not going to be able to get a figure on projected roads, but that (b) frequency was going to be driven by human population increase, which was known to be 4 percent per year. An interesting discussion ensued about how reliably we could translate a 4 percent population increase into increased access. It was thought that a 4 percent increase in access was a conservative estimate, as the rapid population increase was due to in-migration of predominately young males in search of jobs. This group also saw themselves as recreation users and were more likely to carry guns when in the back country.

The subgroup developed an equation that reflected their working assumptions and provided inputs to the VORTEX model that quantified the impacts of various access types on grizzly population viability in the CRE:

Mortality = f [(Human Encounter Rate) x (Lethality of Encounter)]

Access types, their attributes, and the rate of impacts were identified. Scenarios were created that captured the current state of affairs and probable and possible futures. Allowance was made for spatial and temporal variation and recommendations for future actions were made.

Other Habitat Uses Subgroup: This subgroup examined the effect of the physical land use component of human activity without considering human use level issues. The problem statement identified was that human-related developments have the potential to significantly affect grizzly bear habitat. Historically, human activities such as fire management, oil and gas development, and logging have been economic drivers of land uses that can strongly influence grizzly bear habitat. Recommendations were made to mitigate potential impacts of these various land use activities by linking recommendations to habitat greenness, rather than demographic parameters such as mortality. The group therefore suggested that core protected areas alone will not sustain viable grizzly bear populations. It proposed that analysis of multiple-use land be completed prior to further developments, especially in order to address the direct effects of timber harvesting, fire management, oil and gas development, recreational, and residential developments.

Human Impact—Secure Areas Working Group

The secure areas group generated a vision for the future: "To establish and maintain a viable population of grizzlies by accommodating individual security needs in high quality habitat with emphasis on the survivorship of adult females." A gap analysis between the probable and possible futures created the categories for future planning, with feedback loops to the baseline data.

A secure area was defined to be the 9 square kilometers of habitat used by a female grizzly bear every day. This area moves with the bear within her home range. Enhancing the home range that contains connected secure areas can enhance female survivorship. Disturbances higher than twenty human parties per week were considered to cause significant behavioral changes in grizzly bears.

The working group recognized the need for a joint management response from the governments of Alberta, British Columbia, and Canada to manage the grizzly bear as a unit. The group agreed that there is sufficient information to act and that socioeconomic pressures require a focused management response to protect secure areas for female grizzly bears. The group then made some general and specific recommendations for maintaining and increasing the number of secure areas in legally protected areas as well as in landscapes not protected.

Workshop Recommendations and Conclusions

On the scientific side, VORTEX provided for rapid communication and increased understanding among the participants concerning information to be used in local management decisions across jurisdictions. It was evident from the workshop results that demographics matter, and so does landscape use by the human population. It was evident that the human population of the Calgary area adjacent to the central Canadian Rockies will grow at a rate of 4 percent per year and that the profile of the immigrants is predominantly young people attracted to the area because of its natural attractions and outdoor activities, such as hunting, hiking, and skiing. It became clear that the impact of these activities on grizzly habitat and population would be at least as high as the rate of population growth. This analysis thus evidenced that the grizzly population would rapidly crash due to unsustainable killing resulting from the increased frequency of encounters between grizzlies and people.

This analysis of the likely impacts of changing human demographics and activities is useful because it points out clearly that a major thrust of management must be to reduce the average lethality (to grizzlies) of human-grizzly encounters. In addition, new insights into the utility of an expanded PHVA were provided by the addition of complex spatial models of human-induced alterations of grizzly bear habitat. On the one

hand, extensive GIS data exist for the CRE, including standard vegetation classifications and spatial distributions of human land-use patterns,
such as mean road density. On the other hand, because data on grizzly
demographics and habitat use have been collected over the past twenty-
five years, a significant opportunity exists for a productive synthesis of
these two sets of information. However, at the time of the workshop these
data had not been synthesized in the context of a quantitative population
risk assessment. During the course of the workshop, a team of participants developed an algorithm for using GIS and animal telemetry data
to predict grizzly bear mortality risk as a function of selected map variables. While perhaps intuitive, these potentials for new results nevertheless demonstrate the power of combining GIS and demographic data to
derive functional relationships between the spatial characteristics of a
given habitat, use of the habitat by humans, and the population dynamics of the wildlife that use that habitat.

Perhaps the greatest potential for this technique lies in its direct application to wildlife population risk projections. Given some data on how the
landscape will change over time—primarily due to human use—one can
rather easily incorporate future changes into the logistic model and the
GIS maps. For example, if we estimate that road density will increase by
20 percent in the next decade, this process can identify where road building will most likely occur, the types of habitat it will influence, and consequently how the risk of grizzly bear mortality will change through time
and space. This function can then be entered into a population viability
model to more realistically simulate metapopulation dynamics in a landscape increasingly modified by human disturbance.

An important issue for grizzly conservation that emerged was that of
scale. Framing the conservation issue according to different scales has
important implications. It appeared critical for grizzly conservation to
think in terms of greater ecosystems. However, a problem of larger scales
was jurisdiction overlap, which makes it more difficult to get a better grasp
of the problem and to be effective with developing strategies. The inclusion of the perspectives and buy-in from the provinces of Alberta, British
Columbia, and the federal government was recognized as being critical.

Network Learnings

The grizzly bear PHVA yielded a number of learnings for the Network:

Expanded stakeholder processes: This workshop was the only one in the Network's study with a large number of industry participants, and we looked
with considerable interest at the impact of this expansion of stakeholder

groups on the PHVA process. Altogether, participants at the workshop were open, flexible, and adaptive as they went along and eager to know what their role was so they could provide the kind of information required for the modeling groups; participants collaborated not only within their own group or when synthesizing information during plenary sessions, but also between groups as well. In our perception, the intensive facilitation by Network members helped this process, as well as the presence of a convenor such as Herrero, with whom all participants had already established relationships.

The effective use of the bubble diagram: The Network also felt that the use of the bubble diagram as a simple heuristic tool worked well to communicate the Network's project and concerns. This avoided some of the process pitfalls created by the "double agenda" of PHVA and Network experiments, which seemed to plague the mountain gorilla workshop (chapter 6). It also restored VORTEX to a central focus of the workshop, which helped to create convergence among the working groups (see chapter 4), consolidating hard data, building scenarios, fostering communication, and getting participants focused.

The need for local analysis of human population dynamics: As noted in this chapter's introduction, we did not anticipate needing a human demographer because we thought human population pressures would not be an issue in this part of Canada. As it turned out, human population played a pivotal role in grizzly bear mortality, due to the particular population growth and mix in the Calgary area. This convinced us again that macro models of human demography must be combined with more local, micro models, even if those are based on qualitative data, if we are to integrate human demographic pressures into species risk assessment (see chapters 13 and 14).

Limiting factors on the use of VORTEX: The complexity of a model such as VORTEX is both its strength and its weakness. A problem was raised by some participants about the rejection of modeling information when used as evidence during court cases. Evidence based on scientific models has been rejected because judges could not understand it. If managers and decision makers cannot make sense of the models, they too, will not trust or rely on the information generated. The need to articulate modeling technology in language that decision makers can readily understand was raised as an issue, a skill at which scientists are not particularly adept.

Sensitivity analysis has several limitations. The primary one is that much of the interpretation depends on the assumptions implicit in the stochastic simulation model. Second, the confidence limits were based primarily on the expert opinions of workshop participants who may have under- or overestimated the certainty about particular population parameters. It was clear from this workshop that the VORTEX bubble diagram itself is difficult because filling in the arrows is not automatic. The need for a tremendous amount of information increases the complexity of the model, which forces the scientific experts to put their guesses down on paper and to challenge ideals. Future use of the model will depend on the way risk is understood, which also varies with individuals and groups.

The key role of boundary spanners: This workshop also evidenced the central role of boundary spanners. Key boundary spanners—individuals with diverse backgrounds and capable of communicating across disciplines and issues, as well as between scientific and management issues—contributed to positively influencing both the convening of the workshop as well as the workshop process itself. Such boundary spanners were able to facilitate discussion, interpret information, and act as bridges between data sets and working groups.

Recognition of multiple jurisdictions: A critical aspect in the workshop was the need to think in terms of greater ecosystems, where understanding jurisdiction overlap was critical to getting a better grasp of the problem and to developing effective strategies. The inclusion of the perspectives and buy-in from the provinces of Alberta, British Columbia, and the federal government was recognized as being critical especially in those working groups where representatives from across jurisdictions were not present (see chapter 12 for elaboration on this issue).

Efforts to include First Nations and industry: As said earlier, the Network saw this PHVA as a privileged opportunity to include stakeholder groups not included in other cases: First Nations and industry groups (Peterson 1992, 2003; Peterson and Westley-Esquimaux 1992). However, our efforts were not fulfilled. As for First Nations, although groups, tribes, and individuals had been invited through letters, the workshop was unsuccessful in attracting and retaining them. This may be have been caused by the need to build relations and resolve past conflicts and build trust, as we will explore in chapter 15. As the caribou PHVA case study revealed, cultural issues were important to the aboriginal participants, while the biologists defined the problem as a biological one. In that case study, the aboriginal representatives did not participate in the modeling group (see chapter 8). Increased understanding of cultural perspectives and the joint development of working protocols for their involvement in future workshops needs to be strived for.[2]

On the other hand, industry participants saw the PHVA as an extension of their ongoing participation in the ESGBP steering committee. They recognized that the PHVA was an excellent opportunity for the academic community to discuss and present their findings. Despite their presence, however, Network members remained frustrated by our failure to convince industry stakeholders to bring the data required for their more active participation in the analysis. The possible reasons for this are discussed further in chapter 16.

Aftermath of the Workshop

Approximately two years after the workshop, two University of Calgary graduate students, Saundi Stevens and Mary Bennings (2000), carried out personal interviews with a number of the workshop participants in order to assess their impressions after the fact. Representatives of five different stakeholder groups—academic researchers, federal government, provincial governments (Alberta/British Columbia), industry, and conservation—were asked whether the grizzly bear PHVA workshop was effective (for their agency/group) in (1) pooling information and increasing their understanding of grizzly bear population dynamics; (2) identifying information gaps; (3) motivating action toward management and conservation of grizzly bears; and (4) aligning research needs and management action agendas. The interviews yielded the following feedback:

Researchers: As lead researcher and head of the ESGBP, Herrero's views were particularly important. He felt the PHVA process was tremendously useful. He recognized that a few stakeholders thought some of the data was forced to make the situation look worse than it was. However, collectively the researchers felt there were some limitations to the data they had collected and decided it was better to use consensus data from a whole range of grizzly bear populations, which they thought provided a more adequate sample. He concluded that they have since been vindicated in that approach because current data subsequent to the workshop more or less confirms the original hunches they had.

Motivating stakeholder management actions or policy changes in grizzly bear conservation was not an objective of Herrero's when convening the PHVA workshop. However, Herrero felt that the PHVA process improved communication and fostered effective working relationships among the researchers and the expert modelers, a collaboration that has continued since the workshop.

Federal government: Parks Canada participated in the PHVA workshop to contribute toward agency and interagency knowledge and management and education goals. In the opinion of Parks Canada workshop participants, this workshop was not very effective, as the process tried to do too much with too little data. They felt the process could have been improved if much of the analysis and preliminary models had been available prior to the workshop and needed only "fine tuning" during the workshop.

Parks Canada's commitment to grizzly bear conservation issues has continued since the workshop. Parks Canada has held three management plan implementation sessions addressing grizzly habitat effectiveness, security, and mortality rates since the PHVA workshop. Parks Canada's rather tepid response to the workshop possibly resulted from the fact that much progress was made by the agency in the previous few years on grizzly bear issues. This resulted in a perceived need on their part to consolidate where they were rather than once again bring in many stakeholders and use a "crisis management" approach, which was how Parks Canada perceived the PHVA process.

Provincial governments: Two provincial governments also participated in the workshops, Alberta Environment and the British Columbia Ministry of Environment. Both have responsibility for the grizzlies, who range over lands with multiple jurisdictions. When interviewed, both groups felt that some parts of the workshop were very effective, particularly those portions related to population modeling under various scenarios of mortality and productivity. They thought other parts of the workshop and document were weak and of little use for managers. Alberta Environment representatives thought that the greatest contribution of the workshop was that it brought together many people with varied backgrounds (scientific, industry, government) and presented a great deal of information on grizzly ecology and population viability issues. The British Columbia Ministry of Environment felt the workshop was effective in terms of bringing grizzly managers, researchers, and advocates together. They also suggested, on the other hand, that there was not a lot of fresh information discussed at the workshop that would be useful in the management of grizzlies and their habitat in British Columbia.

Industry: Industry participants included representatives from the oil and gas and forestry sectors along with the Alberta Cattle Commission, Canadian Pacific Railway, Ski Louise Resort, and the Calgary Zoo. Industry representatives felt the grizzly bear PHVA workshop was an excellent opportunity for the academic scientific community to discuss and present their research. However, a few industry participants mentioned that they could not help but feel that there was a "hidden agenda" and that the "goals in play" were not necessarily shared with the whole group. Specifically, they believed that some of the data was manipulated or forced to make the situation of grizzly bear viability look worse than it was. They mistrusted the use of numbers based on consensus data from a wide range of grizzly bear populations in North America.

Conservation: The conservation sector included representatives from Canadian Parks and Wilderness Society, the World Wildlife Fund, Alberta Conservation Association, and the Yellowstone to Yukon Initiative. These groups have, for many years, focused their energies on assuring long-term persistence of biodiversity in the CRE. These groups regularly argue that one of the keys to maintaining biodiversity is maintaining the connectivity of viable habitat for the whole complement of natural species. In understanding and assessing this need, the status of secure habitat for grizzly bears plays an important role as a proxy for that of many other species, as well as for the concept of wilderness itself.

Representatives from these groups reported to Stevens and Bennings that the PHVA collaborative process contributed most to their mandate for a greater degree of cooperation between jurisdictions and among land managers, industry, and nonprofit interest groups.

In sum, an unusual opportunity for data gathering resulted in a mixed picture of the appreciation of the workshop, with scientists and conservationists the most satisfied and other stakeholders more equivocal about the use of the workshop results. These questionnaire responses highlight the difficulty in building lasting trust between stakeholder groups.

---

[1] We would like to express our appreciation for Karen Peterson's work on organizing the grizzly bear workshop and for her contribution to early drafts of this chapter.

[2] The identified issues are being followed up and addressed in Karen Peterson's dissertation (Peterson, forthcoming). She may be reached at *karenpeterson@shaw.ca*.

 **Chapter 11**

A Special Concern: The Wolves of Algonquin Provincial Park, Ontario

GEORGE FRANCIS

At the Network meeting in Banff, Alberta, November 1997, members discussed the need to incorporate more societal considerations into the Population and Habitat Viability Assessment (PHVA) designs. Such elements include interpreting human impacts as refinements in the mortality and survival data used for population modeling of "focal" species; obtaining expanded stakeholder involvement and considering the institutional context of the situations being explored in order to enhance understanding of various human stresses on animal populations, habitat-related issues, and possible conservation actions; and identifying criteria for judging the relative success of PHVAs, such as the relative specificity and feasibility of recommendations generated by the workshops and evidence that recommendations are carried out.

On the same occasion, the Network identified the wolves of Algonquin Provincial Park, Ontario, to be a potentially good case study situation. A wolf workshop would provide an example of conservation planning in a developed nation with low population growth and high institutional capacity. In addition, societal considerations would be crucial yet quite focused on a local, relatively contentious issue. In winter, wolves often left the park to prey upon deer in naturally sheltered "deer yards" nearby. A few people in local communities killed the wolves for various reasons, and a small group of scientists who were studying the wolf packs in Algonquin concluded that this human-caused mortality was threatening the continuing viability of wolf populations in the park. Conservation groups were calling for conservation zones to be designated for areas adjacent to the park, in which the killing of wolves would be banned. Some people in local communities interpreted these proposals as attempted imposition by outsiders and the resulting controversy at times became heated.

Since the wolf research group had compiled data on the human-caused mortality of the wolves they were studying, it seemed to the Network that a PHVA workshop could be helpful in two ways. First, the mortality data could be incorporated into VORTEX models to examine more carefully the impacts humans were having on the wolf population. Second, if the right mix of people participated, the workshop could help initiate a conflict resolution process focused on a relatively small geographic area. It was agreed (at the Banff meeting) to explore this idea with the wolf research group.

Unknown then was that the provincial minister of natural resources would, over the next year or so, establish an Algonquin Wolf Advisory Group (AWAG) of stakeholder interests to advise the minister on matters relating to the conservation of wolves in Algonquin Park in particular, and in Ontario in general. As events unfolded, the Algonquin wolf PHVA was cosponsored by the AWAG as a significant contribution for their deliberations. Before proceeding with a description of the workshop and its outcomes and Network lessons, some background on Algonquin Park as a setting and on the Algonquin wolves is in order.

The Context of Algonquin Wolf Conservation

Algonquin Provincial Park was established in 1893 as the first and, with several additions to it over the years, one of the largest parks in Ontario. It now covers some 7,725 square kilometers of the highlands formed by a raised dome at the southern edge of the Canadian Shield between Georgian Bay and the Ottawa River (figure 11-1). At the time of its establishment, the park area had already been heavily logged for mature white pine, with associated forest fires, during the latter decades of the nineteenth century. Nevertheless, it was viewed as a wildlife reserve that also protected the headwaters of several major rivers flowing from the park and agricultural settlements (which had sprung up to the south and east) were excluded from the new park area.

The park lies within the Algonquin–Lake Nipissing Ecoregion (a Canadian ecological land classification), in a transition zone between the boreal forest and the Great Lakes–Saint Lawrence forest region. There is extensive deciduous hardwood forest cover over its western dome area and pine forests grow in the more sandy areas of the park's eastern portion. The many lakes and connecting streams provide extensive canoeing, fishing, and camping opportunities. The main access to the park is from a 56-kilometer stretch of provincial highway that crosses the southern park region and there is intensive recreational use of areas immediately

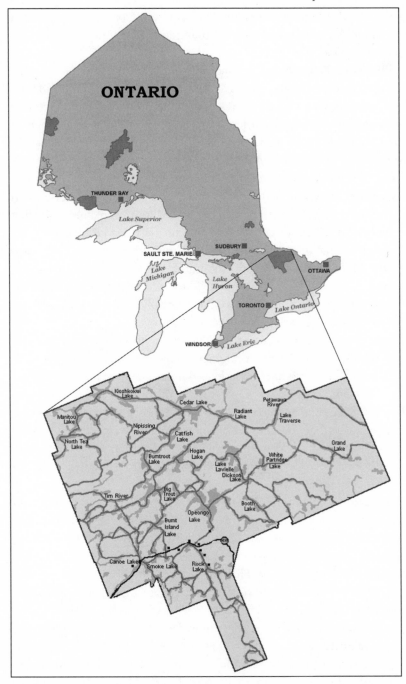

Figure 11-1. Map of the province of Ontario, eastern Canada, showing in detail the location of Algonquin Provincial Park.

adjacent to this highway corridor. The number of park visitors is about one million day visitors annually, including some 220,000 visits to the visitor center alone.

Although called a "park," Algonquin is managed as a multiple-use area similar to a United States National Forest. Management responsibilities are shared between Ontario Parks (under the provincial Parks Act) and the Algonquin Forest Authority, a special-purpose body established in 1975 to manage the about 75 percent of Algonquin Park that is allocated under a park master plan to forestry and that provides saw logs for processing to some seventeen local sawmills outside of the park. About seventy smaller sites within the park are set aside as nature reserves and the research camps for forestry, fisheries, and wildlife were among the first established in the province. Recreational uses are served by over 1,500 kilometers of designated canoe routes, a number of hiking and cross-country ski trails, and about three thousand campsites, most of which are along or close to the highway corridor. The Authority, through a large network of logging roads, maintains other controlled access. Proposals for new access roads are a regular source of controversy.

Areas immediately adjacent to Algonquin Park have similar forest cover, but they are much more fragmented by small farms and other private land holdings and have experienced more intensive resource extraction. This land constitutes a patchwork of private and state-owned ("Crown") lands and also creates the kind of landscape that supports coyote populations that now come into contact with the wolves from the park. Such contact could increase the extent of hybridization between these two rather closely related animals and threaten the distinctiveness of the wolves in Algonquin Park.

The wolf populations of Algonquin Park have been the subject of field studies dating back to 1957 and are among the most studied wolf populations anywhere (figure 11-2). The two most intensive studies were those of Pimlott, Shannon, and Kolenosky (1969) and Theberge and Theberge (1998). Work directed by the Theberges since 1987 resulted in a number of scientific publications and reports (e.g., Forbes 1994; Forbes and Theberge 1996; Theberge and Theberge 1997) and popular articles. An historical overview of wolves in Algonquin Park is provided by Voigt and Strickland (in Ewins et al. 2000). The provincial government has sponsored a review of the status of wolves and coyotes in Ontario and of current management policies for them (Buss and de Almeida 1997) and has compiled anecdotal information from interviews about deer and wolf movements around Algonquin Park (Quinn and Inglis 2000).

Wolves from this park have long been classified as a subspecies of the gray (timber) wolf *(Canis lupus lycaon)* because of their distinctly smaller size, which is intermediate between the gray wolf *(C. lupus)* and the brush wolf or coyote *(C. latrans)*. More recent studies, however, have suggested that *C. l. lycaon* might be closely related to an ancestral red wolf *(C. rufus)* that was once thought to be widespread in eastern North America but has been extirpated from almost all of its former range. This raised a recurring debate among biologists about the taxonomic validity of the red wolf as well as a new debate about whether the Algonquin wolves are possibly one of the last extant wild populations with this ancestry (e.g., Theberge and Theberge 1998). Taxonomic revision could change the conservation priorities assigned to threatened or endangered species. It can be noted in passing that the Conservation Breeding Specialist Group (CBSG) had addressed issues concerning the red wolf in another PHVA (Kelly, Miller, and Seal 1999).

The wolf population in Algonquin Park fluctuates seasonally and from year to year, as well as over longer periods of time associated with the relative abundance of their main prey. In recent years, the numbers are generally thought to have been in the average range of about 200 to 240 animals, in some thirty to thirty-five packs by late summer following the recruitment of young pups. A main issue is the mortality rates of both pups and adults and the consequences for the maintenance of the park's wolf population. It has been noted that park rangers for a long time killed

Figure 11-2. Algonquin Provincial Park wolf *(Canis lupus lycaon)*. (Photo courtesy J. and M. Theberge.)

about 50 to 60 wolves a year in the park. This was stopped in 1959, yet the wolf population did not increase. A conclusion was that the wolves have both a high natural recruitment and a high mortality rate, and that human killing of them is compensatory (rather than an addition) to natural mortality so this killing may not have a great impact on the population. During the 1960s, however, there were major declines in the deer population that may also have limited wolf numbers.

This view about the small effects of human-caused mortality was challenged by Theberge and Theberge (1998) and others. First, it implicitly assumed that the current average population level is sufficient to maintain a viable population over the long term, a question which the PHVA workshop could address. Second, it overlooked questions about the disruption of wolf pack structures and behaviors that may compromise the animals' long-term survival. Third, depletion of wolves may also have opened the park to incursions from coyotes living in the surrounding areas, which could lead to increased hybridization and "gene swamping." Recent genetic studies show some introgression of coyote genes in the Algonquin wolf population. Theberge and Theberge (1998) also contended that the human-caused mortality could be replacing natural selection pressures on the population to the extent that the adaptive responses may be leading toward the evolution of a "human wolf." Conservation groups (e.g., Wildlands League 1997) have called for a 10-kilometer protection zone for wolves immediately adjacent to the park boundaries.

Algonquin Park has a strong appeal to many people in southern Ontario and elsewhere who have had occasion to experience it as a "wilderness" with great aesthetic value and a sense of tranquility. In travel advertisements, in addition to the scenery pictured, these values are symbolized by the cries of loons and the howls of wolves. Interpretive staff from the park's visitor center have sponsored "public wolf howl" outings on summer evenings since 1963, and more than 110,000 people are estimated to have attended over the years. These events, along with media publicity about them, are thought to have influenced many people's attitudes very favorably toward wolves. In this sense, Algonquin wolves are special and the place they are in is special for a very large number of people.

Preworkshop Activities

Network members met in mid-January 1998 with core members of the wolf research group in Peterborough and the Golden Lake area just outside of Algonquin Park. The desirability of holding a PHVA was accepted, but questions of timing and support for it had to be resolved. At the time, the wolf research group preferred to wait a year or so to complete some studies on the genetics of wolves surrounding Algonquin Park for comparisons with wolves inside the park, and the Theberges were in the midst of writing a book on their work and experiences scheduled for publication in just a few months time. The World Wildlife Fund (WWF/Canada) agreed to contribute to workshop costs, but on the condition that the Ontario Ministry of Natural Resources (OMNR) contribute as well. In early February 1998, CBSG outlined an agreement and plans for a workshop.

In May 1998, the minister of natural resources established the Algonquin Wolf Advisory Group (AWAG) to provide recommendations "on an Adaptive Management Plan to ensure the long-term conservation of the eastern (Algonquin) gray wolves of Algonquin Provincial Park and surrounding areas." Terms of reference for the AWAG were approved in September 1998, revised in March 1999, and a PHVA workshop had been tentatively planned for August 1999. However, in the spring of 1999, the provincial government called an election and everything was put on hold, including planning for the workshop, until the election was held in June and the government was re-elected to a second term of office. Once the AWAG was under way again in August 1999, interest in sponsoring a PHVA workshop to assist the group's deliberations resumed. In late November 1999, at the invitation of the OMNR, CBSG met with members of the AWAG and others who were to plan the workshop to discuss details of process, logistics, the role of VORTEX and PHVA, and the desirable range of participants. The Network's emphasis on broad stakeholder involvement was included by CBSG at this preparation stage.

As this sequence of events unfolded, the potential of a case study for the Network became clearer. As noted, the extent of the direct killing of wolves by humans was documented and could be used in VORTEX models. In addition, there were statistical data on human demographics and socioeconomic attributes available for the general areas around the park, although very few people were actually killing wolves. The advisory group was a well-designed recipient for the recommendations from the workshop; its emerging dispute resolution role was one that a workshop could encourage without overloading expectations or the workshop process. While several Network members attended preworkshop planning meetings, only one Network member beyond CBSG staff attended the workshop itself. As a result, the Network played a less active role than in some previous workshops. The Network's primary role was participation in the workshop's planning stages and analyzing institutional and governance issues prior to the workshop, as well as observing the course of the workshop process in a markedly different context from previous cases.

Human Demographics and Socioeconomic Considerations

Some published data on the human demography and on local social and economic characteristics were examined prior to the workshop. They indicated that areas to the southeast of the park, where the human killing of wolves has been documented, remain a resource- and farm-based economy. Local cultural values are quite supportive of hunting, trapping, and fishing. The census statistics for the period 1986–96 for four townships and three towns to the southeast of Algonquin Park show a total population of about 10,335 people, a 3.9 percent increase over 1986, with average annual incomes of Can$19,700 (72 percent of the provincial average) and an official unemployment rate of 17.4 percent (about twice the provincial average).

In contrast, the four main cities/towns to the west of the park had a total population of 93,400 in 1996, a 21.4 percent increase from 1986, with average annual incomes of Can$23,000 (84 percent of the provincial average) and an unemployment rate of 10.8 percent (just slightly above the provincial average). In addition to some resource- and farm-based economic activity, local communities to the west of the park provide services along a main transportation corridor (Highway 11) that links the Greater Toronto Area with the Trans-Canada Highway. This area also includes higher-end recreational and tourism developments, including destination resort complexes and many newer residents who moved in to enjoy what local officials deem to be "life-style options."

Governance

The policies and management that constitute the relevant governance over the area were also summarized pre-workshop. The first land-use planning exercise for Algonquin Park was launched in 1968 and resulted in a master plan in 1974. This was subsequently updated in 1979 and again in 1989–90. Public consultation processes conducted by the Provincial Parks Council for the plan's periodic major updates became increasingly lengthy as they attracted an increasing number of participants. Disputes among user groups in Algonquin arise from the continuing and long-standing provincial government policy that seeks to reconcile continued logging with outdoor recreation activities and "wilderness" values. Forest management planning has been conducted as a separate exercise under the broad guidelines set by the park master plan. Policies for forest management have evolved over the past twenty-five years and they are currently oriented towards restoring forest conditions closer to what they may have originally been, subject to maintaining the supplies of saw logs to the surrounding mills. Currently, the Algonquin Forest Authority has an updated strategic Forest Management Plan for 2000–20 and a more immediate operational plan for 2000–05. About 2 percent of the park is subjected to harvests each year, mainly through selective cutting of individual trees, or cuts over small areas.

Forest management practices, including fire suppression, have a significant impact on the ecosystems of the park. The replenishment of browse vegetation through regeneration in cut-over areas provides critical food supplies for deer and beaver (the favored prey of the wolves), and for moose. Reductions over the years in the areas of hemlock and other conifer stands, which provide winter shelter for moose and deer, may be related to the subsequent movement of deer out of the park in winter. Current management of hemlock and some other conifer stands is directed to restoring some former deer yards within the park.

In the areas around the park, much of the trapping, snaring, and shooting of wolves has been localized within three rural townships to the southeast of the park. In 1993, the OMNR imposed a seasonal regulation in these three townships to prevent the hunting and trapping of wolves from December 15 to March 31.

Issues of native title have also been raised. In 1983, the Algonquins of Golden Lake (now the Algonquins of Pikwakanagan First Nations) submitted a formal claim asserting unextinguished aboriginal title to some 34,000 square kilometers of lands in the Ottawa River valley that included virtually all of Algonquin Park (and also the National Capital Region). Negotiations on this claim were started by the previous provincial government in 1991, followed by the federal government in 1992. By August 1994, the three parties signed a Framework for Negotiations and Statement of Shared Objectives (Ontario Negotiation Bulletin 1994). The next stage in this process, substantive negotiations over specific elements of a settlement leading ultimately to an Agreement-in-Principle (which then has to be ratified by the three parties), is continuing (Ontario Negotiation Bulletin 1998).

At the start of these negotiations in 1991, the minister of natural resources suspended application to native people of the hunting and trapping prohibitions and the fishing regulations in Algonquin. Agreements were subsequently reached with the Algonquin Management Circle of First Nations representatives to limit moose kills in the park and to leave the wolves alone. First Nations have argued that they should take over the control and administration of Algonquin Park. This, along with their unrestricted access to the park has generated local controversy, including potential legal challenges to the legitimacy of their claim. The Pikwakanagans are not involved nor implicated in the wolf kill controversies. Wolves have symbolic cultural and ceremonial values for the First Nations.

Workshop Dynamics: Content and Process

The workshop was held on February 15–18, 2000, at the Leslie M. Frost Natural Resources Centre near Dorset, Ontario, a conference and training facility maintained by the OMNR. These were excellent facilities, with an auditorium for plenary sessions and several smaller meeting rooms for use by the working groups. More than seventy people had been invited to the workshop and about sixty attended, most of them for the entire period. There were several biologists present, renowned for their work on wolves, and others currently involved with field studies in Ontario and western Quebec (where the lycaon wolf is thought to occur also). Managers were mainly from the OMNR, and those present had responsibilities for different aspects of Algonquin Park itself, including the Algonquin Forest Authority, or for resource and wildlife management in administrative districts adjacent to the park. There were representatives from the Pikwakanagan First Nations and from local or provincial organizations involved with hunting and trapping, livestock raising, natural history, wilderness protection, university-based wildlife research, and advocacy on behalf of Algonquin Park. Several members of the AWAG attended, including (when he could) the chair of the group. The other two sponsors of the workshop (WWF/Canada and Canadian Wildlife Service) also had representatives.

Process Notes

The briefing books were available only upon arrival at the Centre, and other papers and reports were distributed during the meetings. This led to a few informal comments, one from a local citizen representative on the AWAG, for example, about feeling overwhelmed with information that was not always easy to digest (e.g., the genetics of lycaon). There was some frustration, commented upon by people who had been closely associated with the Algonquin situation, that too much time had to be spent explaining things that were already well covered in the background material. One constructive suggestion was that organizers should prepare an executive summary or broad briefing statement summarizing the main issues, should identify where in the background materials the relevant facts could be found, and should get this information out to participants before a workshop.

The format for the PHVA was followed closely, starting first with a welcome from the superintendent of Algonquin Park, self-introductions of participants, and comments on the workshop process and expectations by CBSG. Background presentations were then made on the history of wolves and studies of them in Algonquin Park, on recent and current research on population dynamics and genetic studies that are raising basic questions about the wolves' taxonomic status, and on the use of VORTEX as an integration tool.

A plenary theme-mapping exercise was conducted on the first afternoon and succeeded in identifying many particular topics which could be broadly organized into groups. Volunteers from the audience helped facilitators do a first sorting of these as a basis for working groups. The five workshop groups were taxonomy; population dynamics; habitat and prey species; landscape ecology (and "interconnectedness"); and public values. Participants self-selected into these groups, which met concurrently for much of the time, reporting back to plenary sessions convened on the second and third days.

CBSG personnel and representatives of the sponsor organizations met informally each evening to review their perceptions of what had gone on and to note points to be brought back to the working groups the following day. The mandate for the workshop was reinforced at times by the facilitators through reminders about the terms of reference set for the AWAG. For example, the advisory group was to come up with recommendations for the conservation of the wolves regardless of the taxonomic status scientists might eventually assign to them, so suggestions about having the latter as a prerequisite to the former were not deemed particularly helpful. The AWAG was also to develop an adaptive management plan, but other than a few dismissive remarks about adaptive management from one participant on the first day, the workshop had not addressed this during the first two days. At the informal evening meeting with sponsors and facilitators at the end of the second day, the chair of the advisory group requested that this question be taken up. This resulted in the population dynamics group debating the issues among themselves and in a plenary session on the final day without coming to closure. They recommended instead that the design of an adaptive management strategy be the subject of a separate workshop for the advisory group to convene and offered to help out with it.

Discussions during the working groups and at plenary sessions were generally continuous and relaxed, to the point that informal comments were heard about how amiable it all was and whether this would last for the whole workshop. The discussions of the taxonomy and population dynamics working groups overlapped considerably, as did those of the groups for landscape ecology, habitat and prey species, and public values. The latter three groups felt that their discussions were somewhat general or speculative because their input would all depend on some specification of conservation needs and strategies, which they assumed the first two groups were trying to sort out.

A number of participants were aware of deep underlying disagreements, because these had been aired at community meetings many of them had attended over the last few years (and were also reported quite vividly in Theberge and Theberge [1998]). Facilitators commented that participants occasionally declared that they "knew where 'X' [a speaker] was coming from" and that they continued to disagree. There were, however, no noisy confrontations. This was due to skilled facilitation, which nipped potential angry exchanges in the bud, and to repeated assurances by facilitators that all views would be heard and recorded not only on flip charts, but also in the final report of the workshop.

The differences of views did surface in their own ways. While reflecting contrasting values, they were mainly expressed in the different interpretive contexts that the range of participants brought to the discussions. While wolves were the declared focus of the workshop, issues associated with them were defined or interpreted in the contexts of (a) the local agricultural and resource-based economies adjacent to Algonquin Park, which embrace hunting and trapping as an integral part of people's way of life; (b) the managed forest ecosystems in Algonquin Park that supply saw logs to mills outside of the park, but do so in carefully managed ways that create a forage base for the prey species of wolves through vegetation regeneration cycles; (c) self-regulating predator-prey systems with periodic prey-switching, which should just be left alone; (d) anticipated increasing recreational demands on Algonquin Park from southern Ontario (and the Greater Toronto Area in particular), with a resulting increase in the service sectors of local economies and an enhanced symbolic value of maintaining wolves (e.g., wolf howl experiences); and (e) park or wilderness values, which look to a future where all extractive uses and direct management actions in Algonquin Park are phased out and in which wolves will adapt to old-growth forest ecosystems with their inherent cycles of natural disturbances.

Workshop Recommendations and Conclusions

The main findings or conclusions of the workshop were focused specifically on aspects of the biology and status of the species. These included the following:

• The wolves in Algonquin Park are a distinct taxon, but the extent of their geographic distribution is unclear. There is evidence of hybridization, especially with coyotes, but otherwise this taxon is most closely associated with the red wolf, and not the gray wolf. Scientific arguments to support this conclusion are being submitted for publication in the primary scientific literature. Nevertheless, "the gold standard lycaon is in Algonquin Park" (quotations from participants), the park should serve as an "anchor" or "benchmark" for them, and questions about metapopulations of lycaon wolves need more study. Any official change in nomenclature of these wolves has to be a judgment call by taxonomists and falls outside of the role of a PHVA workshop. A scientific meeting of North American wolf taxonomists should be organized to address this in the near future.

• The data on population dynamics and human-caused mortality presented by John Theberge and his colleagues are "as good as it gets" for information about wolves. Evidence for population decline may not be strongly demonstrated statistically, but the park population is clearly at risk of extirpation given the extent and variability of mortality rates. VORTEX simulations with these mortality rates all resulted in extirpations at different time periods over the next century. With declines in human-caused mortality, VORTEX indicated that the population could persist and even become a "source" (rather than a "sink"), which is the appropriate conservation role for a park. The goal should be to maintain the genetic and ecological integrity of the wolf population and a minimum viable population level on the order of at least two to three wolves per 100 square kilometers.

- The possible occurrence and/or effects of genetic swamping of the wolves in Algonquin through hybridization with coyotes in zones along the park boundary needs further investigation and monitoring. (As noted informally in passing, coyotes are impressive "evolutionary units," itself a phenomenon of biodiversity interest.)
- There are a number of possible management actions that could, in principle, be taken to generate more preferred habitat for the prey species of wolves (e.g., through modifications in forest management or harvesting practices) or to reduce wolf-human contacts (e.g., forest road access issues), both within the park and in adjacent townships or wildlife management units outside of the park. However, the outcomes for predator-prey systems cannot be predicted and the many concerns some of these measures could raise in local communities would have to be addressed. A wildlife management plan for Algonquin Park (long called for by the park's master plan) should be completed.
- Long-term adaptive management programs to assess protection or management issues associated with the conservation of wolves as an experimental learning process need to be designed carefully for implementation over a number of years, with appropriate monitoring built into them. There are several possible approaches to this. The AWAG should convene a specialist workshop to address the options.
- Meanwhile, some control over wolf killing should be exerted by declaring wolves to be a game animal (subject to regulations) under the new provincial Fish and Wildlife Conservation Act of 1997, which came into force on January 1, 1999. Regulations, including no-kill restrictions, would have to include all areas adjacent to the park (rather than just focusing on a few townships) and involve all the affected stakeholders in their design. Rules governing compensation for loss of livestock to predators need to be changed. The geographic extent of these provisions needs to be decided (i.e., would they extend outward to township boundaries or include the larger provincial wildlife management units adjacent to the park?). A focused media approach would have to be developed to inform all parties of the new regulations and rationale for them.

While there appeared to be no serious objections to the general thrust of these conclusions, some reservations were expressed informally by participants during plenary sessions. One concerned the PHVA's goals, asking should ecosystem management be wolf-driven or should broader questions of ecological integrity be the main issue to address? Some direct management actions to augment prey species in the park, such as larger clear-cuts to foster revegetation or clearing areas along streams to enhance food sources for beaver, were said to be a "non-conforming" use in the context of the park master plan; such management was also noted to be inconsistent with wilderness concepts of letting nature take care of itself. The regulatory approach of "managing the users and not the resource" upset some local users and raised issues of compliance and enforcement.

Network Learnings

Human impact on wolf mortality: The data gathered on human-caused mortality of wolves was particularly useful for the VORTEX simulations. The implications of this source of mortality, and the sensitivity of the wolf population to plausible reductions in it, were clearly demonstrated. The availability of these data was fortuitous, since the data are time-consuming and often difficult to gather.

The census and other statistical data on human demographics and socioeconomic attributes were probably too coarse to be other than suggestive in this particular situation.

Wolf killing is done by relatively few people, most of whom are known to the local communities. It is an acceptable part of the local culture associated with an agricultural and resource-extraction economy. The direct killing of wolves is a practice that might just fade away if the economic base of nearby communities changes and provides more secure livelihoods. This hypothesis might be tested with a comparison between the southeast and western areas adjacent to the park, where the latter's local economy has a more prosperous recreational tourism component.

Taxonomic status of wolves: The question about the wolves' taxonomic status was interesting in its own right, but was viewed by many participants as secondary to (or even somewhat of a distraction from) the main questions. It was hypothesized during the workshop (and repeated in the AWAG report) that this taxon might be quite wide-ranging (from Quebec to Manitoba), hence the sense of urgency for taking strong, special measures for the Algonquin population might have lessened. Whether or not this helped the advisory group reach a broader consensus is not known, but conservation groups are still following the issue closely.

The Algonquin Wolf Advisory Group (AWAG): This group did reflect a range of stakeholder interests. In governance terms, it served as a "bargaining system" that was intended to lay an appropriate "regime" for wolf conservation in the spatial domain consisting of the park and adjacent lands. The workshop did reveal some areas where common ground might be reached, and a number of participants were perceived by others to have "given" a little (at least personally) on some points of contention. The AWAG seems to have been able to build upon this.

Greater ecosystems concept: The issues raised by this case study are similar to many other situations of "people and parks," especially where larger mammals range outside of the park in question. In the context of the "ideal type" situation of protected areas in industrialized countries (see chapter 12), the concept of a "greater ecosystem" to promote conservation objectives in a larger regional context has not evolved in the Algonquin Park case, although the geographic extent of such an ecosystem could readily be defined by the park, township boundaries, or the surrounding provincial wildlife management units.

The AWAG made one reference to "the Greater Algonquin Ecosystem," without suggesting its geographic extent, in one recommendation (number 22), that the OMNR continue to review broad-scale resource planning in the areas surrounding Algonquin Park. More generally, the ministry should consider issues of habitat fragmentation, road densities, and habitat linkages among wolf populations that would help assure the continued presence of the lycaon wolves "across a large portion of central Ontario." Barriers to adopting this approach in the Algonquin situation merit further exploration, possibly in relation to the primary use of the park in maintaining the resource-extraction economies of neighboring communities (and the cultural values associated with this) and uncertainties for governance occasioned by ongoing native land claims negotiations.

Success?: As an event, the PHVA workshop went well. Two features merit recognition as factors associated with success. First, the VORTEX modeling with the field data on human-induced wolf mortality was able to confirm the view that human impacts were "additive" (not "compensatory"), an issue in dispute for several years. Second was the importance of having a stakeholder group (AWAG) already established and underway (in this case under a ministerial directive); this meant that the workshop could assist the group and the group was attentive and receptive to findings from the workshop in which they also participated. Ultimate success, of course, goes beyond both of these evaluations and will be confirmed largely by the fate of the wolves.

Aftermath of the Workshop

The AWAG met almost monthly during a period of just over one year (1999–2000) and the PHVA was acknowledged to have provided significant input for their deliberations (AWAG 2000). The group took up one specific recommendation and had selected population data from the The-berges' work reanalyzed by two wolf specialists who participated in the PHVA modeling group; their analysis concurred with the view that the Algonquin population has quite likely declined (Vucetich and Paquet 2000). The AWAG also reviewed the findings from the PHVA workshop (Ewins et al. 2000) to arrive at their own recommendations. The "prin-ciple recommendation" from the AWAG was:

> We recommend implementation of a long-term Adaptive Manage-ment Plan for the wolves of Algonquin Provincial Park to significantly reduce the risk of population decline arising from human-caused mortality, and to maintain the wolf population at a level which is con-sistent with the long-term carrying capacity of the Park.... The plan would feature modest regulation of trapping and hunting of wolves, provide for the management of wolf prey and their habitat as is fea-sible, and improve public education and communication that would raise the focus on wolves and their management needs. (AWAG 2000, 14–15)

There are twenty-four specific recommendations in the AWAG report, most adopted unanimously by the advisory group. One recommendation about restricting hunting and trapping in the thirty-three townships adja-cent to the park (number 18) was apparently the main source for con-tention within the group; decision in this case was reached by a voting procedure rather than by consensus.

The AWAG report was submitted to the minister of natural resources in December 2000 and it was then made available for a two-month public comment period in January 2001. In November 2001, the minister announced a thirty-month moratorium on the hunting and trapping of wolves in all townships immediately adjacent to Algonquin Provincial Park; this was welcomed by conservation NGOs, some of which have since campaigned to make the moratorium permanent. The wolf research group's paper on the status of the eastern Canadian wolf (*Canis lycaon*) has been published (Wilson et al. 2000) and the Theberges decided to conclude their work in Algonquin. In May 2001, the Committee on the Status of Endangered Wildlife in Canada (COSEWIC) listed the eastern Canadian wolf as a species of "special concern."

PART FOUR

Understanding and Integrating the Dynamics of Human Systems

 **Chapter 12**

Governance for Conservation

GEORGE FRANCIS

Once the biological feasibility of survival for a focal species' threatened populations has generally been confirmed by a Population and Habitat Viability Assessment (PHVA) or other conservation plan, issues of governance for conservation become central. There is a need to identify the structure and functioning of the governmental agencies and other organizations that overlay the threatened populations' habitats and through which the appropriate conservation actions would have to be enacted. A PHVA helps specify what is needed in this regard. The challenge becomes how to best proceed.

There are several different definitions for the word "governance," but they each refer to interpretations of institutional arrangements extending beyond just government. Various private sector and other nongovernmental organizations (NGOs) and groups, as well as the rule systems under which these different organizations operate, must also be included. Most definitions restrict the term "governance" to alliances of organizations that collaborate to achieve mutually agreed upon purposes, for example, "Governance ... is the complex art of steering multiple agencies, institutions and systems which are both operationally autonomous from one another and structurally coupled through various forms of reciprocal interdependence" (Jessop 1997, 575).

This chapter is in three sections. The first outlines a broad perspective on three general kinds of societal contexts within which governance for conservation has to be addressed. Each of these contexts exhibits recurring governance-related issues. If designers of PHVAs are alert to these contextual backgrounds, they might anticipate issues likely to arise surrounding the particular questions they want to address. They could also use reports in the relevant literature to inform themselves about the experiences of others in dealing with such issues. The second section sketches a conceptual framework drawn from the social sciences to analyze the current state of governance for conservation in a particular situation. It focuses on long-term processes associated with the emergence and recognition of domains, the formulation of regimes, the development of full-fledged new institutions, and, in some cases, the gradual displacement of institutions by new domain-regime formations. Conservation issues associated with a particular focal species often have to address earlier phases of institutional development, either to augment or modify what is already in place, or to more generally begin developing institutional capacities. The third section discusses more-hypothetical approaches to assessing the overall sufficiency of organizational arrangements constituting governance for conservation.

"Ideal Type" Patterns in the Societal Contexts for PHVAs and Conservation

Part 4 in this volume discusses an array of social dimensions (or the "human ecology") that should be considered in designing PHVAs. As noted in the case studies, most attention has gone toward more accurately quantifying the human impacts on mortality or survival of the focal species of interest. However, such approaches can also point, at least implicitly, to important conservation actions that might be taken to lessen these negative effects. Implementation of the conservation actions will also be affected by the human ecology of the situation. The various social dimensions described in this volume tend to cluster into relatively few "ideal type" (or "generic type") configurations in terms of recurring conservation issues. Such ideal types (Weber 1978) provide an analytic tool for grouping many individual cases into clusters that approximate the pattern embodied in the ideal type; it is recognized that individual cases will sometimes exhibit elements of more than one type.

There are at least three ideal type situations discernable from recent literature on people and protected areas. The reason for reviewing this literature is that conservation strategies called for by PHVA analyses will almost inevitably entail issues about the establishment or management of protected areas in which viable populations of a given focal species can thrive. The three ideal types are protected areas in industrialized countries, protected areas in "developing" countries, and protected areas on Indigenous peoples' lands. They show similarities to the groupings used by Ness (chapter 13) to differentiate major demographic profiles and are summarized below.

Protected Areas in Industrialized Countries

In the context of industrial societies, relatively well-developed sets of protected areas are administered by specialized government agencies, supplemented by private conservation initiatives on privately owned lands. Resource management agencies and private sector organizations also own and manage lands under a relatively clear structure of property rights. With the growing awareness of issues concerning biodiversity, the limitations of having to rely upon the existing protected areas to conserve biodiversity have become widely discussed. In part, this is a matter of the relatively modest size of protected areas relative to the needs of certain biota (e.g., Wallis de Vries 1995; Clark, Paquet, and Curlee 1996). There may be contradictory policies for wildlife that inhabit parks as well as surrounding lands, and research conducted only in the parks may be inadequate for formulating cooperative management policies at a larger regional scale (e.g., Wright 1999). In addition, some landforms and their associated habitats and biota are not well represented in protected areas. As a result, "gap analyses" to identify additional areas to be protected are being pursued in a number of jurisdictions (e.g., Hudson 1991; Baydack, Campa, and Haufler 1999).

"Threats" to the existing protected areas are commonly defined as human-induced stresses of sufficient magnitude and duration as to cause an undesirable change in the structure and function of ecosystems in the protected area. While human demographic properties such as overall population growth and transient (tourism) or permanent immigration into the vicinity of the protected area therefore play a role, the more important impacts are in industrial and commercial usages. In Canada, for example, the main threats to national parks are thought to come from toxics and pollutants (such as acid precipitation, ground-level ozone, pesticides, or heavy metals); habitat change (from transportation and utility corridors, resource extraction activities close to the park, and from park infrastructure and management practices); direct impacts on wildlife (poaching, vehicle kills, human disturbances); and from exotic (non-native) species of plants, animals, or microorganisms (Woodley and Sportza 1996). Only some of these threats can be dealt with effectively from within the protected areas. Collaboration with neighboring landowners, or the effective implementation of international environmental agreements, is the key to removing other stresses.

Faced with these inadequacies in the number and sizes of protected areas for conserving biodiversity, and the threats to those areas which do exist, scientists and conservationists have promoted the notion of "greater ecosystems." These are regional-scale perspectives that view protected areas in their larger surroundings and raise questions about how land use by multiple owners or managers of some greater ecosystem might be better directed towards conservation purposes. In part, this may require some "redesign" of the landscape to help buffer core protected areas from certain human influences and to link the cores together so that enhanced connectivity would meet the needs of selected focal species' metapopulations. Comprehensive design guidelines for how isolated protected areas could be buffered and interconnected through natural landscape corridors have been spelled out (e.g., Grumbine 1994, 1997).

In North America, some of these greater ecosystems are being viewed at a continental scale in the Wildlands Project (Foreman et al. 1992). The implications for an ecosystem vision have been explored for particular regions and species. For example, Clark et al. (1996) discuss the requirements for conserving large carnivores in the Rocky Mountains from the legal, economic, cultural, and management perspectives as well as from the application of conservation biology. Other examples of conservation guided by greater ecosystem perspectives include the Southern Appalachian Wildlands (Newman et al. 1992), Crown of the Continent Ecosystem (Miistakis Institute; see *www.rockies.ca*), and the Greater Yellowstone Ecosystem (e.g., Baden and Leal 1990; Primm and Clark 1996).

The collaborative partnerships needed to realize these larger-scale schemes may entail new societal arrangements for conservation purposes. The difficulties encountered in the conflicted Greater Yellowstone Ecosystem demonstrate this. A series of more manageable pilot projects viewed as learning opportunities may be the best way to proceed (Brunner and Clark 1997). "The challenges in this world-famous region are contextual (rapid change, growth, pluralism, complexity, state/federal conflicts, and lack of common perspective), institutional (multiple organizations with overlapping authority and control and disparate mandates, uneven leadership, lack of creativity in problem solving, and resistance to change), and human (diverse perspectives and values and epistemological limitations)" (Clark 1999, 393).

Clark (1999) also suggested other approaches along with pilot projects, including "capacity building," leadership, and staff development training workshops; case study analyses for policy learning; and joint problem-solving exercises and seminars involving government staff, NGOs, and citizens, directed toward long-standing management issues. The experience of trying to apply the concept of a biosphere reserve could also be helpful (e.g., UNESCO 1996). It can be noted in passing that there are many examples of interorganizational collaborations to enhance governance for other purposes, such as the management of urban regions or watersheds; these cases are not reviewed here.

The case study of grizzly bears in the central Canadian Rockies (chapter 10) is an excellent example of the desirability of thinking in terms of greater ecosystems for a region that has a complex mosaic pattern of entrenched property rights and institutions. It is discussed further in Developing Governance for Conservation, later in this chapter. Two other case studies, for the muriqui in Brazil and wolves in Ontario (chapters 7 and 11) also exemplify this kind of situation. The key questions for conservation purposes relate to:

- the mosaic of human boundaries (jurisdictional, administrative, and proprietary) that overlay the geographic area of interest;
- the property regimes that pertain to each of the units in the mosaic (state-controlled or "public," private, common property, or open access with no property regime);
- the numbers and kinds of local organizations whose activities affect the survival of a focal species, or that would be the most affected by conservation measures, and their general attitudes toward the species or toward conservation measures needed on its behalf.

Protected Areas in Developing Countries

Over the past two or three decades, and in response to concerns about the loss of biodiversity or of renewable resources such as forests, wildlife, or water supplies, governments throughout the world have been establishing various kinds of resource reserves or protected areas. Most international and national conservation organizations continue to promote the expansion of systems for protected areas within each country and the strengthening of agencies that administer protected areas. Often in the past, protected areas were imposed by central authorities upon landscapes long occupied by farmers, pastoralists, swiddeners, and hunter-gatherers who were directly dependent upon local resources for their own survival. Sometimes whole communities were moved out of the new protected areas and resettled nearby, and some or all of the resource-use practices residents had previously engaged in became prohibited.

Protection for most protected areas was, and in many cases continues to be, based on some police or militialike operations. This "fences and guns" approach to conservation may be changing in favor of collaborative management, at least in the more participatory and less repressive governmental situations. Nevertheless, at local levels, antagonisms and conflicts can fester between protected area organizations and villagers in surrounding communities. Issues of social justice associated with such situations have garnered increasing attention, often through the efforts of local rural development organizations.

A sampling of recent writing about these issues follows. Of particular importance is the manner in which protected areas have been established. When a protected area is proclaimed from the top-down, without local consultations and accompanied by eviction of villages from inside the new protected area, the levels of local hostility can remain high. Hough (1993) cites cases of revenge by arson, and Fiallo and Jacobson (1995) report confrontations with protected-area staff. Kothari, Suri, and Singh (1995) note how expulsions and restrictions on uses of protected areas are often arbitrary, with little evidence of how local resource use threatens biodiversity. Community distrust is intensified if protected-area staff are seen to be taking out frustrations on local villages while accepting bribes from stronger commercial interests whose activities degrade the protected area. Commentators on these kinds of situations may denounce administrators of protected areas, sometimes in quite harsh terms (e.g., Guha 1997). Attitudes can also turn negative toward protected areas (although not necessarily wildlife) in cases where few benefits are forthcoming from conservation, but where expectations were once high (e.g., Newmark et al. 1993). Certain species of wildlife also generate negative attitudes when they are believed to be destructive, even if preventive measures to avoid things like livestock predation are at hand (e.g., Oli, Taylor, and Rogers 1994).

Conversely, Hartup (1994) describes a situation in Belize where a community voluntarily established a protected area on their own lands and received some benefits from visiting ecotourists. Villager attitudes toward the protected area remained positive, despite growing economic pressures on agricultural land use and the lack of alternative economic opportunities. Tacconi (1997a,b) describes a situation in Vanuatu where a participatory identification, assessment, and establishment process that recognized customary property rights and local people's needs proved successful for conserving protected areas. Badola (1998) notes that local people were supportive of the concept of forest conservation in an area linking two national parks in northern India, but nevertheless continued to use them in customary ways because they had no alternatives. Antagonisms toward a forestry department were apparently reduced when the state government involved local communities as partners and provided them usufructuary benefits in return for forest regeneration and protection. Similarly, people near a tiger reserve in India retained positive attitudes toward the reserve despite crop and livestock damage in villages adjacent to it because they were able to continue using the reserve for fodder and fuelwood; in addition, their cultural and religious attitudes toward wild animals were positive (Sekhar 1998). De Boer and Baquete (1998) report a similar situation among people living near an elephant reserve in Mozambique, in which people accepted damages from wildlife so long as they could continue to hunt, fish, and use plant materials from the reserve.

The intensity of human-wildlife disputes depends in part on the relative densities of populations. Hoare and Du Toit (1999) note that coexistence between people and elephants in northwestern Zambia was dependent on human settlement patterns. Coexistence was possible up to a human population density of fifteen to twenty-five people per square kilometer, which was associated with land-cover transformation of woodland to "settlement coverage" from 40 percent to 50 percent; beyond this threshold, elephants moved out. This example hints at a possible greater-ecosystem approach that, if combined with economic benefits associated with, for example, domesticated elephants and other wildlife, might be a viable conservation strategy in appropriate locations. Kiley-Worthington (1997) argues, with examples, that the semi-domestication of some wild large mammals would be a key for their survival. Trained elephants kept in or near reserves with their handlers might be contracted out part-time as draft animals in local communities.

One other major challenge is to take a wide variety of measures that serve to "decouple human needs from wildlife harvest" (Barrett and Arcese 1995: 1081) and from the need for other extensive resource extraction from protected areas. Wainwright and Wehrmeyer (1998) describe a situation in Zambia where hunting of wildlife was culturally important and could not be "bought off" with economic benefits from other activities. In Cameroon, hunting pressures in a reserve were heavy in areas near villages adjacent to the reserve, and wastage of animals caught in snares was high in areas farther away, which were visited less often; the hunters were men between the ages of eighteen and thirty-five who would have to be involved in developing designated zones with limitations on hunting in the reserve (Muchaal and Ngandjui 1999).

Swanson (1999) notes the essential importance of finding ways to generate economic returns for the "reserves" sector as well as the economically "productive" sectors of national or regional economies as a prerequisite for the conservation of biodiversity (including the implementation of the Convention on Biological Diversity). Solutions to problems associated with communities near protected areas have been sought through incorporating rural development projects into the conservation programs associated with the protected areas. It was believed that if economic and other benefits could be linked directly to protected areas, the social injustices associated with the manner in which protected areas had been established could be addressed, and local support for conservation strengthened. The result has been a large number and variety of Integrated Conservation and Development Projects (ICDPs), discussed in the following paragraphs.

"The distinctive feature of ICDPs as conservation strategies is that rural residents are induced to surrender access to, or to curtail illegal offtake of, native species and their habitats in exchange for alternative sources of income and sustenance, or for the provision of direct compensation, infrastructure, or social services associated with an improved standard of living. Such exchanges are sometimes contractual, but whether formalized or not, the basic notion of an exchange of access for material consideration is central to ICDPs" (Barrett and Arcese 1995: 1074).

There seems to be general recognition that conservation should be linked to socioeconomic needs of people affected by conservation policies (e.g., Alpert 1993) and that, in at least some cases, ICDPs can work (e.g., Matzke and Nabane 1995). ICDPs have sometimes been seen as promising initiatives (e.g., Wells and Brandon 1992; Alpert 1996) but not yet proven, given the relatively long time it can take for some rural development projects to generate noticeable improvements in people's lives. ICDPs may best be viewed as medium-term strategies to resolve park and people conflicts (e.g., Alpert 1996). In addition, they can only work in situations where the threats to a protected area come from just a few local sources and where the planned development projects can provide sufficient benefits to the right villagers or households to remove incentives to further encroachment on a protected area and its biodiversity. There has been a tendency to conceive of local communities as if they were small, homogenous, and passive social structures so that critical factors of ethnicity, gender, landowner classes, and customary rights of subordinate groups were ignored (e.g., Neumann 1997; Leach, Mearns, and Scoones 1999). It is instead important to recognize "communities of interest" that are particular subsets of a local population. It is also important, whenever possible, to understand the subtleties of household economics and how the resources of a protected area fit into an overall strategy of income and sustenance (e.g., Coomes 1995; Coomes and Barham 1997).

The main criticisms of ICDPs concern either the planning or management of individual projects, or the failure to think about them in relation to larger-scale issues such as political commitment to large regional development programs. Some of the projects have been described as "marriages of convenience" (Alpert 1996) that overlook the rich experience available in the rural development literature about factors involved in successful rural development (e.g., Wells and Brandon 1992, 1993). In the planning stages, such projects may ignore critical ecological and economic assumptions about their feasibility, likely impacts, or sustainability (Barrett and Arcese 1995; Gibson and Marks 1995; Shyamsundar 1996; Vandergeest 1996). They may also underestimate management needs because conservation agencies that initiate ICDPs often lack experience in rural development (e.g., Wells 1996). Some lack the capability for working and communicating with local groups (Pandey and Wells 1997). Others fail to provide for back-up research and monitoring of the projects (Alpert 1995). There may also be fundamental mismatches between the scale of ICDPs and the scale of the issues to be addressed to resolve the perceived problems (e.g., Ghimire 1994; Southgate and Clark 1993; Wells 1996).

The preliminary conclusions about ICDPs are that, like other kinds of development projects, they can only work under certain circumstances even if they are otherwise well planned and managed. Wells and Brandon (1992) suggest that "project preconditions" include "serious political commitments to the project, legislation conducive to the achievements of ICDP objectives, realistic institutional arrangements for project management, compatibility with regional development, systematic attention to land ownership and other resource access rights of the project's beneficiaries, and commitment [by conservation organizations conducting ICDPs] to institutional reorientation." Alpert (1996: 853) concluded that "... ICDPs can be a viable medium-term strategy for conservation and development at a limited number of sites that are biologically rich, aesthetically attractive, economically poor, geographically isolated, sparsely populated, and culturally traditional."

The need for collaborative management for protected areas is now widely recognized and regularly called for, whether or not ICDPs are the focus for the collaboration. Besides adopting guidelines for effective collaborative processes (e.g., Borrini-Feyerabend 1996), other contextual factors have to be considered, such as the different power and knowledge bases of the participants. Like ICDPs, there may be situations where the collaboration can be undermined or overwhelmed by other factors. Examples include lack of clarity over the holding of rights or tenure and what these entail for different community groups (e.g., Andersen 1995); fundamentally different perceptions of the environment and resources among diverse groups (e.g., Kaus 1993); contradictions between central government policies for modernization that create local dependencies on the cash economy and community-based management of resources sanctioned by taboos or the sense of sacredness surrounding particular places and species (Horowitz 1998); economic incentives at various jurisdictional levels that serve to drive the overexploitation of resources (e.g., McNeely 1993); instances in which conservation measures lead over time to unexpected ecological changes (e.g., Khan 1995); political practices whereby important transactions or land-use decisions leading to resource exploitation take place "outside the law" (e.g., Richards 1996); and outbreaks of civil strife (e.g., Hart and Hart 1997). Local groups are the least powerful when it comes to developing local conservation policies; Agrawal and Gibson (1999) propose that agencies and nongovernmental groups direct more support to them, perhaps through some kind of federated structure that can network and negotiate against arbitrary actions by governments.

The mountain gorilla case study (chapter 6), as well as others cited in chapter 14, represent the range of problems often associated with reconciling basic human needs and conservation within traditional use areas. The tree kangaroo case study (chapter 9) illustrates the use of rapid rural appraisal methods (e.g., Chambers 1994) to better understand community resource uses, including their probable effects on tree kangaroos in the immediate area. The key questions for conservation purposes relate to:

- the local history of establishing (or trying to establish or maintain) protected areas associated with the focal species of interest;
- impacts on neighboring communities or on particular groups arising from restrictions on access to, or use of, certain resources formerly extracted from the protected area; and
- the relative success of local community development projects, especially ICDP kinds of initiatives.

Protected Areas on Indigenous Peoples' Lands in "Frontier" Regions

There is recognition that biodiversity may also be associated with cultural diversity and that important areas of biodiversity are associated with Indigenous peoples who have maintained traditional ecological knowledge associated with hunter-gatherer cultures. Such people's relation to the landscape is embedded in a religious or mythical context, rather than a proprietary or instrumental one, and the concept of protected area is best expressed through recognition of sacred sites and the special economic or cultural values of various plants and animals.

The imposition of some formally recognized protected area by an external authority representing a dominant culture can only be done effectively if the Indigenous peoples' traditional rights to the areas they occupy are also acknowledged, if cultural resource-use practices are allowed to continue, and if Indigenous people retain a significant control over decisions affecting the area. Otherwise, the result may be little more than a "paper park." Wells (1998) notes how state interference undermines local resource management capability without being able to replace it with effective administration and he comments that the dynamics of interactions among all the actors that affect biodiversity may be as complex and poorly understood as biodiversity itself.

The situation may be further exacerbated when a state takes control over wildlife or protected areas and tries to administer them remotely. Local communities come to realize that these are no longer community resources for them, and so will take to poaching or other forms of exploitative use.

> Wildlife degradation results from the systematic alienation of Indigenous people from wildlife by creating ecological islands and denying them access to wildlife as a resource. It is also due to removing responsibility for managing wildlife and ownership from Indigenous people and making wildlife the responsibility and property of the state; degradation results because the state has failed to effectively manage by remote control this vast estate in communal areas in light of shrinking resources. The failure of the state creates characteristics of "open access." The state fails because it is far removed from where the action is. (Sibanda 1995, 85)

As we saw in chapter 8 and will touch on again in chapter 15, co-management agreements between Indigenous people and the external, dominant culture, set in some larger agreement about land rights, are one way of resolving this issue.

Stevens (1997) reviews the literature and some case examples from throughout the world on "indigenously inhabited protected areas." He notes several distinctive types of such areas in terms of the extent to which recognition is given to Indigenous peoples' settlement and subsistence practices and their involvement in resource management and protected area administration. From the perspective of "alliances for conservation and cultural survival," the major policy issues include full recognition of Indigenous peoples' land tenure and subsistence practices, and institutional arrangements for comanagement leading at some point to full Indigenous management of the protected areas.

Key concerns of Indigenous peoples are about "... sovereignty, self-determination, land tenure, resource use rights, fulfillment of agreements, culturally sensitive enforcement of protected area regulations, development, coordination with national and international NGOs and governments, and tourism management" (Stevens 1997, 280). A more recent review by Beltran (2000), also based on examination of case studies, comes to similar conclusions and develops a set of "principles and guidelines on protected areas and indigenous/traditional peoples." Future establishment of comanaged and locally managed protected areas as the standard practice in homelands of Indigenous peoples will require changes in practices long associated with national parks in a number of countries. The cross-cultural challenges can often be difficult (e.g., Gadjil et al. 1993; Cox and Elmquist 1997). Stiles (1994) outlines possibilities for developing sustainable commercialization of renewable forest products in the context of cultural survival and Indigenous rights. Decher (1997) outlines the conservation potential inherent in giving formal recognition to traditional sacred groves and sites in Ghana, and from offering management help if village councils so wished. Protected areas established in Indigenous regions for other reasons, such as to reduce border tensions and evolve "common security zones," may be more problematic (e.g., Weed 1994).

The Peary and Arctic Islands caribou case study (chapter 8) was situated in a "frontier" region long inhabited by Indigenous Inuit peoples, in which a comanagement regime is being established. The case study of tree kangaroos in Papua New Guinea (chapter 9) was situated in an

Indigenous peoples context, but governance issues were not addressed. The key questions for conservation purposes relate to:

- whether there has been clear recognition by the state of Indigenous peoples' rights, and if so, the extent to which rights and responsibilities have been devolved to Indigenous communities;
- the kinds of comanagement regimes (if any) that are in place; and
- other issues associated with Indigenous peoples' self-government or autonomy that can be expected to arise and possibly dominate discussions in a PHVA.

Analyzing the Current Status of Governance

An analytical framework is described below to help discern the current status of governance for conservation in particular geographic areas of interest. It makes use of the concepts of institutions, actors, domains, and regimes, which are introduced briefly.

"Institutions" can simply be defined as rule systems that specify acceptable social practices (Knight 1992); this is contrasted with organizations, which are goal-directed and work-oriented and are usually specialized to carry out certain sets of tasks. The recurring patterns of behaviors that are constantly being created and re-created by the actions of people following rule systems encoded in laws, tradition, and custom, appear as social organization or social structures, in large measure because they are predictably repetitive most of the time (Giddens 1984). A change in basic rules can change recurring patterns of behavior. "Actors" is a generic sociological term that refers to an array of different organizational forms (i.e., agencies, corporations, associations, commissions, clubs) and sometimes to key influential people within them. Often, the word "institution" is used to refer to rule systems that also have well-established organizational arrangements for monitoring and enforcing the rules (e.g., criminal justice systems). From this perspective, the processes of institutionalization refer to the development of organizational capacity to foster compliance with particular rule systems.

A "domain" is a "social space" as perceived and defined by the actors who share it. The focus of a domain can be a geographic area, a social or economic sector, or certain kinds of problems and issues. As a social construct, a given domain may have no firm boundaries because as actors come together within it, their perceptions of what should be included by it can change. Domains arise when actors within them become aware of their interdependence with similarly situated actors. This begins to occur when actions taken by any one of the actors affect the organizational environment (as defined in organization theory) of a number of others. As the numbers of actors increase or the scale of their activities enlarges, the domain develops a kind of "turbulence" (Trist 1983), where everyone in the domain seems only to be able to react to the events and conditions created by everyone else. At some point, given sufficient turbulence, the domain will start to self-organize through processes suggested in Developing Governance for Conservation section, below.

The term "regime" is used, especially in the international relations literature, to refer to a governance system intended to deal with a limited set of issues or a single issue area (Young 1994). The many examples of regimes at the international level are expressed in the form of treaties or conventions among nation-states to address issues that individual countries cannot deal with on their own. Regimes can carry out a variety of institutional tasks: they may define regulatory codes of conduct in regard to some shared resource, agree on operating procedures for resource allocations, engage in joint collaborative projects, or develop shared understandings and agendas from which actors can then work together more closely (Young 1997). Regimes can exemplify the two definitions of institutions noted above: generally accepted rule systems either with or without organizational capacities to foster compliance.

Within nation-states, governments are able to exercise their sovereignty through administrative bureaucracies and police powers, so that it might first appear that all rules can be enforced. However, regimelike arrangements can be found within nation-states as well as among them. The number and variety of common property regimes (CPRs) is one major example (Berkes 1989; Ostrom 1990) where corporate, communal, or other kinds of group ownerships set rules of use for their members while excluding outsiders. The similarities between CPRs and international regimes have attracted recent scholarly attention (e.g., Keohane and Ostrom 1994). Scale issues become important because there could be insurmountable problems for CPRs with too many participants, considerable cultural diversity, and complex interlinkages among different CPR systems. A key requirement for CPRs is that national governments and their policies remain supportive of local self-organization; protecting institutional diversity related to how diverse peoples cope with CPRs may be an important contribution to sustainability (Ostrom et al. 1999). Regimes have also evolved for multiple municipalities in larger urban regions and for watershed planning and management. The argument for considering governance within nation-states in terms of effective regimes is strengthened from the recognition that many governmental "rules-in-the-book" do not translate into "rules-on-the-ground" and that those actually implemented may be ecologically incoherent (Sproule-Jones 1989).

Critical Questions

To address governance issues in the context of PHVAs, or other conservation planning initiatives, the following questions might be posed with reference to the focal species of interest.

What is the current geographic scope and organization of the perceived domain for the focal species?

Some kind of domain will exist because the range-country sponsors of a PHVA would be part of it. So the question becomes one of the domain's geographic scope as perceived by the actors within it and the degree of its organization. The geographic scope of the perceived domain appeared to be sufficient in the case studies for the Peary and Arctic Islands caribou, the mountain gorillas, the grizzly bears, and at least one population of muriqui (chapters 8, 6, 10, and 7). It was not sufficient for the Algonquin wolves (chapter 11). The organization, in terms of the key actors involved, appeared to be well developed for the caribou and for one of the muriqui populations, in disarray because of civil strife in the areas inhabited by the mountain gorillas, and evolving for the grizzly bears.

Does a conservation regime exist?

This question explores the extent to which there are some generally agreed upon sets of social rules that provide for the conservation of the focal species across the geographic extent of the relevant domain. Some parts of the regime may be institutionalized as special administrative agencies, for example, those having responsibilities for wildlife management or national parks. The case studies show considerable variety here. The conservation regime for the Peary and Arctic Islands caribou is in place, but it is evolving into a more balanced comanagement regime between the Inuit and Euro-Canadian governance systems. There appeared also to be a regime for tree kangaroos (chapter 9) embedded in some mix of private land ownership and customary rights for certain resource uses by community groups and it was able to mobilize conservation actions when the PHVA workshop made apparent the critical need for such actions. The conservation regime for grizzly bears was a rather complex regional mosaic within which grizzly conservation applied fully, somewhat, or hardly at all in the different parts of the region. Different institutional configurations for governance were in place for different components of the regional mosaic.

Is there a fully institutionalized system for conservation in place?

"Fully institutionalized" refers to administrative and enforcement capabilities such as those associated with national parks. The main question is the sufficiency of these arrangements either to cover the relevant domain, or to carry out the conservation actions required. Most of the case studies point to some weaknesses or shortcomings from what would be most desirable. The national parks for the mountain gorillas were sufficiently large, but were either very weak or overwhelmed by events surrounding them. The provincial park for Algonquin wolves did not cover the essential winter range, and the state park for one population of muriqui apparently experienced extensive poaching.

Developing Governance for Conservation

Domains and regimes can be interpreted as phases in institutional development processes that continually evolve. Domains become organized into regimes and regimes can become fully institutionalized over time (e.g., Young 1994; Harvey 1995; Brunne and Toope 1997). The process does not just stop with a fully institutionalized structure. The social and ecological context for the challenges institutions were meant to address may themselves change in ways that the institutions either cannot perceive, or to which they fail to adapt. In due course, because of their widely perceived "management failures," the institutions lose their effectiveness and legitimacy. Gunderson, Holling, and Light, (1995) postulate an institutional cyclelike phenomenon associated with ecosystem dynamics at regional scales.

The human agency components associated with institutional development have been described through use of concepts such as:

- epistemic communities, "... a network of professionals with recognized expertise and competence in a particular domain and an authoritative claim to policy-relevant knowledge within that domain or issue-area" (Haas 1992, 3);
- bridging organizations or functions (e.g., Westley and Vredenburg 1991);
- referent organizations, a term analogous to referent groups in social psychology and referring to organizations that play a key role in cultivating

"domain-based inter-organizational competence (that) enables the organizational life of society to be strengthened at the domain level in ways
that are self-regulating rather than becoming imperial or remaining
ineffectual" (Trist 1983, 270–271);

- bargaining systems; and
- institutional transformation associated with social learning (e.g., Michael
  1993).

There are overlaps among these categories; for example, epistemic
communities may serve a bridging function, referent organizations may
foster negotiation systems, and social learning could give rise to new epistemic communities. Human agency associated with each of these phenomena may also be vision-led, planning-led, or learning-led networks of
actors (Westley 1995). Each kind of network has both strengths and weaknesses of which actors should be aware.

The relevant questions for a PHVA are whether or how the above phenomena are occurring in the societal contexts associated with the focal
species of concern. The PHVAs themselves are meant to catalyze some
movement in these directions. Arguably, PHVAs do this in part because
the core expertise drawn upon, especially for the modeling components,
constitutes a kind of epistemic community to whom the other participants
are expected to provide relevant data. The workshop process is also meant
to provide some bridging function among scientists, managers, and other
stakeholders.

The Algonquin Park wolf case study (chapter 11) dealt with a contribution to a formal bargaining system already set up—the Algonquin Wolf
Advisory Group of stakeholders who, with the stimulus provided by the
PHVA, developed a set of negotiated recommendations to the provincial
government. These were subsequently accepted, thereby addressing at
least some of the main conservation concerns about the wolves, including taking steps towards an appropriate regime for the relevant domain
lying just outside of the provincial park. In the mountain gorilla case study
(chapter 6), the PHVA stimulated the formation of the Mountain Gorilla
Foundation, which may become a kind of referent organization. The
caribou case study (chapter 8) exemplified a strong overarching bargaining system, at work to develop a mutually acceptable and effective
comanagement regime for the five metapopulations of caribou.

The grizzly bear case study is more organizationally complex (chapter
10). Box 12-1 notes six interorganizational groups or networks associated

Box 12-1.
Interorganizational groups or networks associated
with grizzly bear issues in the central Canadian Rockies.

Central Rockies Ecosystem Interagency Liaison Group (CREILG): Established in 1991 by Parks Canada and agencies of the Alberta and British Columbia governments. Share information on the theme of ecosystem management and work on issues of mutual concern. Management of grizzly bears was one issue they addressed in the context of the Central Rockies Ecosystem. Prepared the *Atlas of the Central Rockies Ecosystem* (White and Scott-Brown 1995).

Long-Term Strategy for Large Carnivore Conservation in the Rocky Mountains (WWF): Established 1995. Promotes carnivore conservation areas and associated measures for ten species in the Rocky Mountains, from Yellowstone to the Peace River country in northern British Columbia. Conducts landscape analyses and GIS mapping of habitats and identifies important human-use factors.

Rocky Mountain Ecosystem Coalition: Established 1993. The coalition promotes principles of ecosystem conservation and the conservation of ecological integrity and biodiversity in the Canadian Rockies through applied research, education, advocacy, and "aggressive litigation." Investigated effects of roads and habitat fragmentation on grizzlies along the eastern slopes of the Rockies.

Rocky Mountain Grizzly Bear Planning Committee (RMGBPC): Established in 1997 by Parks Canada and agencies of the Alberta and British Columbia governments. Exchanges information and practical tips on methods for grizzly bear conservation and management. Can make policy-level recommendations (but had not done so as of January 1999).

The Miistakis Institute for the Rockies: Established 1995. Promotes transboundary ecosystem management for a Crown of the Continent Ecosystem focused on the headwaters of the Flathead River and Glacier-Waterton International Peace Park. Provides GIS and other information analyses and services in part through a Dynamic Online Reference Information System (DORIS) for use by the ESGBP, the WWF project, Crown of the Continent/Y2Y, and other conservation groups.

Yellowstone to Yukon (Y2Y): Established 1993. Network of more than 105 organizations promoting broad-based biodiversity conservation strategies that protect wildlife habitat, wildness, and the "mountain assemblage of large carnivores." Mobilizes public information and understanding of the landscape-level biodiversity conservation concept. Region extends from the Greater Yellowstone Ecosystem to the Mackenzie Mountains in the Northwest Territory.

with grizzly bears in the central Canadian Rockies, in addition to the Eastern Slopes Grizzly Bear Project (ESGBP), which cosponsored the PHVA. While more understanding of these groups is needed before drawing conclusions, each appears to be performing some bridging function, albeit in quite different ways. They also exemplify different styles associated with the evolution of domains into regimelike formations, with the CREILG being mainly learning-led (like the ESGBP itself), the RMGBPC mainly planning-led, and the Y2Y mainly vision-led.

Not surprisingly, all of the PHVA case studies identified ways in which conservation actions might be strengthened for the focal species of interest. A broader question concerns the sufficiency of governance arrangements to provide the degree of conservation needed to allow the focal species not merely to survive, but to thrive. This question can be explored at the national level and at the domain level.

Ness (chapter 13) introduces the idea of assessing the "political-administrative system strength" (PASS) that could identify governance prerequisites for carrying out sophisticated policy measures at a national scale. Components of PASS include the existence of a strong center for political decision making over the geographic area of interest, the administrative capacity to monitor socioeconomic and environmental change, and a strong commitment to sustainable development. In reflecting upon the muriqui PHVA, Ness notes that PASS increases environmental protection in Brazil.

An appropriate PASS could increase the capacity to develop a conservation strategy for other biota and to carry out the necessary suite of conservation actions, which are to control direct and indirect human-caused mortality, secure key habitats and sites, rehabilitate degraded areas and secure habitat connectivity at the landscape scale (where necessary), and promote a sustainable regional economy (that redirects human economic activities away from exploitation of particular biota or their habitats). The question then is whether this capacity is evident at the regional or domain level and is effectively back-stopped by the overall PASS capabilities at the regional and national levels.

The situation at a regional or domain level might be assessed with another approach, applied here to the grizzly case study as an example. With reference to the main categories of conservation actions that may have to be considered (see box 12-2), critical information includes relevant programs of the federal and provincial governments, special measures being taken under intergovernmental agreements, initiatives taken

Box 12-2.

Main categories of conservation actions for grizzly bears.

Control direct and indirect human-caused mortality
    Develop and enforce hunting regulations
    Disperse human activities and facilities over time and space
    Compensate for livestock depredation

Secure key habitats and sites
    Create zoning provisions within publicly owned lands
    Purchase lands at market prices
    Purchase, or acquire by donation, covenants or conservation easements

Rehabilitate degraded areas no longer used for resource extraction
    Reforest old logging roads and seismic lines
    Remove unused facilities (e.g., campsites, generators)

Secure habitat connectivity at the landscape scale
    Maintain or redesign corridors
    Modify zoning to provide buffers around core sites
    Foster interagency coordination and multi-stakeholder agreements

Promote a sustainable regional economy
    Market wildlands and charismatic species (ecotourism)
    Identify costs of unsustainable resource use
    Maintain regular contact with local communities

---

by the corporate and nongovernmental sectors, and special collaborative arrangements associated with the conservation of grizzlies. These organizational systems would then have to be assessed for their scope and depth. Scope addresses questions about the extent of formal agreement for grizzly bear conservation over the entire range (domain) of four subpopulations and the extent to which organizational systems are in place to carry out the main categories of conservation actions. Depth addresses questions about the extent to which organizational systems for conservation have been fully developed to include policy commitments, with funded programs having standards, benchmarks, timetables, and periodic evaluations made publicly available. Gaps revealed by this analysis would become the focus for developing or strengthening governance for the conservation of grizzlies.

Priorities for dealing with the gaps could be set by relating the overlaying mosaic for governance to the regional analyses of critical habitats for grizzlies. The PHVA was able to relate landscape analyses of secure areas and high-habitat quality for grizzlies to the VORTEX model. The workshop also provided examples of specific actions that could be taken at sites in or near Banff National Park to reduce negative impacts on the grizzlies. Geographic Information System (GIS) mapping at the same scale should be able to overlay the human boundaries on the landscape mosaic for the entire domain. It then becomes a strategic decision about which gaps to attend to first, given the relative importance of different gaps for the grizzlies and the relative feasibility of encouraging human uses of landscapes that could reduce the more serious impacts on grizzlies.

Conclusions

It is clear that the introduction of human dimensions into PHVAs adds a measure of complexity and only some of the considerations can be integrated directly into VORTEX models simulating the population dynamics of the focal species of concern. Once the biological feasibility of allowing for the continued survival of some population is judged to be positive, usually with the proviso that certain critical conditions relating to population mortality and survival rates are dealt with, then the human dimensions critical for conservation measures become central.

It is suggested here that some of the human dimensions can be anticipated given the general "ideal type" contextual situations a PHVA will be addressing, and that understanding the institutional arrangements for governance will be important in order to identify what needs to be done. Some of this might be addressed in the design of a PHVA background inquiry prior to a workshop and as part of the attempts to involve some of the "right" people. While follow-up must be done by the relevant stakeholders in the range country, it may be possible to identify particular aspects or local areas where some collaboration or conflict resolution would be the most helpful for conservation purposes. At this point, a PHVA can leave off and some strategic thinking must begin.

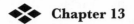

Chapter 13

Human Population Dynamics and Integrative Action

GAYL D. NESS

Dealing with the human population as a challenge to biodiversity conservation involves addressing two general questions: what is happening with the human population? and, what can be done? This chapter focuses attention primarily on the first question, examining human population dynamics. Brief attention is given at the end to the question of what is to be done. The two strategies suggested here, organizational and methodological, are treated extensively in the rest of the book.

Understanding human population dynamics involves telling two stories; one is of general forces, the other concerns location specificity. The general forces at work can easily be identified. These include the past two centuries of rapid human population growth, closely associated with the rise of urban industrial society based on fossil fuels. Here is where human numbers multiplied by human activities have placed enormous burdens on the natural environment, destroying both habitat and species at rates hitherto unknown and unimagined. But this global revolution has also produced two major demographic regimes, each with its own distinct set of challenges to biodiversity protection.

In addition to the easily identified general forces, however, there is another immense and complex diversity of challenges that are highly location-specific. They require both analytical and programmatic approaches that focus on the complex conditions of any specific area in which we wish to promote biodiversity conservation.

Demographic Dynamics: General Forces and Location Specificity

Let me begin with a brief review of the general underlying forces that are at work and move progressively to the kinds of different local conditions that pose very different challenges. We must understand both the general and specific conditions before we can address the question of what is to be done.

General Forces: Population Growth and the Rise of Urban Industrial Society

Figure 13-1 shows a portion of human history that is well known. The last millennium opened with a continuation of the long, slow growth of the human population. Perhaps as many as five hundred millennia ago, the human species emerged in Africa and expanded to fill the entire globe, though at annual rates only just above 0 percent. A millennium ago humans constituted perhaps no more than the population of today's United States, about a quarter of a billion. In the long fifteenth century, when the West discovered the seas and tied the world together in one highly interactive ecosystem, growth rates rose slightly, especially in Europe, Asia, and possibly Africa, while the European impact on the Americas produced a huge demographic collapse. (I date this from 1415, when the Portuguese conquered Ceuta in North Africa and began the great explorations that brought the Europeans to the Americas and Asia. It ends in 1523 with Magellan's circumnavigation of the globe.) Then in the eighteenth century, world population growth rates began to rise to near 1 percent per year, followed by the much more rapid exponential growth after 1945 that the figure shows so clearly. This dramatic increase in numbers and rates of growth mark our era as highly distinctive, and also highly destructive of the natural environment and biodiversity.

There is no doubt that this huge increase in the human population has placed immense and destructive pressures on the natural habitat of many species and has resulted in extensive species destruction. The human population today is a major challenge to biodiversity conservation (Harrison and Pearce 2000).

But it is not only the number of humans that poses the challenge—it is also what they do to the natural environment. In general terms this can be seen in figure 13-2, which shows the close association of rising numbers with a transformation in human social organization from rural

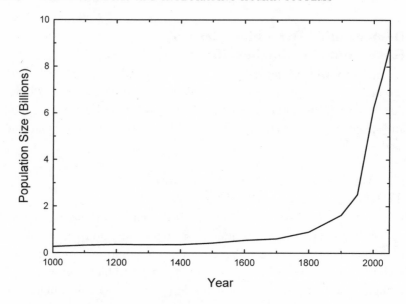

Figure 13-1. World population growth, 1000–2050.

agrarian to urban industrial society, based on the revolutionary development of fossil fuels.

Fossil fuels produced a new type of society with a voracious appetite
for natural resources. Forests were felled to produce timber for construction and to clear land for agriculture. The earth was turned over in the
search for minerals, from precious metals to coal and oil to exotic metals
and finally to the powerful and toxic uranium. Probes drilled deep into
the earth for waters stored over millennia to slack the thirst of the crops
that would feed the growing population. This appetite for resources
wreaked havoc on the natural habitat that supported the earth's great biodiversity. But the appetite included the species as well. Long before the
revolution to this new society, when humans first moved across the Bering
Strait, it took perhaps a millennium for them, along with other forces, to
eliminate the great mammals of North America (Gibbon 1998). With
the new type of urban industrial society, it took only a few decades to
bring the North American Bison or the Passenger Pigeon to near extinction. And so it has continued, with species extinctions now approaching
unprecedented rates.

Even here, however, the development of this new type of society and
its voracious appetite was not evenly spread across time and space. It

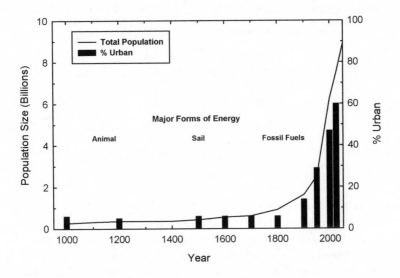

Figure 13-2. Human population growth, energy and social transformation. (Source: Ness 1993 and Ando 1997.)

began in the North Atlantic community from the late eighteenth through the nineteenth and into the middle of the twentieth centuries, usually leaving the rest of the world far behind, still locked in a rural-agrarian society that had roots extending back many millennia. Only in the mid-twentieth century has the rest of the world experienced this social-technological revolution. This has produced two different demographic regimes, which present very different types of challenges to the conservation of biodiversity.

To understand this uneven distribution of change and the different regimes it has produced, we turn to a simple but powerful concept, the demographic transition.

Two Demographic Transitions

The rise of urban-industrial society has brought a remarkable transformation simply but powerfully designated as the demographic transition: the movement of a society from high to low birth and death rates. Most important, however, is that there has not been just one, but two major demographic transitions, one in the past and one currently in progress. This timing of demographic transitions has given us today two different regimes, one of low growth and one of high growth, with very different

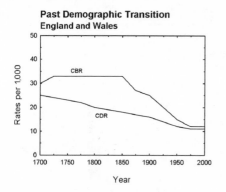

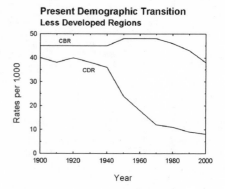

Figure 13-3. Two demographic transitions: past and present. The demographic transition is the movement of a population from high to low birth and death rates. The past demographic transition occurred more gradually from lower original levels, bringing slower rates of population growth. The present demographic transition occurs more rapidly, from higher original levels, bringing higher rates of population growth. CBR = crude birth rate; CDR = crude death rate.

challenges for biodiversity conservation. The two transitions can be seen in crude outline in figure 13-3. (Material in this section is developed more fully in Ness and Golay 1997.)

The past demographic transition is illustrated here by the experience of England and Wales. Every currently industrialized society has gone through roughly the same transition, beginning about 1700–1750 in England and Wales and ending with Japan in 1950. Birth and death rates moved from high to low in the process of the development of urban industrial society. The process differed somewhat from society to society. In Europe, France has one of the earliest fertility declines, Holland one of the latest. Japan was the last of the major industrial societies to complete the transition, as mortality began to fall slowly in the late nineteenth century and fertility began to decline slowly around 1920, with rapid declines coming in the aftermath of defeat in World War II. In all cases, however, the process was a relatively gradual one, with mortality and fertility starting at moderate levels, mortality declining gradually with a general rise in the standard of living, and fertility falling later as urban industrial society demanded a different reproductive strategy for families. Both mortality and fertility declines came without major medical developments in the control of either mortality or fertility. This past transition has produced a distinctive demographic regime today throughout the major industrial world: slow to negative rates of population growth with great wealth and exceptionally high standards of consumption.

There is another demographic transition underway today in the rest of the world, in what the United Nations has called the "Less Developed Regions," though some of these today, such as South Korea, Taiwan, Hong Kong, and Singapore, rank with the wealthy urban-industrial societies of the West (UN 1966). Here we begin with higher rates of mortality and fertility than in the past, for reasons that are still hotly debated and not fully understood (Livi-Bacci 1992; Caldwell 1976; Ness and Golay 1997). Developments in medical technology especially stimulated by World War II brought new medical capacities to control mortality. Another development, the spread of large-scale international organizations, played a supporting role, as the new mortality-controlling technology could be carried by the new international organizations rapidly to all parts of the world. The result was a massive and rapid decline in mortality following 1945–50, clearly evident in the right side of figure 13-3. Mortality declines that had required a century or more in the past could now be accomplished in a decade or so. But with high original levels of mortality, the continuing high fertility following the mortality decline brought exceptionally high rates of population growth. The high growth rates of the past demographic transition barely reached 1 percent per year; those in the current transition have often risen above 3 percent per year. The present transition is not only more rapid, however; it involves far greater numbers than in the past. Both the speed and magnitudes present new and powerful challenges.

Fertility is now beginning to fall in much of the less developed world as well, especially after 1965–70, which saw the peak of world growth rates. Again, new technological and organizational developments, as well as policy changes, have been a major feature of the transition. New contraceptive technologies have been developed and, since 1965, have become widely available. The spread of markets, changes in government policy from pro-natalism to anti-natalism, and the spread of governmental and nongovernmental organizations for fertility limitation, or family planning, have brought rapid fertility declines in many parts of the world (Ness and Ando 1984). But the fertility decline that marks the end of the present demographic transition is not evenly experienced throughout the world.

Most of East Asia (China, South Korea, and Hong Kong) has achieved the low fertility that is now common in the more developed regions. The process is also well advanced in most of Southeast Asia and Latin America. South Asia is more mixed, with great advances in Sri Lanka and some of the states of India (Kerala, Tamil Nadu, and the Punjab). Bangladesh is making good progress, as is the very poor and geographically constrained Nepal. But Pakistan, parts of northern India, and the rest of what can be called West Asia still show high fertility. Most of Latin America has experienced fertility declines, but the region as a whole has not yet completed the transition, and there are pockets in Central America and the Andes where fertility remains high. Africa is now the major area of high fertility, with a future that leaves many questions.

These two historical stages of the demographic transition have given the world two very different patterns of growth, which can be seen in figure 13-4. The graph shows the average annual percent growth of the world and three major regional categories used by the United Nations: the More Developed Regions (MDR), or the wealthy nations that experienced the past demographic transition; the Less Developed Regions (LDR), or all of Africa, Latin America, and Asia except Japan, which are now experiencing the demographic transition; and the Least Developed Regions (LeastDR), the poorest countries of Africa and parts of South Asia.

Here one can see the rising growth rates that came with the rapid reduction of mortality after 1945–50. Moreover, most of the world experienced a baby boom after the war, but by 1950 growth rates began to diverge. The MDRs experienced an early decline in growth rates, and by 2025 the rates are projected to be negative. Growth rates in the LDRs continued to grow through the period 1965–70, when they peaked at

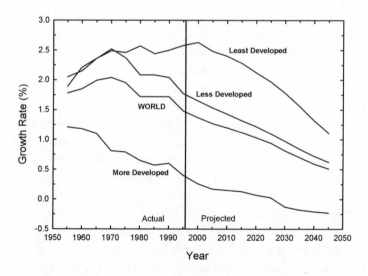

Figure 13-4. Population growth rates for the world and its major "development" regions. (Source: UN 1996.)

about 2.5 percent. By that time the LDR population stood at 2.7 billion, or 73 percent of the world's population. Although the LDR as a whole began to show a decline in its growth rate after 1970, the LeastDR rate continued to rise as mortality declined with no change in fertility. In 1970, the LeastDR held 300 million people. Today the LDR's 4.9 billion account for 80 percent of the world's population, while the LeastDR population has grown to 658 million, just over 10 percent of the world's population. It is projected to rise to 1.6 billion, or 17 percent of the world's population by the year 2050 (UN 2001).

Two Different Demographic Challenges to Biodiversity Protection

For today's problem of biodiversity protection, this two-stage history of the demographic transition has produced two major demographic regimes with considerably different challenges.

On the one hand are the major industrial societies of the world, with low to negative rates of population growth. They are also wealthy and have highly developed state administrative capacities. The challenges they pose come from their voracious appetites for natural resources, with great implications for the less developed regions, and their rapidly rising appetites for nature as an item of domestic consumption. Here the challenges come not from sheer numbers of human beings and their rates of growth, but from their wealth and their demands for natural resources and for nature itself.

Domestically, these wealthy countries not only show slow population growth, they have also experienced a century or more of increasing governmental efforts to protect the natural environment and biodiversity. Parks and forests have been set aside and protected from destruction, and individual species have often been given legal protection when they appear threatened. (It is also fair to note that in some of the more developed countries, such as Holland and Belgium, there is rather little "pristine nature" left to conserve.) Both the wealth and the administrative capacities of these countries have meant that designated protected areas are usually in fact protected. At the same time, the natural environments of these countries are themselves under pressure from the wealth they generate and from the increasing demand for "nature" as an item of popular consumption. The demand for housing produces an urban sprawl that eats into the natural environment. The rising demand for recreational housing, or summer homes, adds to the pressure. Finally, the rising demand for nature itself places increasing pressures on protected areas that are designated protected and recreational areas. Though the wealthy nations have developed effective administrative capacities to make protection work, they also harbor internal forces of large and wealthy extractive companies and "development forces" that consistently threaten to undermine such protection.

The wealthy world's demand for natural resources also produces other pressures far from home. It has an especially marked impact on the less developed parts of the world, which are rich in natural resources and poor in the political and administrative capacities to protect or husband those resources. Tropical rain forests are cleared to provide timber and pasture for wood and beef for the wealthy nations. Oil and mineral deposits in the less developed world are targets of the large corporations that feed the demand of the wealthy world. The exploitation of these resources often produces great environmental destruction both through direct extraction and through the conflict generated by these wealthy outsiders in countries with weak political and administrative systems.

There is one additional population dynamic in the industrialized societies to which we must refer, though its implications for conservation are mixed and unclear. As wealthy nations, these are powerful magnets for immigrants from the poorer countries. In addition to legal migrant streams, there are increasingly large illegal streams, some of which are now run by global organized crime. Many of these migrants find their way to urban areas where they may find employment, and thus do not directly constitute population pressures on protected areas or on vulnerable species and habitats. At the same time, they give rise to significant social problems, such as crime and unemployment, and everywhere are fueling rising "right wing" political movements. Such movements tend to be less tolerant of, or place less importance on, environmental conservation, and could constitute forces undermining the protection that is available in the rich countries. Attempts to control this migration have been notably unsuccessful and it is unclear what the long-term implications will be for environmental conservation.

In any event, it is clear that the challenges found in the wealthy, low fertility nations come not from the sheer press of more people, but from the press of people with more wealth. There remains, however, a large portion of the world's population that continues to show high rates of population growth. And here the pressure of sheer rising numbers of people poses most serious challenges to biodiversity protection.

The LDRs in 2000 (including China) contained a population of about 4.9 billion, and though their growth rates have declined, their large numbers will mean that they may add some 75 million persons per year over the next decade. The numbers are unprecedented and frightening from the perspective of biodiversity protection.

If the numbers of new people to be added in the LDRs are frightening, those in the LeastDRs are even more so for two reasons. First, though their estimated population (UN 2001) was 658 million in 2000, less than a fifth of the overall LDR population, it is still growing at 2.5 percent per year. Thus it is projected to add nearly 186 million people over the next decade. To make matters worse, this is also the region with some of the world's richest areas of biodiversity and in which the threats to biodiversity are the greatest. A recent study by Population Action International has identified 25 biodiversity hot spots around the world (PAI 2000). These are areas especially rich in biodiversity and are now seriously threatened with habitat and species destruction. In eleven of these areas, where the total human population is 472 million, population growth rates are 2 percent per year or more. These are areas where mortality has declined but fertility remains high. They have moved into the demographic transition, but are not projected to close the transition for another few decades. Here the pressure of sheer numbers on biodiversity is especially powerful and destructive.

If the numbers alone were not enough, there is even greater pressure from the changing age structure that accompanies rapid population growth. When mortality declines, large numbers of young people are added to the population, greatly increasing their proportions and reducing the overall age of the population. Perhaps the most critical of the age groups for biodiversity protection is the young males, or those of 15–19 years. (This five-year group is used because it is part of the standard age/sex classification by which population data are presented. The United Nations Population Division, for examples, publishes estimates of sex and age distributions for all countries of the world, for the period 1950–2050 [UN 1998].) This young male group represents a volatile and often highly mobile population (Daly and Wilson 1988; Mesqueda and Wiener 1996, 1999). It is often found at the center of urban riots and "ethnic cleansing," or in frontier regions. It is the group that infantry sergeants most prefer because its members are energetic and fearless (largely due to lack of experience). It is also a group that provides fodder for demagogues, as we saw with grisly details in the Khmer Rouge and Hutu-Tutsi mass killings. In wealthy societies, the energy of young males is channeled into productive activities by education, sports, and the prospect of future employment. But when schools and sports are not available and when unemployment is their common lot and expected future, young males can be especially volatile and destructive, especially of the natural environment.

Throughout the LDRs this young male population has been growing rapidly, and it continues to grow for 15–20 years after fertility begins to fall. Figure 13-5 shows how this volatile population can grow rapidly, and continue to grow even after fertility begins to decline. The figure compares Sri Lanka and Kenya. With roughly comparable population sizes in 1950, Sri Lanka experienced a major fertility decline starting in the 1950s, while Kenya's fertility actually rose and did not begin to decline for another generation. Both had just under 300,000 young males in 1950. Sri Lanka's numbers grew to almost 1 million and leveled off there by 1980. With continued high fertility, Kenya's young males climbed to 1.2 million by 1990 and are estimated at near 2 million today. Though Kenya's fertility and growth rates are declining, the young male population will continue to grow to nearly 4 million before it starts to decline in 2040. This presents Kenya with a series of demands for education and jobs that it cannot now satisfy, portending even greater potential for instability for the next generation.

Much of the population in the LDRs will show growth patterns similar to, or more rapid than, those of Kenya; and these are the areas where we find many of the world's biodiversity hot spots. They are also the poor-

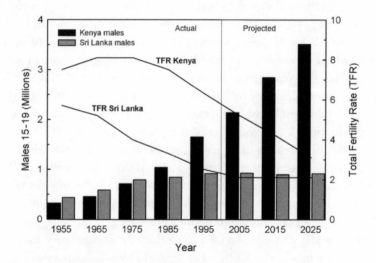

Figure 13-5. Fertility and young male populations in Kenya and Sri Lanka.
TFR = total fertility rate. (Data source: UN 1994.)

est areas of the world, with little in the way of schools, jobs, or land available for this rapidly growing and highly volatile population. It is not difficult to predict increasing pressure on the already vulnerable areas of biodiversity.

Location Specificity

General processes are precisely that, general. They are historically and geographically broad and do indeed have a pronounced impact on biodiversity. To engage in specific activities to protect biodiversity, however, we must note that each specific situation of habitat or species destruction has distinct conditions that must be considered. Here is where location specificity dominates the analytical process.

A first approach to location specificity can be made through a brief retrospective view of what has gone just before. We saw many cases where population projections are made for the next fifty years. This is one of the powers of demography: its capacity to make projections about future changes in mortality, fertility, and population growth. The United Nations Population Division has been making projections for much of the last half century. Every two years a new set of estimates is made of both past and projected movements. Looking at the world as a whole, these projections have been quite accurate (Ness and Golay 1997). For many relatively large populations, projections for the next twenty years can usually be quite accurate. But the limitation of such projections of "relatively large populations" is that they often do not work for smaller populations. Earlier projections of total world numbers for 1990, for example, have been off by less than a percentage point or two. But for Africa they were underestimated by almost ten points, and for Latin America overestimated by about the same amount.

The problem becomes more difficult when we deal with the relatively small areas and populations where a specific habitat or species is to be protected. This is quite clear in the analyses of most cases in this book. In these relatively small areas, population projections can be overwhelmed by uncertainty. It is very difficult to make accurate projections of human population movements for small areas without gathering specific field data (see chapter 14). Too many different processes are at work besides the broad general forces of human mortality and fertility decline. And for small areas, a very small movement of population, in- or out-migration, a mild epidemic, or a very good harvest can overwhelm more general forces. Consider just two of the cases detailed more fully in earlier chapters, the mountain gorilla of Uganda and muriqui of Brazil (chapters 6 and 7 respectively).

Brazil has completed the demographic transition. Mortality began to fall rapidly after 1945, while fertility remained high. Then fertility began to fall in the late 1960s, and finally came into line with the lower mortality by about 1995 (UN 2000). This was driven in part by Brazil's massive urbanization. By 1970–75, the rural population began an absolute decline that continues today (UN 1999). Consequently, nowhere is the muriqui habitat threatened by simple human population growth. In all of the critical habitats, the rural population is declining; low fertility and urbanization have reduced the human population in and around all critical habitats. While this relieves the general pressure on the habitats, it says nothing about specific threats. In one area critical to the muriqui, forest clearing for pasture, despite population decline, threatens the species' habitat. The programmatic solution that seems most feasible is for conservation groups to buy up lands to protect them from clearing. In another area, illegal cutting of palm hearts threatens not the habitat but the monkeys themselves, as the palm harvesters kill monkeys for meat during their illegal forays. This occurs in a national park that is supposed to be a protected area. But the weakness of the state precludes effective protection. Here the programmatic solution suggests that moving to plantation production of palm hearts could divert demand from natural to controlled land systems. The point of these illustrations cannot be overemphasized. Specific locations produce different population-environment dynamics: population-environment dynamics are location specific.

In Uganda, Rwanda, and the Democratic Republic of the Congo, both the mountain gorilla and its habitat are protected. But the protected area was completely overwhelmed with the ethnic warfare that broke out in the 1990s. With hundreds of thousands of refugees fleeing ethnic slaughter, no habitat could be protected. Nor could a species that might provide protein for those refugees. This was, to be sure, an extreme case of population movement overwhelming a small area, but we are likely to see more of these ethnic wars in the near future and they are especially threatening in some of the world's richest areas of biodiversity.

Small areas, like the upper Amazon gold fields, can be overwhelmed by even small migration streams in a nation where fertility and population growth rates have dramatically declined. And small areas can either attract new people, or send them away, threatening or relieving specific habitat and species in ways that defy prediction.

What Is to Be Done?

Given this considerable complexity on the demographic side, what can be done to address the challenges they pose to biodiversity protection? This book focuses precisely on that question. It describes the work of an experimental network of scientists from different disciplines, some of whom are currently engaged in ongoing efforts of species and habitat protection, specifically the Population and Habitat Viability Assessment (PHVA) workshops. In the Network meetings, the group struggled to envision how we could integrate knowledge from demography with that of population biology to best influence policy. At the November 1997 Network meeting, conceptual developments from past population policy work (Ness and Ando 1984) provided a process model that could be translated into the PHVA processes.

Figure 13-6 describes a path analytical model used to examine population policy formation, implementation, and impact in Asia. The process identified here flows from a policy decision to limit fertility, through implementation, usually in the form of a national fertility limitation program, to impact in fertility decline. Each of these stages can be quantified. The policy decision is measured in time (years), on the assumption that the earlier the policy decision, the stronger the policy. Program strength has been extensively quantified as the strength of the family planning program. And rates of percentage fertility decline over time offer the

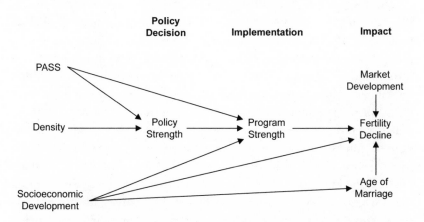

Figure 13-6. A path analytical model of population policy and fertility change. All paths are positive. (Source: Ness and Ando 1984.)

quantification of the impact. The model proposes three major determinants: political-administrative system strength (PASS), population density of the country, and the level of socioeconomic development. These require brief discussion.

PASS is a concept developed from empirical examination of the political and administrative conditions surrounding Asian population policy formation. It includes three major components: the strength of the political center, the capacity to monitor social and economic conditions, and the level of commitment to economic development. Each of these components was quantified using a five-point grounded scale, giving weights to various qualitative statements running from stronger to weaker conditions. Population density is the simple measure of persons per square kilometer for the country as a whole. Socioeconomic development is composed of three measures: per capita gross national product, infant mortality rate, and female school attendance.

The model was estimated using data from twenty-three Asian countries for the period 1952–80, with very good results. Each of the three multiple regression equations explained 60–90 percent of the variance. A further step was devised to examine deviant cases. Outliers (countries) were identified for each of the three equations and their political and social histories and conditions were examined to provide further understanding of the process. In essence, this strategy allowed us to move from quantitative, multiple regression coefficients to the richer qualitative analysis of individual cases.

This suggested that the PHVA process might be examined as a similar process, with five identifiable steps: (1) institutional contexts; (2) PHVA process; (3) PHVA recommendations; (4) institutions and management plan; and (5) the outcome in species conservation. The institutional contexts included two major components: the sociological/demographic and the biological conditions. These identify both the broad conditions that affect species risk of extinction and the data that will be needed in the PHVA workshop. The workshop itself includes both qualitative discussions of conditions and the more quantitative VORTEX modeling that is used to assess risk of extinction. A typical PHVA workshop, six of which are described in part 3 of this book, develops specific recommendations that are agreed upon by various stakeholders who participate in the workshop. Those recommendations then face various probabilities of being implemented and having the desired impact as they move through the

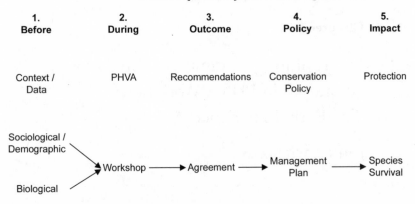

Figure 13-7. A model of the PHVA process.

larger social conditions of the affected area and the specific management plan developed after the workshop. The final outcome of this process is the variable condition of species and habitat protection. Figure 13-7 shows our initial version of a model for the PHVA workshop process.

Placing the PHVA process in this larger social process allowed us to ask questions about the conditions that affect the success of a PHVA workshop itself and that determine whether that degree of success will actually result in the protection and continued viability of the effected species. Much of this will be described in the following chapters of this book.

◆◆◆ **Chapter 14**

Incorporating Community Population
Appraisals in PHVA Workshops:
The Early Experience

JOHN S. WILLIAMS

This chapter explores a methodology for collecting and integrating location-specific demographic data into Population and Habitat Viability Assessment (PHVA) workshops. As such it offers a complement to the large-scale models explored in chapter 13. While they have direct implications for the work of the Network, the cases reported in this chapter predate the cases that are explored in part 3 of this book by some five years. In effect, the work reported in this chapter anticipated that done in Papua New Guinea (chapter 9) and is offered here to give broader context to that case in particular and to offer a blueprint for future efforts to include local landowners in PHVAs in general.

In the fall of 1993, in preparation for the Population Habitat and Viability Assessment (PHVA) workshop on the Indian/Nepali rhinoceros, the Conservation Breeding Specialist Group (CBSG) set for the first time an explicit workshop objective to consider "the impact of human demographic change on rhino populations" (Molur et al. 1995). At that time, World Conservation Union (IUCN) headquarters, located in Gland, Switzerland, had a program in population and environment as part of its Social Policy Service. (Funds for a demographer in this department were provided by the U.S. Agency for International Development under the Population Fellows Program, administered by the University of Michigan.) The CBSG invited me, the IUCN's Population and Environment Fellow, to participate in the workshop, charged with bringing the science of human demographics into the workshop process and specifically providing input to the modeling of the viability of the rhinoceros population.

Of all the social sciences branches, demography seemed most compatible with the vision and approach of scientists involved in the PHVA process. Demography works with a set of models that are used to forecast future population size and age distributions. These models are based on empirical, numerical data, as collected by census and vital statistics. The growth of human numbers in areas inhabited by endangered species was perceived to have a negative impact on many species. At CBSG's annual meeting in 1993, the staff was encouraged to include an analysis of human population impacts as part of its workshop processes.

Prior to this workshop, I had provided support to the IUCN projects in five Asian and African countries. These conservation and community development projects were concerned with seeking cooperation between protected-area managers and local communities, for the benefit of both. More specifically, I trained staff working in the IUCN's conservation and development projects in participatory research techniques and accompanied the staff into villages that were located in the immediate vicinity of protected areas. I was particularly interested in the ways in which the population dynamics affected how local communities manage and use local resources.

Participatory Population Appraisal

One of the tools that proved useful was the taking of a participatory census in each village, in the course of which we documented recent demographic history and probed how the village and the surrounding environment had changed over the previous half century. The data collected locally were analyzed along with any available on the locale from national census and vital statistics sources. Using the demographic projection model DEMPROJ (The Futures Group 1990), I prepared projections of population (size, age structure, and households) over the next thirty years. These projections, when presented to village groups, stimulated thoughtful discussions of the impacts of recent demographic change. Where were additional houses built and why? What additional land had been needed and used? What happened to land-use patterns and use of local resources as the village changed? And, even more interesting, given the changes over the last twenty years, what was expected over the next thirty years given the population projections?

The Participatory Population Appraisal is one of a family of participatory approaches used in community research. The term "Participatory Rural Appraisal" (PRA) describes an approach to the use of these techniques in which community members take an active part in conducting and interpreting the community study. Community members make maps and models; they investigate and interview; they present information; they plan. The outsider does not extract the information for external analysis, but serves as facilitator to establish rapport. Following a PRA, the facilitator then seeks to support the community in its attempt to implement its plans and projects, obtaining external assistance if desired. For a good discussion of participatory methods, see, for example, Case (1990) and Chambers (1992). Such exercises have the following basic characteristics:

Participatory: They require the full involvement of local community members; respect for them as subjects, not objects; respect of local perceptions and choices; sympathy for local problems; humbleness on the part of external researchers; use of visual rather than written material; and spending enough time in the field, nights included, for establishing rapport with people and obtaining genuine feedback.

Relevant: They focus on in-depth information, validity and reliability of data, systematic and structured procedures and recording of notes, and triangulation of information collected.

Flexible: They mix a variety of techniques (a "basket of choices") to adapt to the local situation.

Rapid and low-cost: They involve a trade-off between quantity and relevance, accuracy and timeliness, and a maximum use of local resources and reaching "optimal ignorance" or "appropriate inaccuracy."

Empowering: They are oriented toward social change. Such PRA exercises include the collection and systematization of local views, thus enhancing their visibility; they provide incentives to local leadership and local institutions; and they promote local dialogue and cooperation.

Most of those working with participatory methods have eschewed quantitative analysis in favor of in-depth interviews and qualitative information (e.g., Chambers 1992). However, I found preparing an analysis of population extremely useful in looking at patterns of change in the environment over time. It helped add historical perspective and a long-term vision for the future.

It was quite natural that this model, utilized by the Social Policy Service in support of other projects, was selected for bringing demographic expertise to the rhino PHVA workshop held in Jaldapara, India. This basic model was also later used in preparation for the Papua New Guinea workshop (chapter 9).

The model for a Participatory Population Appraisal within the PHVA workshop process involved the following elements:

1. Staff support from an in-country group capable of participatory research in the local communities;
2. The gathering and analysis of existing census and vital statistics data for the area under study;
3. The taking of a participatory census of households and population, mapping of household locations and land use;
4. An analysis of household patterns of harvest of forest products (i.e., firewood, other plant products, animal products);
5. A participatory analysis of the impacts of past changes in the size, movement, and characteristics of households.
6. A projection of households and population for the next 30 years and likely impacts on agricultural lands, forest lands, wildlife, and water; and
7. A participatory analysis of policies that might improve living standards for local people while conserving local resources.

The Rhino PHVA Workshop in Jaldapara, West Bengal, India

In the Jaldapara area, there were no ongoing community development projects that could be used as a basis for a research effort. We had no project staff and no existing base from which to conduct research. More in question, what would we have to offer the local communities in asking for their cooperation in our research efforts? In the case of community development projects, there are funds available and the belief that staff could follow up on the community assessment to provide tangible benefits to the local communities. However, in Jaldapara, we would be in no position to follow up the visits. An interesting debate ensued at IUCN headquarters, in which some thought that we should try and support the workshop, others argued that it was not ethical to conduct research and have no resources to support the community, and everybody felt that more time was needed for planning. There was particular concern that the work in the community would seek to extract information for the workshop and would not be an active research undertaking by and for the community. Nevertheless, the decision was made that if we could obtain in-country support, the assessment would be done (Williams 1994).

Calls were made to several groups in India, including Development Alternatives, a nongovernmental organization (NGO) in Delhi whose mission was to promote sustainable development through appropriate technology and environmental management. They were interested, and there was at least the possibility that they would be in a position to follow up with the community after the workshop. A small grant of US$2,500 was provided to Development Alternatives out of Social Policy Service discretionary funds to send a team to the Jaldapara area. They would conduct the research with support from the IUCN on demographic statistics and participatory research techniques. The Development Alternatives team had experience working in communities, but lacked training in participatory research methodology and demographic analysis.

The agreement with Development Alternatives was reached only three weeks prior to the workshop, without discussion with or input from the workshop organizers in India. When the organizers heard that an American demographer was going to visit the field to look at local population dynamics, there was immediate and very strong resistance. The message came back that the American demographer was to be forbidden from doing any research in the field. Permission was reluctantly given to allow the team of two women from Development Alternatives to visit villages in the Jaldapara area. Staff from the Ministry of Environment and Forests arriving at the workshop were appalled that a group had been out talking to people in the villages prior to the workshop. The department was seriously considering relocating several villages from key areas near the Jaldapara Wildlife Sanctuary in order to increase the size of the park. Forestry staff felt, probably rightly, that the research work undertaken would make such a task more difficult.

The Jaldapara Setting

A hundred years ago, the floodplains and savanna of northeastern India's Brahmaputra River were home to more than 100,000 of the great Indian one-horned rhinoceros *(Rhinoceros unicornis)*. Extension of tea cultivation and waves of human migrants in the area resulted in extensive loss of rhino habitat. The agricultural and settlement patterns served to fragment a larger habitat into isolated pockets. This, coupled with overhunting in the first decades of the twentieth century and later poaching, brought the rhinoceros close to extinction during the 1970s.

In 1941, an area of 116 square kilometers became the Jaldapara Game Sanctuary (later renamed Jaladpara Wildlife Sanctuary). It is one of eight protected areas in India where there are still rhinos in the wild. Currently, about thirty-five rhinoceros inhabit the sanctuary. When the sanctuary was first formed, the human settlement areas along the river were excluded from the area under protection. The result was a park that was shaped like a pair of trousers, with the area between the legs inhabited. These areas were initially sparsely settled. However, having excellent soil and water characteristics, they attracted in-migration.

By 1970, the area could be described as fully settled; most families cultivated 3–5 acres of land and maintained small homestead gardens near their houses on the higher ground of the village proper. Flooding during the 1960s had ravaged the lands. In the village of Jaldapara, it had claimed 20 percent of the cultivated land. Of this, some was lost in the changing river course and the rest was converted to a dry and pebbly wasteland.

Even though remote from large urban centers, this area was faced with large-scale movements of people. Conditions in neighboring Bangladesh were deteriorating. During the decades preceding the workshop, many migrants from Bangladesh moved into the area, greatly increasing the rate of population growth. Many of the newer migrants had sufficient resources to buy land, which escalated in value ten-fold between 1970 and 1993. Those who live in these villages have no ancestral attachments to the land. The earlier settlers arrived from 1930 to 1960, the others during the last thirty years. Half of the growth in population since 1970 is attributed to in-migration.

The Development Alternatives team arrived ten days prior to the workshop, reviewed census records, historical records, health and mortality studies, data on land use, migration, and other information (both published and unpublished) pertaining to the area. Interviews were held with local park officials, other local officials, and health workers to fill in the background. The statistical information was incorporated in a population model to analyze the population changes over the previous two decades and to project trends from 1991 to 2031.

The field survey focused on interviewing forest officials and local people to identify issues and areas of conflict important for conservation of the rhino and its habitat, identifying three representative village clusters for field work with the help of the forest officials and local people, and conducting field work at the community level using Rapid Rural Appraisal (RRA) techniques in the selected villages.

There are some thirty-three villages that surround the sanctuary, with a 1993 population of about 200,000. The villages differ considerably in history and ethnicity. An analysis was made of three village clusters, Jaldapara, Sidhabari, and Laksmandabri. These villages were selected because they were adjacent to a part of the sanctuary that is rich in wildlife and has ideal rhino habitat. The area is also known as a corridor for rhino movement. The other most important selection criterion was the fact that people in these villages are highly dependent on the sanctuary's natural resources for their income and livelihood.

Jaldapara village is near the geographic center of the park, between the "trouser legs" of the sanctuary. Sidhabari depends upon resources from the eastern leg, Laksmandabri, from the western leg. Park managers have been talking for years about relocating those living in Jaldapara village and incorporating that area into the sanctuary. These villages have poor access to transportation, communication, and health facilities and have a high dependency on the sanctuary's natural resources. Villages depend on four primary sources for livelihood: cultivation, work in the tea estates, income from seasonal work outside the area, and the sale of resources extracted from the reserve. While not wholly representative of all thirty-three villages in terms of population change over the last two decades and general patterns of land use, these three villages do not appear to be atypical.

Population Characteristics and Change

The population immediately surrounding the sanctuary approximately doubled during the twenty years from 1972 to 1992. The population of Jaldapara totaled about 8,800 in the 1991 census, which is up from some 4,500 in the 1971 census. The population growth was as much a result of migration as of high fertility and declining mortality. Tables 14-1 and 14-2 describe recent and projected population size using the population projection model DEMPROJ, given alternative assumptions about migration and the prevalence of family planning.

Migration

There are no Indigenous people in most of the villages surrounding Jaldapara. The immediate area around Jaldapara was sparsely inhabited prior to 1930. Those who did live there, the Totos, now comprise less than 1,000 people and living in the village of Totopara on the northern boundary of the sanctuary. Villages to the south know little about them.

The migration patterns have shaped the area's settlements and land-use and continue to be a significant force. A typical village around Jaldapara has early settlers from Nepal, Bihar, and Assam who arrived more than forty years ago and more recent settlers from Bangladesh.

In-migration from other countries (Bangladesh) has been illegal in recent years, but it continues. While some in-migrants remain in the area as landless laborers, many from Bangladesh arrived with sufficient resources to buy land and settle down. Poorer families with land may be forced to borrow money (usually from the wealthier Nepalese landholders) to meet expenses (such as to cover medical costs), using their land as collateral. Inability to repay the loan results in the need to sell the land. Thus, they too become land-poor or even landless. The state of being landless is new; no families were landless prior to 1970 and there were no traditions or patterns to mark such families.

The migration patterns over the next few decades will be a major determinant of the future population of the area. Using the DEMPROJ model, future population projections were made given alternate migration assumptions, in combination with the median projection for fertility and mortality. The first assumed a continued in-migration of 50 to 60 persons per year in a base population of 8,790 for the next two decades. The second assumed a comparable out-migration over the same period. The resulting projections demonstrate that the future population of this small area is extremely sensitive to minor changes in migration (keeping in mind that future migration patterns in local areas are difficult to predict).

As Table 14-1 shows, low levels of in-migration will result in a population increase of the Jaldapara village area from the current level of 8,790 to 14,080. On the other hand, low levels of out-migration will lead to a lower increase of 11,260. If low-level net in-migration continued for an additional two decades, the population would be 50 percent higher than with out-migration. A population policy that encouraged low levels of out-migration could have major impacts on reducing local population pressure on resources. Experience in other countries suggests that increasing education levels will usually stimulate such out-migration.

Table 14-1.

Population projections showing effects of variations
in migration for Jaldapara village (including
Phradhanpara and Natunpara), 1981–2031.

Migration Assumptions	Year					
	1981	1991	2001	2011	2021	2031
In-Migration	6,630	8,790	11,360	14,080	16,710	19,200
No Migration			10,750	12,670	14,440	16,000
Out-Migration			10,140	11,260	12,170	12,800

From 1981 to 1993, in-migration averaged about 0.5%. In-migration assumptions include a continuation of in-migration of 50–60 persons per year. Out-migration assumes an out-migration of 50–60 persons per year. Projections based upon median fertility and mortality assumptions.

In 1993, the Development Alternatives team thought it likely that there would be a shift to out-migration from the area, but they were mistaken. Since 1993, substantial in-migration has continued from Bangladesh, resulting in more rapid than expected population growth. (See discussion in chapter 13 concerning location specificity, a more theoretical note on making demographic projections for small areas.)

Fertility

The infrastructure in the vicinity of Jaldapara in 1993 for child/maternal health was exceedingly poor. The closest health facility was 10 kilometers to the south, but it was poorly equipped. Support for family planning was virtually nonexistent and villagers reported no use of contraception. However, there was a government program promoting sterilization and the total fertility rate of the village had declined over the previous twenty-five years. The sterilization campaigns reduced the total fertility rate from a level of about 5.0 in 1970 to between 3.6 and 4.0 in 1993.

Table 14-2 shows the high, medium, and low population projections for Jaldapara village area, with the variations in population size determined primarily by fertility rates. The data indicate that changes in the fertility rate do not make a great deal of difference in population size for about two decades, but the differences are cumulative and increase with each decade. The low projection shows that the reduction to a replacement level of fertility by the year 2011 combined with a modest level of out-migration could result in a stable population of 11,500 in two decades. That is less than 50 percent higher than the current population. On the other hand, the high projection leads to a doubling of population (to 16,500) by the year 2021, even assuming that fertility will be reduced to 3.3 in 2021; and the population would continue to grow rapidly thereafter.

In the years since the workshop, there has in fact been action to bring better health benefits and family planning to these communities. It is believed that the workshop and its recommendations played a part in making this happen (Sanjay Molur, CBSG India, personal communication with the author).

Mortality

From 1971 to 1985, the health conditions in the vicinity of the Jaldapara reserve improved. Primary health clinics had been established and there was an increase in vaccination programs for children. Most women in the three villages that we visited now go to the hospital in Falakata to give

Table 14-2.

High, medium, and low population projections for Jaldapara village (including Phradhanpara and Natunpara), 1981–2031, and corresponding total fertility rates.

Fertility Assumptions	Year					
	1981	1991	2001	2011	2021	2031
TFR	4.0	3.8	3.4	3.3	3.3	3.1
High Projection	6,670	8,810	11,330	14,030	16,500	19,200
TFR	4.0	3.7	3.3	2.8	2.3	2.1
Medium Projection	6,670	8,810	10,860	12,030	12,861	13,323
TFR	4.0	3.6	2.8	2.1	2.1	2.1
Low Projection	6,670	8,810	10,430	11,250	11,500	11,500

TFR = Total Fertility Rate
Note: Projections made using DEMPROJ (Futures Group 1993). High population projection includes in-migration and higher mortality rate than medium and low projections.

birth. Those hospital facilities have expanded their services and are increasingly prepared to provide emergency care. Deaths from childhood disease and maternal mortality have steadily declined.

These improvements in health have been hampered in recent years by a rapid increase in population (with which the existing health facilities could not cope); increasing poverty, also associated with increasing population (greater number of landless/land-poor); and a lethal form of malaria that has become prevalent in the area.

In the Jaldapara village area, fifty children below the age of five died during the two years prior to the workshop. We concluded that there was an actual increase in the child mortality rate during the three to five years before our visit. The local health clinics did not even have a microscope for diagnosing malaria from blood samples.

Nevertheless, the population projections assume a rapid decline in child mortality during the decade following the workshop. The type of mortality found can certainly be greatly reduced through the provision of maternal and child health services; and we understand that such services have improved for local communities, as recommended by the workshop.

Population and Land Use

The villages around the sanctuary have relatively good quality agricultural land. Farmers practice traditional agriculture and efficient crop rotation, which helps to retain soil fertility. On a typical plot, three crops are grown each year: two crops of paddy and/or jute and one crop of vegetables. Villagers consider that three acres is the minimum that a family must have in order to grow most of the food that it needs. In addition, a small homestead garden with diverse and useful trees and plants is a necessity. Most of the households in the area have a homestead garden of this sort if they have sufficient land. By 1970, the land area of the villages in this study was almost fully occupied; most families had 3–5 acres and virtually no families were landless.

Inevitably, when there is an increase in the number of households and a decrease in the amount of land available for cultivation, the average land available per household decreases. In the three villages, there were 1,400 households in 1971, a number that increased to 2,700 in 1991. The average landholding (under cultivation) per household declined from 3.69 acres in 1971 to 1.84 acres in 1991. This did not mean, however, that most families had half the land they used to have. We undertook a mapping exercise of the agriculture lands used by each family and were totally surprised to discover that those families with agricultural land generally had at least 3 acres, while the poorer families were essentially landless. Any increase in the number of households results in an increase in landless or land-poor households. We would not have learned this in the absence of our fieldwork in the community.

The number of households by size of landholding (1971–91) is shown in figure 14-1, which pictures the area's dramatic increase in the number of land-poor and landless families in the recent past and near future. The number of households with at least 2.5 acres has remained roughly constant over the last twenty years and will continue to be rather stable over the next few decades. This despite a large expected increase in number of households. Those forming households for the next twenty years have already been born, and hence household formation during that period is independent of the fertility rate. The number of households, as noted above, could be altered by changes in migration. The proportion of households with less than half an acre will rise quite dramatically, much more rapidly than the increase in the population itself. At this time, there is virtually no local employment that is independent of the use of land and local natural resources. Landless households will have to scrounge for a living using available resources. The only local resources that are available are found in the sanctuary. A continuation of this pattern of population growth and economy would likely doom the wild species living in the sanctuary. People do come first.

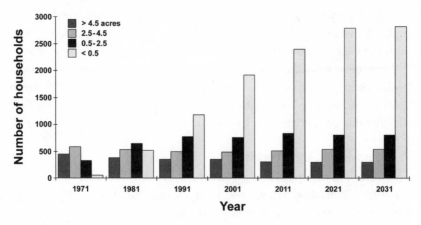

Figure 14-1. Number of households by landholdings, 1971–2031, Jaldapara, West Bengal, India. Selected villages adjacent to Jaldapara Wildlife Sanctuary, including Jaldapara (with Pradhanpara and Natunpara), Sidhabari, and Laksmandabri. The 1991 census showed a population of 2,800 households (15,400 people).

Migrant households moving into the area often came with greater financial resources than many existing families. There was a pattern by which formerly landed families became landless while the in-migrants purchased the land. This resulted in an increased level of conflict between older and newer residents. Economic disparities between landholders and those without land are likely to widen in the future.

Dependency of Villagers on the Sanctuary

Until recent years, villagers' dependency on the sanctuary consisted of items for their own consumption. These included firewood; grass for thatching roofs, for grazing, and for mats/baskets; and timber for furniture and construction.

As the number of landless has increased, many families have turned to the reserve for collecting resources that may be sold in town as a means of livelihood. Once again, firewood is the single commodity most actively collected from Jaldapara. The women of the poorer households go out in groups of ten or fifteen. The forests of this area are dangerous places; those who enter may encounter elephants, wild boars, tigers, or leopards. During the week before our field trip, a leopard attacked a group of women from Jaldapara village; one woman was badly mauled and was hospitalized. The women collecting wood may go to sites as far as 5 kilometers from the village, and carry back bundles weighing up to 30 kilograms.

Dependency on the sanctuary for income is inversely related to the landholding per household. The prevailing situation of small landholding and absence of nonfarm employment opportunities leads to a considerable dependency on the adjoining natural habitat for livelihood.

Participation in the Workshop

The workshop took place within Jaldapara Wildlife Sanctuary in December 1993. Workshop conveners, in particular the Forest Department administrative leaders, were antagonistic to the arrival of a research team that wanted to bring their community-based information into the workshop deliberations. From these conveners' point of view, the communities lay outside of the park, and residents had no legal rights to park resources. There was a willingness to consider issues of trade, focusing on the illegal rhino horn trade and ways to obtain cooperation from local communities in stopping the killing of rhinos for that trade. Issues concerning the population and health of local communities were considered outside of the scope of the workshop and were entertained with high reluctance. A trade and human impact working group was formed.

In order to actively involve the community in using the research from the village, the Development Alternatives team invited representatives of the communities where we had worked to observe the workshop proceedings. The workshop organizers were not in the least interested in the villagers' presence and saw their participation as an obstacle. They were allowed to sit in the back, but there was no attempt to translate the proceedings for their benefit, nor were they made to feel welcome. They left in the course of the first day, not to return. On the first day, it was unsure whether or not the Development Alternatives team would itself be given standing to be part of the workshop. It could be argued that the presence of the team represented the interests of the local communities. However, the villagers were denied true participation in the workshop. The later workshop in Papua New Guinea (chapter 9) gave one demonstration of how useful such participation could be.

The conservation biologists used the VORTEX model to look at the long-term viability of the rhino population; they were pleased to have the additional insights and information derived from the demographic analysis of the local communities and they attempted to consider it within the modeling process. (See chapter 3 for an analysis of how the modelers struggled with using the inputs from the social scientists.) Within the working groups, the results of the community appraisal were discussed and taken into consideration in preparing recommendations.

It is evident from the foregoing sections that the adjoining human population's dependency on sanctuary resources is one of the major factors that will affect the future viability of the sanctuary. However, the impact of the community research effort on this workshop was small. There was a failure to integrate the human demographic community appraisal with the workshop process, though the information was of interest to the modeling group.

Participatory community appraisal activities looking at population dynamics and local resource use provide useful input into conservation planning in areas where wildlife managers must work with local people. Local people have a great deal of knowledge concerning local circumstances and that knowledge must be used. If local park managers are to obtain cooperation with local people, they must manage their resources in a manner that sustains both local communities and wildlife. In Jaldapara, the workshop recommendations to bring health and family planning services to the local community (noted below) were later implemented, for the benefit of both people and wildlife. However, the Jaldapara experience shows how difficult it is to graft broader perspectives onto a workshop at the last minute.

As a footnote, the workshop made recommendations to provide better maternal and child health facilities and family planning to local communities. The local protected-area staff was more interested in the observations of the Development Alternative team than other Forest Department officials, and clearly saw their relevance. Several of these local officials talked at length with the team and added their observations to the problems facing local communities. A few years after the workshop, the recommended health facilities were established with the active support of local park officials.

The Java Gibbon PHVA Workshop in Bogor, Indonesia

Population growth in local communities is one of a number of human dynamics that may affect a local habitat. In parts of the world, including South America and large parts of Southeast Asia, the rural population has either stabilized or is in some cases decreasing. Even in areas where the local population is increasing, the use of local resources by local people may not be a major factor in degrading the habitat or endangering species. The number of gibbon spp. *Hylobates lar, H. pileatus,* and *H. agilis* in Thailand and *Hylobates moloch* in Java have declined in recent years; PHVA workshops were held in these two countries in May 1994.

The gibbon workshop in West Java was held immediately after the Thai gibbon workshop, and there was no opportunity for any participatory fieldwork prior to the workshop. However, several community development NGOs were represented and so there was good participation of those that had deep knowledge of community uses of local resources.

Prior to the West Java workshop, data from 1980 and 1990 were obtained from the Bureau of the Census in Jakarta on districts adjacent to Halimun and Gede Pangrango National Parks, which held the most significant remaining habitat for the dwindling population of Java gibbons. For the Cisolok district adjacent to Halimun, the human population had grown from 63,400 to 75,600 during the decade 1980–90, a growth of about 20 percent. This information was brought to the workshop and considered. Workshop participants were concerned about the effects of local population on the parks.

The gibbons were threatened in Halimun Park primarily by the loss of habitat and continuing encroachment on land. As they formed, new households would attempt to clear a forested area for agricultural use. In recent years, park officials have made such encroachment more difficult and have succeeded in stopping it in some locations. This has resulted in anxiety for the local villagers, who are only now beginning to see that they no longer have expansion areas. The problem of landless people is in its early stages. An increase in landless families could become a severe problem if new households are prevented from encroaching on forested land.

At the start of the workshop, human impacts were considered with the support of several NGOs that had been working to support community development. However, the representatives of these organizations at the workshop were more interested in community empowerment and the funding of community development programs than in developing conservation programs to protect gibbon habitat. In fact, several participants were quite hostile to the modeling process and to the emphasis on conservation.

The recommendations of the working group on human demography focused on increasing cooperation between park managers and local communities, notably on how park managers could assist communities in their community development issues. The chair of the working group was the head of Halimun and Gede Pangrango National Parks and he brought with him a paper that he had written entitled "Promoting the Role of Local Communities and NGOs in the Management of Indonesia's Halimun National Park."

The working group, with input from both the park director and members of NGOs working with local communities, drew up a profile of the villages in the vicinity of Halimun Park. Halimun is a large park of 40,000 hectares affected by some one hundred communities adjoining and within its boundaries. Villages to the south of the park are highly traditional in lifestyle and culture, and are less accessible to the larger towns. Sirnarasa village, for example, is located 18 kilometers from the nearest medical clinic. Families there have three to four children, higher than typical in West Java. To the north of the park, more accessible to the towns, villages are less traditional. These areas have received more in-migration over the last two decades, with migrants generally coming from elsewhere within the district. The number of households in the peripheral villages had approximately doubled over the previous twenty-five years.

The current growth rate of villages around the park is estimated to be between 1.8 and 2 percent, with a fertility rate below the average of rural West Java. The fertility rate has fallen sharply during the last twenty years.

Community Resource Use

Populations in communities adjoining these two parks are to a considerable degree dependent upon the parks for food, fuel, timber, and wildlife poaching, particularly birds. But the greatest single long-term threat to the parks comes from land encroachment. Such encroachment is for housing, crops, grazing, fish ponds, mining, and roads or paths. The increase in number of households over the previous twenty-five years had put additional pressure on land. This was particularly true for villages that were located entirely within the park. However, at the workshop, we had no quantitative information on the projected growth of households in these communities or the likely impacts on Gibbon habitat.

The human demography working group's recommendations were based on several objectives: to enable park rangers and staff to work more effectively with local communities, to increase welfare of surrounding communities, and to improve awareness in local communities of the value of conservation and sustainable management of local resources.

With regard to human population issues, the group's recommendations were quite general and included the following: "Population growth in the areas around the park continues, and places additional pressure on park resources. Efforts should be made to discourage the migration of people from other areas into these peripheral areas. In addition, consideration should be given to encouraging low levels of out-migration from the villages around the park. One way of accomplishing this would be to increase the levels of education and job skills of people living in the villages" (Supriatna et al. 1994).

These recommendations were adopted at the workshop with little discussion or controversy, although they were sufficiently vague as to have little chance of being implemented. Further, the demographic analysis using only census data was unable to give the modelers information that would assist them in looking at the impacts of the human populations on the gibbons of the park. Regional census data examined in the absence of local environmental impacts gave only a background demographic trend divorced from actual impacts on the ground.

Field Trip to Halimun National Park

Following the workshop, BCCI, a community development group working with villages located inside of Halimun Park, made possible a participatory research appraisal of the impacts of local population change on local resources. These tools included individual interviews with community leaders, group interviews, a transect walk through the village, and a participatory census of the village.

A map was prepared on the floor of the village head's house, showing all houses, the numbers of adults and children in each house, and the land plots that each household used for rice production. Then we asked that all houses built during the previous ten years be identified. This map specified relatively precisely the pattern of encroachment on the park resulting from new households taking land. There was also recognition of a predictable squeeze on available resources should the park management stop all future encroachment. The mapping exercise indicated that for each new household, approximately 2 hectares of land were cleared. The encroachment occurred primarily along the watercourse in areas that were not overly steep.

One of the overall workshop recommendations was to encourage low levels of out-migration. The village headman reported that only two men had permanently left the village during the previous twenty years, though a few others left from time to time for temporary employment in town. Village elders, however, were very uncomfortable with the idea of their children leaving and moving elsewhere; there was the belief that life in the village was better than elsewhere. In general, the households had three to five children each, and the villagers participated in a rough analysis of how many new households there would likely be over the subsequent ten years. The information on numbers of new households, and the land encroached upon by new households, would have provided useful information to the VORTEX modelers in projecting loss of habitat.

It is well-known that park management intends to curtail all new encroachment. A group discussion developed surrounding this policy and the need for additional land for new households in the village. This discussion, without any probing on my part, led directly into a consideration of family planning. The headman complained that the birth control that was being practiced was ineffective and that there were many pregnancies where family planning had failed. "The pill has not been effective in this village. Our women take the pill, and they still have children. How can we make family planning more effective here?" Despite the complaint, it seemed clear that the ideology of wanting smaller families was well developed and that future population growth within this village would slow.

Fieldwork undertaken prior to the workshop would have provided quantitative specificity to the impacts of local demographic trends; and it would have provided needed input in practical and specific recommendations for the planning process.

The Thailand Gibbon PHVA Workshop at Khao Yai National Park

The Royal Thai Forest Department sponsored the gibbon PHVA workshop to develop management strategies for wild and captive populations of gibbons in Thailand. The workshop goals set up by CBSG with the forest department were:

1. to make a preliminary analysis of gibbon distribution and status so as to develop long-term management strategies for the remaining wild population;
2. to establish a managed captive breeding program for gibbons; and
3. to link these two management programs by developing rehabilitation and reintroduction protocols for returning captive gibbons back to the wild.

Unlike the Jaldapara workshop, there were no funds available at IUCN to obtain the participation of a local community development group in undertaking data collection in support of the workshop. However, the

Population and Community Development Association (PDA), a Thai NGO, had been working over the previous decades to provide support for family planning and community development in rural areas of Thailand. One of PDA's projects had directly supported communities adjoining Khao Yai National Park, home for a significant gibbon subpopulation and the site of the PHVA workshop. The project had provided family planning services and community development support.

PDA's programs have been marked by extensive villager involvement, not only as beneficiaries, but also as planners of their own development. Suptai was one of the communities bordering the park that PDA had assisted. What happened in this village provides an illustration of how participatory appraisal and community development activities may reduce the pressure of a local population's use of resources in and around a protected area. PDA enabled me to visit Suptai prior to the workshop.

Suptai Village—Population, Migration, and Resource Use

When PDA's project started in the 1980s, Suptai was an isolated rural community, marked by strong dependency upon wild resources from the forest and an agriculture based primarily on the cultivation of rice. It was isolated in the sense that there was no all-season road to provide access. It had no local health care facilities, high fertility, and a rapidly growing population. Incursions into the park for collecting firewood and hunting were having a negative impact on the park.

One of the first activities of PDA was to provide a community-based family planning service, a model that they had developed that has worked successfully within rural areas of Thailand. By the time of my visit, the fertility rate in Suptai had dropped to a replacement level.

When I arrived in Suptai village in 1994, I saw an orderly looking village, with electric wires going to each house. What was most striking was the absence of young people. There were children present in the village, but few families have more than two children. And what surprised me was the feeling that the village was elderly. Simply reducing fertility does not reduce the growth of households for a generation. What had happened here in the course of a single decade?

In their work with the village, PDA responded to the specific concerns for each village. Suptai's priorities were the construction of an all-weather road by which supplies could be transported to and from the village and an agreement was reached by which the village obtained a road and each household received free electricity and an electric pot. In return, the villagers agreed to refrain from collecting wood or game or making incursions into the park. The park boundary marked a clear delineation between village lands and parkland. From all accounts, the villages were keeping their agreement, and the previous degradation of that region of the park had ceased. I was concerned that the increased prosperity and the completion of the road might serve to stimulate migration into the village and result in more rapid population growth, a pattern I had see in Cameroon and elsewhere.

I sat down with several villagers and had them discuss what had happened to their community and their local resources. They said that living conditions in the village had really improved in recent years. The availability of electricity had made cooking fires largely obsolete, eliminating the tedious collection of firewood from the forest. The village economy had developed a self-sustaining agriculture based on improved varieties of rice and improved agriculture practices. The raising of poultry and other farm animals had largely displaced the need for hunting. Older village students now go to school via motorcycle, and several of the men of the village go by road to town as temporary laborers.

The average household size had dropped from 4.7 in 1980 to 4.2 in 1990, reflecting both a reduction in fertility and the out-migration of young adults. The children in the village now receive more schooling, but when they complete school at 15 or 16, most of them depart for the city. I asked the group why this had happened, which stimulated quite a discussion. In the end, the general consensus was that television played a major part in stimulating the migration. Most houses now had television, which provided a view of urban life in Bangkok, with its many economic and cultural opportunities. Wage paying jobs did not exist in Suptai, and most of the younger generation, males and females, departed. While there has been some return migration of older adults coming back to the village, the predominant trend has been out-migration. The road, which might have facilitated in-migration of new households, instead facilitated the departure of the young. The out-migration was the unanticpated consequence of higher education, information via television on life in the city, and lack of local employment. The dynamic economic growth of Thailand very much increased the pull factor of the urban areas. Such out-migration might not occur with stagnant economies.

Not everybody was happy with the changes. Some of the older people in the village felt that change had come too fast, that there was no longer anything to hold young people to the area. There is something depressing in the out-migration of the young. I have noticed that places that are growing tend to be proud of that growth. However, everyone agreed that conditions in the village were better than a few years earlier, and no one suggested removing the electricity and the road. Thanks to the out-migrations, there was, for sure, a reduction of pressure of the local population on resources.

While population growth in Suptai had essentially ceased, census data show that over the recent past there has been an increase in the human population around protected areas, but the growth has slowed, particularly in the southern half of Thailand where many of the gibbons live. In 1994, there was evidence that the out-migration of school graduates to urban areas was not confined to Suptai, but was a more general phenomenon, reducing the pressure of human populations on forested habitats such as Khao Yai National Park.

Before visiting Suptai, I had obtained census and vital statistics data for 1980 and 1990. These data failed to show the dynamics of local population change. Data from such sources as census provide a good starting point for a participatory census, but do not substitute for a participatory population appraisal. A participatory census in a smaller community (less than 2,500 people or 500 households) may usually be accomplished quickly and accurately and provide detailed qualitative as well as quantitative information on local patterns of migration. Such information on migration patterns is critical in making local population projections.

The Workshop

At the workshop, the human dimensions working group was chaired by the manager of one of Thailand's national parks; we found high interest and support from the park managers, who were interested in working with and gaining greater cooperation from the villages adjacent to the parks. Demographic analysis was readily admitted into the workshop process.

However, population growth and the use of natural resources by nearby communities did not appear to be a major factor affecting park habitat or the gibbon population. Of much greater concern was the pet trade. Demographic tools were enlisted to make a detailed analysis of the pet trade, with analysis of available data on the trade and the numbers of pets being released into the wild or turned over to zoos. Gibbons as pets are satisfactory for about the first seven years of their life, but as they mature they become aggressive, difficult to manage, and dangerous. Those who own pets often get rid of them, overwhelming zoos or groups that wish to protect the gibbons. Releasing domesticated gibbons into the wild is dangerous. They are not afraid of humans and, in fact, can readily injure humans.

Using demographic tools as part of an analysis of the pet trade was a useful component of the workshop and was incorporated into other workshop processes. For this workshop, a community appraisal of local community population dynamics was probably not particularly useful for the workshop processes, primarily because local population dynamics did not lead to degradation of the local habitat. Indeed, analysis of the way that local human communities use resources is not always an essential component of a PHVA workshop. Other approaches to looking at the manner in which human activities affect species are necessary. The Network attempted to expand the view of how social science might inform the workshop activities. While the information on gibbon biology was carefully elaborated at the workshop, the human population census data that was brought to the workshop contributed little. Further, the absence from the workshop of groups that had experience with local communities and the absence of representation from the local communities limited the workshop focus to the animals themselves.

Implications for Later Workshops

These three first attempts to incorporate human demography into the PHVA workshop process raised some insights and set the stage for the later activities of the Human Dimensions Network. It has not been easy to incorporate human demography as a useful component of a workshop process that focuses primarily on the protection of an animal species. It remains difficult to attract demographers and social scientists to workshops concerned with endangered species, as such workshops fall outside of the their discipline's mainstream issues. Most community development groups are more interested in seeking resources to improve the welfare of local communities than in looking at the human impacts on local endangered wildlife. Members of the communities themselves have indigenous knowledge, but are easily disaffected, as at Jaldapara, by a workshop process that seems to consist mostly of outside scientists and land managers.

We have learned from these exercises that the simple display of numbers, even numbers that show population growth, have little meaning at the local level unless they are directly tied to the manner in which that population uses local resources. In Jaldapara, participatory research methods proved to be useful tools in such an analysis, as later seen in the

Papua New Guinea workshop (chapter 9). National population data are virtually useless to the workshop process. The size and growth of the human population is only one of several factors affecting how populations use and abuse local resources. However, the local population figures are enormously useful in preparing community development and conservation plans for an area.

The experience of community participation in Jaldapara was not replicated in the Thailand or Indonesia workshops. It was possible in Jaldapara only because there was an in-country NGO with community experience that had an existing tie with the IUCN, had the interest and personnel to support the workshop effort, and was able to carry out a research trip into a remote community with no advance notice or preparation. It is no wonder that this experience has been difficult to replicate. To institutionalize such an approach would require the commitment of the workshop planners, advanced planning, and allocation of at least modest resources. In Thailand, those organizing the workshop had no interest. The Indonesian workshop followed immediately after the Thailand workshop and, though there were in-country groups that might have been interested in facilitating community involvement, there were insufficient resources and time to act on those interests.

In Thailand, there were two community development organizations invited to the workshop, both of which had worked in communities located near gibbon populations. One of these groups provided support for a one-day visit to a community adjoining Khao Yai National Park prior to the workshop. However this organization was unable to participate in the workshop. The other organization appeared during the first day of the workshop, but quickly departed after the opening plenary session. Given the pressure of local populations on local resources and the low projections of future population growth, population growth was not a major concern of the Thai workshop. The one village visited had reduced its use of forest resources and appeared to pose no particular threat to the gibbon. Demographic analysis is certainly pertinent to an examination of the impacts of population growth on local habitats, but such an analysis was not needed in this area of Thailand. The biggest threat came externally, from the urban pet trade markets. As a result, other social science skills might have been of greater use.

In Indonesia, the workshop needed better information from the local communities as an input for modeling. A community appraisal of local population dynamics and use of environmental resources would have enriched the workshop. Several NGOs with community knowledge participated in the workshop, but their knowledge of the local communities contributed little. Looking back on that workshop, I believe that if properly approached, they might have contributed more. In Papua New Guinea, where the team had familiarized themselves with the issues facing local villagers, we were able to draw the NGO representatives into a useful dialogue. The difference may have been that in Papua New Guinea, we had village leaders who were able to provide highly useful information, supported by the NGO representatives (see chapter 9).

The experience of the Jaldapara rhino workshop informed preparation for the workshop in Papua New Guinea and preparation of a population appraisal at the community level prior to that later workshop. A major difference, however, was that in Papua New Guinea the effort supported fuller workshop participation by community representatives, addressing a significant weakness of the Jaldapara workshop.

In Papua New Guinea, the Network sent out its own team to undertake an appraisal of local population and resource use. Such an approach is expensive and leaves little behind for follow-up. An alternative would have been the Jaldapara model, by which a small grant could be made to an in-country NGO to conduct a community appraisal. Such fieldwork might best be made (as in Jaldapara and Papua New Guinea) with the support of a person knowledgeable about CBSG processes (either a local workshop organizer or a CBSG staff member). That person could ensure that the information collected was relevant to the workshop. Failure to plan for the inclusion of this type of data means that such input might occur only by happenstance.

Chapter 15

Caveat on Consilience:
Barriers and Bridges for Traditional
Knowledge and Conservation Science

DAVID A. LERTZMAN

Multi-stakeholder processes and collaboration in conservation planning have become increasingly common, yet the involvement of Indigenous peoples remains illusive and often disregarded. Even so, conservationists are becoming aware of the vast storehouse of ecological knowledge and wisdom in the annals of oral traditions. Scientists are coming to recognize the contribution of traditional ecological knowledge (TEK) in gathering baseline data as well as the dynamic abilities of traditional knowledge practitioners within management contexts (Freeman 1985, 1995; Cruik-shank 1981; Scientific Panel in Clayoquot Sound 1995; Duerden and Kuhn 1998; Lertzman 1999). Researchers are less familiar with the cultural protocols or methods that guide the transmission of traditional knowledge, the philosophical foundations upon which TEK is based, and the social institutions through which it is transmitted. This is a distinct stumbling block for understanding, research, and application of TEK in conservation science. Indigenous scholars indicate that "research" is one of the "dirtiest words" in the Indigenous world's vocabulary because it is inextricably linked with European imperialism in a global history of colonialism (Tuhiwai Smith 2001, 1).

In his book on the unity of knowledge, E. O. Wilson offers a clarion call for "the greatest enterprise of the mind" in the unification of the sciences and humanities (Wilson 1998, 8). He bases this on interlocking causal explanation across disciplines and terms it "consilience." Wilson gives ample evidence why this is a critical quest for global ecological stability. Imperative and noble as the task may be, practitioners of the natural and social sciences working on case studies in this volume have found it a difficult one. Although these disciplines, as Wilson points out, all share philosophical origins in the Enlightenment, they seem at times to have divergent methodologies and theoretical underpinnings. This is a challenge for the complex, transdisciplinary problems to which conservation sciences are applied. How much more challenging, then, would it be to find common ground between traditions arising from completely dissimilar cultures of origin and historical circumstances based on profoundly different philosophical systems for ordering reality? Yet this is exactly what the attempt to link TEK and traditional Western science (TWS) in conservation assessment and planning tries to do. The caveat for consilience is that conservationists will need to think beyond the philosophical foundations of their cultural paradigms to work successfully with TEK practitioners for conservation goals.

This chapter addresses barriers to bridging traditional knowledge and conservation science and proposes pathways for getting through. It draws on the author's background in working with First Nations in Canada. Although the Network had hoped to bring such a perspective into the case studies reported in this book, it was unable to enlist such expertise until after case studies were completed. Hence, this chapter offers a retrospective examination of the topic in light of other case studies involving Indigenous peoples and TEK.

Working with Indigenous Peoples in Conservation: A Global Context

Sustainable development came onto the world stage in 1987 with the World Commission on Environment and Development report *Our Common Future*. Its authors called upon world governments and their people to take responsibility for global environmental damages and the policies that cause them. Regarding Indigenous peoples, the commission recognized:

> These communities are the repositories of vast accumulations of traditional knowledge and experience that link humanity with its ancient origins. Their disappearance is a loss for the larger society which could learn a great deal from their traditional skills in sustainably managing very complex ecological systems. (115)

At the 1992 United Nations Conference on Environment and Development, several international agreements were adopted by 174 countries framing a global policy context for sustainable development. These agreements, including the International Convention on Biodiversity, Agenda 21, Guiding Principles on Forests, and the Rio Declaration on

Environment and Development, explicitly recognized the role of Indigenous peoples and their communities. Previous and subsequent international conventions have been signed dealing with the protection of Indigenous peoples' intellectual and cultural property.

There are compelling drivers behind such international environmental initiatives. One of the most powerful indicators of human impact on the biosphere is loss of biodiversity. According to E. O. Wilson (1986, 1998) we are witnessing the greatest rate of extinction of species in the last 60 million years due to human intervention in the biosphere. This trend has been accompanied by a similar impact on cultural diversity. Drawing the connection of ecology and culture, research demonstrates a correlation between biodiversity and linguistic diversity (Nettle and Romaine 2000). The greatest "biolinguistic diversity" on the planet is found in areas inhabited by Indigenous peoples, where 4 percent of the world's population speak 60 percent of the world's languages. Most of these languages and the ecosystems their speakers inhabit are threatened or on the verge of collapse. The viability of one is linked to the conservation of the other.

Loss of a language signifies much more than the depletion of words. Languages represent meaning systems, a way for organizing and making sense of the universe. Each embodies an inimitable example of human ingenuity and adaptation to the environment. Along with words, knowledge, and understanding, when a language is lost we lose a way of life and our species is diminished. Indigenous peoples living close to their ecosystems for long periods of time have garnered an enormous degree of descriptive and applied knowledge. Much more than "data," this information characteristically functions within time-tested resource management systems of long-resident peoples. Thus, adaptation to and use of ecosystems by Indigenous peoples offers alternatives for Western science–based conservation and management strategies.

Indigenous elders explain that conservation and sustainability represent core values in their traditions based upon spiritual teachings that translate into ethics of environmental stewardship. Former Scientific Panel for Sustainable Forest Practices in Clayoquot Sound member and Nuu-Chah-Nulth elder Roy Haiyupis explains:

> Respect is the very core of our traditions, culture and existence. It is very basic to all we encounter in life. . . . Respect for nature requires a healthy state of stewardship with a healthy attitude. It is wise to respect nature. Respect the spiritual. . . . It is not human to waste food.

It is inhuman to over-exploit. "Protect and Conserve" are key values in respect of nature and natural food resources. Never harm or kill for sport. It is degrading to your honour.... It challenges your integrity and accountability. Nature ... once broken, will hit back.... (Scientific Panel in Clayoquot Sound 1995, 6–7)

Given a dependence on stable habitat structures and species viability, Indigenous resource users share goals with conservationists. Yet motivations for conservation can be different. Indigenous peoples use the products of ecosystems; this can put them at odds with conservationists. Where traditional customary law and resource management practices have been degraded and local people are experiencing economic hardship, there can be conflict. Variables behind such divergence between Indigenous resource users and Western conservationists must be respected and understood to engage in mutually beneficial endeavours.

Conservationists need also be aware of Indigenous peoples' growing political power. Due to their long-standing history, distinctive identities, relationship to the land, and unique collective rights, Indigenous peoples are not merely another stakeholder in conservation and sustainability processes. In North America, First Nations are increasingly effective at exercising their rights and asserting their constitutional position, becoming more influential players in all levels of society, significantly in access to land and control of natural resources. In geographic areas where Indigenous people are a significant presence, conservation initiatives are unlikely to succeed without their participation.

Indigenous people are an essential element of conservation. Their knowledge and experience is a boon. They have also an urgent need for community economic development. To collaborate with Indigenous people, conservationists must develop skills not typically found in most conservation organizations. The next section offers conceptual tools to help deepen conservationist's understanding of TEK.

Traditional Ecological Knowledge Systems

Traditional ecological knowledge (TEK) is a young term used to describe something ancient. It draws our attention to the knowledge outcome of a complex system of social relations and institutions founded upon beliefs and values mediated by the protocols of oral tradition. Most important for these dynamic, adaptive, and living TEK systems are the ways of life, ecosystems, and people who generate them.

Although various scholars have tried to define TEK, there really is no universally accepted definition (Berkes 1993). Nor is Indigenous knowledge a uniform concept across Indigenous peoples (Battiste and Youngblood Henderson 2000). Some question the value of discussing how such knowledge is constituted (Cruikshank 1998). Others assert that power is the real issue (Nadasdy 1999). Indigenous scholars (Battiste and Henderson 2000) suggest that attempts by Western academics to define Indigenous knowledge are inherently colonial, grounded in a Eurocentric need to categorize and control.

Be that as it may, it is possible to glean common themes from a literature review of TEK (Lertzman 1999). Traditional knowledge includes a spatial aspect (i.e., it is geographically located), has an historical or temporal nature with very long time frames, is socially mediated (i.e., transmitted through social institutions in communities), and is culturally located (it functions within a larger cultural worldview and belief system). Not so prevalent in the literature is a fifth aspect: the "methodological" element of TEK. This pivotal feature refers to the traditional knowledge protocols that govern how TEK is accessed, verified, and transmitted. Boiling down these elements, "TEK systems" (TEKS) refers to the structure of social relations and institutions (social capital), founded upon philosophical beliefs and cultural teachings (cultural capital), mediated by practices and protocols (methods) of oral tradition (Lertzman 1999). All are necessary elements of these knowledge-producing systems and features of TEK. If any aspect is missed, the integrity of the system and its components are compromised.

The *Report of the Traditional Knowledge Working Group* (Northwest Territories 1991) offers a description of traditional knowledge based on elders' input from several backgrounds (Inuvialuit, Inuit, Dene, Metis). Traditional knowledge is rooted in the landscape, offering "a view of the world, aspirations and avenue to 'the truth'" different from the Euro–North American mainstream whose knowledge is based largely on European philosophies.

> Traditional knowledge is knowledge that derives from or is rooted in the traditional way of life of aboriginal people. Traditional knowledge is the accumulated knowledge and understanding of the human place in relation to the universe. This encompasses spiritual relationships, relationships with the natural environment and the use of natural resources, relationships between people and is reflected in language, social organization, values, institutions and laws. (12)

It can be difficult, they suggest, for people to grasp that traditional knowledge is more than mere information. In the above description, one perceives references to the biophysical, social, and spiritual relationships in which TEK is embedded. Rooted in the traditional "way of life," this shifts the emphasis from "knowledge" to context of knowledge: ecologically, socially, and philosophically. Unlike Western science, empirical knowledge is not seen as separate from other beliefs; there is no attempt to separate values and understanding from knowledge.

Many authors have sought to compare Western science and TEK. Cajete (2000), a Tewa Native American, refers to "Native science," the "peoples ecology" which is a lived story of place, kinship, and environmental knowledge.

> Close observations of plants, animals, landscape sights and sounds, changes in wind and humidity—everything surrounding people—is part of Native science as it is in the Western scientific tradition. It is the Native emphasis on participation and experience that embeds the sense of kinship with all nature in the minds, hearts, and souls of all members of the community. (101)

Native science comprises considerable empirical knowledge, yet understanding the empirical relationships of nature "is not enough." According to Cajete the key is in living and nurturing these relationships. Both TEK and Western science share observation and description of the empirical world. Yet the understanding each brings to its experience is based on quite different perspectives.

Traditional Western science (TWS) addresses phenomena that can be measured in time and space and does not generally recognize that which lies outside that domain. With important aspects of TEK outside the researchable realms of science, there are gaps in understanding when it comes to perceiving the origins of traditional knowledge. Most people agree that TEK has a spiritual foundation, often described as holistic in nature, yet descriptions of this by Western scholars tend to be shallow and vague. The standard epistemological account for the origins of TEK is through trial and error over time. This makes sense. Empirical observation and deduction are certainly an important aspect of TEK, as many researchers can attest. Yet this is only a partial account representing but one among other just as important means for creating and transmitting TEK.

Addressing "The Different Origins and Shared Goals of Scientific and Traditional Knowledge," the Scientific Panel for Sustainable Forest Practices in Clayoquot Sound (1995) offers the following discussion:

... consider traditional medicinal knowledge: it is acquired through the rigours and methodology of a vision quest, in which persons isolate themselves and undergo fasting, cleansing, and other ritual activities to receive inspiration and medical knowledge from supernatural powers. Although the methodology of the vision quest is unfamiliar to the modern medical community, the knowledge gained often coincides with that of modern medical scientists, acquired by wholly different methods. (16)

Thus, a variety of intuitive, somatic, and other spiritual modalities are vital to the generating of TEK, including singing, dancing, drumming, dreaming, fasting, praying, purifying, periods of isolation outside of the community, and other ceremony. Partial accounts result in recognizing knowledge outcomes but not the means by which knowledge is generated. This is inaccurate and insulting to TEK practitioners. TEK systems are holistic because they synthesize empirical observation and deduction with other ways of knowing. It is proper to respect the spiritual and philosophical foundations of TEKS along with their empirical observations and management applications.

Traditional resource practices such as hunting, fishing, wild rice and berry cultivation, medicine gathering, cedar or birch bark and spruce root collecting, and so on involve cultural practices that are a means for generating and transmitting TEK. These prayers, ceremonies and songs, and community celebration and collaboration express the beliefs, values, and relationships that are the foundations of TEK. Central to this are traditional knowledge protocols, often entailing extensive consultation among elders, specialists, and learned or involved individuals.

Some may not think of such cultural practices as "methods" in the academic sense. In the context of oral traditions, however, knowledge and its transmission are guided by the rigor of strict rules of learned protocol that are generally replicable and consistent within language areas. These protocols are methodological principles by which oral knowledge is generated, transmitted, and legitimized. One might think of traditional knowledge protocol and consultation among practitioners as not unlike scientific methods and peer review. Protocol is learned from an early age and takes years in which to become proficient. Protocol ensures that knowledge is shared properly, in a manner consistent with and legitimate to the beliefs, teachings, and practices of oral tradition in a given area, language, and community. These learned skills typically require years of mentoring and may involve arduous physical and intellectual training.

Some of these abilities can be passed interculturally, forming the basis for a skills set that has been called cultural literacy (Lertzman 2002).

Bridging TEKS and TWS in conservation entails joining two entirely different worldviews. Both share the same world, but bring very different understandings to their experience of it. TEKS and TWS constitute discrete ways for ordering the universe based on different assumptions about the nature of reality (Tambiah 1990). They represent parallel, potentially complementary knowledge systems. Through the collaboration of fifteen prominent scientists and four Nuu-Chah-Nulth elders, including a hereditary chief, the Scientific Panel for Sustainable Forest Practices in Clayoquot Sound concluded that TEK provides for Western science an "external, independently derived reference standard" (1995). When respected individuals from completely different worldviews reach consensus on fundamental ecological principles and conclusions about the nature and status of ecosystems and their components, it is worth listening. This provides a basis for bicultural standards of verification. It is like saying, "here is a joining of the best of two worlds." The point is not that one way of knowing is better or more valid than the other. Both TEKS and TWS are valid on the basis of their own merit. Drawing on both strengthens planning for conservation (Lertzman 1999). In addition to the epistemological challenges described above, some important social barriers are reviewed below, followed by bridging strategies.

Barriers and Bridges for Traditional Ecological Knowledge and Conservation Science

Many of the barriers to bridging traditional knowledge and Western science are perceptual in nature. Some are historical; others are methodological. There can be legal, procedural, and institutional barriers. There are ecological, political, and economic barriers. Some are endogenous and exogenous to Indigenous communities. Many barriers relate to a lack of familiarity with Indigenous people and cultures, and a general lack of information and culturally appropriate research. The notes below are based on a combination of literature review, historical research, and the author's professional and personal experience. The material is neither exhaustive nor comprehensive; the intention is to stimulate thought and discussion.

Colonial and Historical Barriers

The persisting history of colonialism has created an array of perceptual and structural barriers to bridging TEKS and TWS. The perceptual legacy of this is still prevalent in North America and throughout the world. Old racist attitudes and chauvinistic stereotypes are changing but their impact is lasting. Many individuals are still of the attitude that Indigenous peoples and their cultures are less sophisticated than those societies based on a TWS perspective. The idea that TEK is somehow less credible than TWS-based knowledge, and that Indigenous cultures have been superceded by modernity, are a liability for collaborative efforts in bridging TWS and TEKS.

Legal and Institutional Barriers

These are often slower to change. The Assembly of First Nations and the Inuit Circumpolar Conference (1994) cite lack of legislative authority as the greatest barrier to implementing traditional knowledge, particularly in environmental protection. Legal and political barriers to land claims continue to exacerbate the problem, while the slowness of the courts and other legal processes continue to frustrate efforts to preserve TEK in spite of a growing cultural revival in many communities.

Political and Economic Barriers

Political and economic barriers are enmeshed. Current political and economic barriers to the persistence, research, and implementation of TEK are framed within a colonial history and global economy, including social, institutional, and political outcomes. Some of the greatest political barriers relate to governance regimes and land use. Changing positions and policies toward Indigenous peoples at various levels of government exacerbate these. Weakening of traditional governance and cultural authority (diminishing social capital) and conflict with elected authorities may be a factor. Such conflict may take the form of community-level disagreements over resource use, economic development, and community health needs.

Economic barriers include impacts of the market on TEKS subsistence economies, which are often far more sustainable. Local resource users are rarely isolated from the impacts of the global economy. Socioeconomic factors have a profound impact on many Indigenous communities, entrenching cycles of poverty and exacerbating community health issues. Many communities are overwhelmed or preoccupied with day-to-day survival, making proactive, long-term community development planning a challenge. Traditional technologies and economic structures require a base of biophysical, social, and cultural capital to function effectively. Market-based economies tend not to value these, while valuing less-sustainable practices (e.g., in fisheries, forestry, mining), offering short-term financial gain but disturbing the ecological and social systems upon which TEKS depend.

Cultural Appropriation

Cultural appropriation is an ongoing issue for many Indigenous peoples; sharing of TEK opens this possibility. The Assembly of First Nations and the Inuit Circumpolar Conference (1994) state that TEK is a growing field in which Native people have to subsume their views to the expertise of non-Native analysts and consultants. The result is that knowledge which ought to be at the disposal of Native people and their communities to use according to their priorities and values has been appropriated. A common story is told of the non-Indigenous researcher who benefits greatly from the sharing of TEK but does not give back to the community. The same person winds up a "TEK expert" with more authority than the people whose rightful cultural property this is. There are real and perceived dangers for holders of TEK to become involved in TWS-based research.

Language/Communication Styles

Languages are critical to TEKS; many TEK practitioners speak English as a second language. Language differences can pose time and other constraints. Language is not so much a barrier as is the lack of familiarity with language differences and communication styles (see in particular, chapter 9). Learning languages and different communication styles, especially those culturally based, can be fulfilling and personally challenging. These are specialized skills requiring time, effort, and experiential training.

Bridging Strategies

With proper education, training, and relationship building, many of the barriers and constraints raised above can be overcome; others must be managed and worked with. Either way, bridging TEKS and TWS in conservation assessment and planning takes a special kind of cross-cultural competence. The foundation for this is respectful, ongoing face-to-face experience between the members of differing cultures and the worldviews they represent. This extends the interdisciplinary concept of consilience cross-culturally to bridging knowledge systems. Bridging strategies are suggested below.

Cultural Literacy

Working in such intensely cross-cultural situations can challenge one's perceptions, causing individuals to rethink their ideas, their actions, and to think beyond them. One must learn to think outside one's cultural box, without leaving it behind. Those who acquire the capacity to communicate effectively in a culture outside of their origin have achieved within that society a degree of cultural literacy. Literacy with TEKS protocol must be acquired through direct interaction and mentorship with the holders of such traditional knowledge. Putting this knowledge into action requires diligence, sincerity, respect, sensitivity, and open-mindedness; it can be both challenging and fulfilling.

While cultural literacy is gained through personal learning, it has powerful professional applications. People who are proficient enough in skills of cultural literacy can become functionally bicultural. Being bicultural is the ability to function comfortably and effectively within two cultures, and it means that members of those cultures experience you as an effective, comfortable communicator in their world. It is not about leaving one's culture and replacing it with another. The bicultural person has learned to think and act outside of his/her cultural box without leaving it behind. These people possess specialized abilities that make them invaluable to circumstances of cross-cultural bridging. They are the bridge builders; their role is essential to conservation and other initiatives that bridge TWS and TEKS.

Relationship Building

Quite fundamentally, quality relationship building and ongoing relationship maintenance (stewarding social capital) is one of the most effective antidotes to the barriers presented in this discussion. Yet this takes time, motivation, human resources, and skill. There was a need for greater lead time and relationship building in both the Peary and Arctic Islands caribou and the Papua New Guinea tree kangaroo cases (chapters 8 and 9, respectively). Relationship building is also a key element and output of Population and Habitat Viabilty Assessment (PHVA) workshops. The Papua New Guinea case, in particular, seemed to make gains in this area. These social capital investments are worth stewarding for ongoing conservation efforts. Lack of relationship building contributed to the absence of local First Nations in the Eastern Slopes grizzly bear workshop (see chapter 10).

New Institutional Arrangements

New institutional arrangements and governance regimes can provide excellent opportunities for long-term conservation collaboration (see chapter 12). One of the most powerful examples of this is found in comanagement, which in Canada has been the result largely of land claims. Such institutional configurations rely on power sharing and collaboration, illustrating how self-determination of Indigenous peoples is an asset to conservation. The caribou case (chapter 8) provides an example of a PHVA taking place in this context, setting the stage for co–decision making. Comanagement is an evolving form with local variation.

Economic Development

This remains one of the most urgent issues in Indigenous communities. Conservation practitioners could find significant allies with Indigenous peoples were they able to harmonize conservation efforts with sustainable community economic development. Indigenous participants of the caribou case study seemed to be attempting such a link with "country foods" but this was not explored (chapter 8). While it may not have been within the objectives of the workshop, the occurrence is exemplary of missed opportunities.

Ecosystem Conservation

TEKS cannot function without the ecological bases from which they are born. One obvious barrier to the persistence, research, and application of TEK is the ongoing degradation of ecosystems and their components by current resource, industrial, and other development practices. A mutual commitment to preserve the ecological integrity of natural systems is a central stone of the bridge between TWS and TEKS conservation practitioners. Whatever their differing motivations, this provides inspiration on both sides to support necessary relationship and cultural literacy development.

Strengthening Participatory Design

As discussed in chapter 4, PHVAs place high emphasis on equal participation, participant facilitation, power equalization, and consensus building around recommendations. Both the caribou and Papua New Guinea tree kangaroo cases evidenced that Indigenous participants found the process amenable (see chapters 8 and 9). The moment in the Papua New

Guinea workshop when Indigenous landowners gave their full support to the recommendations after separate internal deliberation represented a high point for everyone. The PHVA offers a bridge for working with Indigenous peoples in a way that top-down processes do not. More could be learned from Indigenous peoples about culturally based participatory and inclusive decision-making processes (Scientific Panel in Clayoquot Sound 1995).

Conclusions

Consilience between Indigenous practitioners of traditional ecological knowledge systems and those of traditional Western science is unlikely to occur based solely on modernist rational empiricism. Yet the barriers to bridging TEK and conservation science are not due to any inherent flaw in Western philosophy. As with any fundamentalism, the problem lies in seeing science as the only path. Yet, these barriers are exacerbated by imbalances of political and economic power, intellectual hegemony, and colonial history. Getting beyond this does not require giving up one's tradition, be it TWS or those of TEKS; it requires diligent experiential learning in a context of mutual respect. This is critically important in order to address the multifaceted challenges of conservation, which require collaboration among the very best that the worlds of TWS and Indigenous peoples have to offer. Such determined cultural interface has an ability to shift people's perceptions and expand their capacities. It is transformational by nature.

There are ways in which the PHVA might be enhanced to be more effective in these circumstances.

• Build a bicultural bridging team of appropriate individuals, including local community representatives, to perform legwork for the project.
• Have this team commence relationship building at the community level months in advance of the workshop. Intensify these activities at least one month prior, with at least two team members including the community representative.
• Create opportunities for the team to sponsor cross-cultural training workshops for both TWS and TEKS participants. Include outings on the land and/or water with local TEKS practitioners.
• Engage TEKS practitioners in conservation processes that draw upon the philosophical foundations and management applications of TEKS beyond their role as a data source.

- Become aware of the development needs of local Indigenous peoples and seek to harmonize conservation goals and planning processes with sustainable community development.
- Facilitate networking between TEKS practitioners from various communities and support their ongoing dialogue with the conservation community.
- Develop organizational capacity to steward the social capital investments made by PHVA process: engage in ongoing, long-term relationships.

For conservationists rooted in a Western science–based perspective, it is important to deepen one's understanding of culture, history, and community needs to work effectively with Indigenous peoples and TEK. Indigenous people who have an understanding and appreciation for TWS may employ it in their own conservation efforts, yet science is not the primary driver of conservation in their communities. There can be spiritual, cultural, economic, political, and other motivations for Indigenous peoples to engage in conservation. Thus, while there may be shared goals and motivations between TWS and TEKS conservation practitioners, there can also be divergent ones. Effective bridging of TEK and conservation science manages such divergence. Listening to Indigenous peoples, as well as respecting and understanding their desires and aspirations will help scientists and others partner more effectively in wildlife and habitat conservation. Similarly, conservation scientists will need to find effective ways to share the wisdom of their tradition. TEKS and TWS offer each other externally derived, independent reference standards that provide a basis for bicultural verification. They represent parallel, potentially complementary knowledge systems with their own methods, philosophical foundations, ordering principles, and communities of respected experts. Including these different ways of knowing generates robust data, enhances assessment processes, and strengthens conservation planning. Based on the cases presented, Indigenous peoples' participation and knowledge is key to developing a sound basis for conservation policy planning and action.

 **Chapter 16**

Strangers at the Party: An Industry Strategy Perspective on PHVAs

HARRIE VREDENBURG

The Population and Habitat Viability Assessment (PHVA) workshop process has the dual objective of surfacing and modeling current species-level data in order to devise effective species and habitat management plans. Further, the Conservation Breeding Specialist Group's (CBSG's) aim for PHVAs is to foster and increase communications between relevant stakeholders in a species' conservation, with a view to effectively implementing such plans. Historically, the workshop process involved scientific experts on the species in question and government-agency personnel who were responsible for managing the habitat. In practice, workshops were comprised of university- or scientific institute–based biologists working with wildlife agency– or national parks–based wildlife managers.

In many conservation contexts this was quite appropriate. Both groups were in a position to contribute in some way to the dual objectives of the workshop, the scientists perhaps more so to the population modeling and devising of management plans and the wildlife managers more to the implementation of the plans. One of the early lessons learned from the workshop process was that there are large disciplinary and cultural differences dividing these two communities. These differences require active workshop facilitative management in order to achieve a productive collaborative workshop. Another early lesson was that wildlife managers are often in a position to contribute quite significantly to the data surfacing and modeling aspect of the workshop, based on the "local" knowledge of species biology and habitat interaction. Again, in order to fully incorporate such local knowledge into modeling and management planning, the workshop process must be actively facilitated to achieve this end. Chapter 4 addresses some of the key process facilitation aspects related to this.

In recent years, nonscientific local stakeholders have been increasingly involved in the workshop process. We are recognizing that scientists and wildlife managers alone cannot conserve species threatened with extinction. Local landowners and communities clearly have a role to play in the implementation of conservation management plans, since their behavioral practices in a habitat area are critical to the success of any conservation plan. Even when a species habitat is a protected area, national park, or nature reserve, there is almost invariably some local resource utilization or extraction in the protected area itself and often considerable utilization and extraction of resources immediately adjacent to the protected area. Some local landowners and community members also may have important species and habitat data to contribute to the modeling process as a consequence of their habitat and species utilization practices as hunters, gatherers, or agriculturalists. The roles of these local landowners and communities have been acknowledged by their inclusion in the workshop process in recent years. The challenge of cross-cultural communication between these groups and the scientists and wildlife managers in a conservation workshop is significant. When local communities are Indigenous or live traditional lifestyles, the challenges of workshop collaboration with scientists and scientifically trained wildlife managers are exacerbated. Other chapters in this book address specifically some of these unique challenges.

Resource utilization and extraction activities in a habitat area are often carried on by local communities or by local landowners. Such activities can range from traditional hunting, gathering, and subsistence agriculture to commercial timber extraction, ranching, and mining operations. These larger commercial enterprises might be owned and operated by communities or by prominent local families or individuals. Local community and landowner stakeholder groups participating in a CBSG workshop usually provide the information and represent the interests of these commercial stakeholders in a habitat area.

There is another type of resource utilization and extraction activity that is often important to the conservation of a habitat area but that is not represented directly by local landowners and communities. This is large-scale industrial commercial activity. This type of activity is differentiated from the community-based and local landowner-based commercial activities by both the relative scale of the activity and the relative importance of nonlocal linkages represented by the activity. Community-based and local landowner–based commercial resource extraction and utilization activity tends to be relatively small in scale as it is constrained by locally available financial capital and technical expertise. It is also constrained geographically by the more locally oriented strategic scope of those making the commercial decisions.

Industrial commercial resource utilization and extraction activities are characterized by close connections to multinational corporations, high financial capital content, high professional and technical expertise, high technical sophistication of capital equipment, close connections to world commodity or consumer markets, the importance of government ministries and regulatory agencies to commercial operations, a high proportion of nonlocal national and international managers and corporate professionals, and close connections to world financial markets. These characteristics differentiate these enterprises from the local nonsubsistence commercial enterprises.

In a world that is undergoing increasing economic integration, this industrial type of resource extraction and utilization activity is becoming more and more important. The global societal pressures on this type of enterprise to operate in an environmentally and socially responsible manner are also increasing. These enterprises rarely own the land on which they are carrying out their activities, but hold the timber, mineral, or other resource extraction or utilization rights. These more limited ownership rights are granted to them by the relevant government authorities and are subject to regulation by relevant government agencies. Although, strictly speaking, habitat-related decisions usually remain the prerogative of the government authorities and regulators, in fact, in almost all jurisdictions there is considerable latitude available to these enterprises with respect to decisions that could have impacts on habitat. Decisions made by these firms can have significant impact on habitat area available (e.g., clear-cut forestry, open-pit mining operations, tourism resort development); habitat fragmentation (e.g., oil and forestry access roads, seismic access lines); habitat quality (e.g., chemical contamination of land and water from mining and oil extraction operations, noise from tourism activities, predator access on roads and pathways developed); and species exploitation (e.g., hunting by nonlocal workers employed by industrial enterprises, "trophy" hunting by international tourists).

Although industrial commercial enterprises have historically not been stakeholder participants in CBSG workshops, their absence is of increasing concern for at least two reasons: due to global economic integration, there is more of this type of activity going on in many parts of the world where critical habitat is at stake; and, due to societal expectations in developed (and increasingly in developing countries), these firms may be more motivated than in the past to be environmentally and socially responsible. One hundred years ago, gold mining companies could come into the Canadian Yukon and leave behind a moonscape of tailings piles with equanimity; fifty years ago tourism developers could come into the Canadian Rockies and build a major ski resort in Banff National Park with equanimity; twenty-five years ago oil companies could "open up" the Ecuadorian Amazon to migrant peasant farmer settlement by building accessible oil access roads while simply discarding polluted water into streams and rivers with equanimity; ten years ago forestry companies could clear-cut rain forests in Brazil and Indonesia with equanimity. While the pressure to develop ecologically sensitive areas for industrial commercial purposes has increased due to globalization, the social license for these firms to act without consequences has decreased. Including representatives of these enterprises in the CBSG workshop process presents a particularly challenging problem.

What Role Can Industrial Resource Developers Play in Conservation?

Industrial commercial resource developers can play a significant role in conservation primarily by taking account of conservation concerns and sensitivities in making development decisions. Resource developers make many decisions that are inimical to conservation because they are simply unaware of the specific conservation impacts of their landscape-based development decisions. They are certainly aware that no development is better for the conservation of habitat and species conservation and have been made continually aware of this fact by many advocacy-oriented environmental organizations. The focus of major environmental groups for so long has been on halting resource development in high-profile situations that it is difficult for many conservation advocates to imagine situations in which development might take place in a manner more consistent with conservation goals.

CBSG, as a science-based organization working for integrative stakeholder-based solutions to conservation problems, is well placed to influence how resource development takes place rather than whether or not it takes place. In situations where government decisions have already been made to allow resource development in ecologically important areas (often in or adjacent to protected areas), the CBSG workshop process can play an important role in opening channels of communication between conservation scientists and resource developers.

A concerned corporate oil production manager or corporate chief forester who is trying to comply with their organization's professed objective to improve the firm's environmental performance does have decisional latitude that can have conservation benefits. Examples of these kinds of decisions include where, exactly, to locate a forestry "cut-block," an access road, or a pipeline pumper station. Armed with conservation biology information about the degree of threat of extinction a particular species faces and the important role of certain habitat types (particularly if these are precisely geographically mapped), an operations manager can avoid significant habitat areas and use less valuable areas for development purposes.

Having operations managers from various firms in the same or different industries present at a workshop can also result in these normally independent and competitive firms collaborating in ways that are beneficial to habitat and species conservation. For example, oil production, mining, and forestry infrastructure development might be geographically concentrated and coordinated and situated in a lower-quality habitat area in order to minimize critical habitat destruction effects. The added costs of building infrastructure on more difficult terrain (for example, on rocky ground away from a river valley) or rerouting around critical habitat areas can be offset or recouped by collaborating with several organizations.

Particularly powerful is the opportunity to include the resource developers' future landscape-based plans in the computer simulation model's extinction scenarios. This way, the operations manager can have a reasonably precise idea of the conservation effect of certain mitigating actions and their expense. Showing a scientifically based estimation of the conservation effect of operational mitigation action is valuable in promoting these types of actions in resource development firms.

Although in most cases resource development managers can be most useful to the conservation process by adjusting their resource development plans to the needs of conservation as expressed by conservation scientists, in some cases they may be in a position to help implement actual conservation plans. By participating in conservation workshops and becoming knowledgeable about conservation problems, resource development managers may be able to play a role in restoring critical habitat or underwriting field research. On rare occasions, in the case of, for example, hydroelectric companies or forestry firms, the company may actually employ field biologists who may be able to supply important species or habitat data to the workshop for modeling.

What Will Motivate Industrial Resource Developers?

Industrial resource developers, as organizations, have different goals than governments, environmental groups, universities, and research institutes. Whereas the latter may have as their primary goals such things as protection of the environment and the creation and development of knowledge, private corporations are chartered to provide products and services for the profit of their shareholders. They are subject to what economists refer to as the "market discipline." This somewhat narrow goal of corporations may be lamented and criticized, but this is unlikely to have much effect on corporate behavior.

Some corporations that are wholly or in part owned by the state may have a broader goal in their charter, including environmental protection or providing an affordable service or product for a large proportion of the population. In effect, these organizations are adopting some of the goals of the state. Increasingly, however, many of these state-owned firms are privatizing and also subject to market forces.

Some privately held companies, that is, firms whose shares are not traded on public stock exchanges but are owned by a group of wealthy individuals or families, have a little more flexibility in decision making and the ability to take a longer-term view than firms with publicly traded shares. The reason for this is that these firms are not as immediately subject to market discipline as are publicly held firms. A privately held firm can sustain a drop in profitability for as long as financial resources are available to continue operations. A publicly held firm, on the other hand, is subject to immediate share price declines as a consequence of poor profit reports; senior corporate management and the firm's board of directors are held accountable for share price declines as reflections of their leadership and corporate governance performance.

This market discipline is the reason why environmentalists' appeals to firms to "do the right thing" with respect to the environment are often unsuccessful. Corporations do not take positive environmental or conservation actions in order to do the right thing. Individual corporate managers may well want to do so. In fact, many people working for resource development companies are personally committed to the environment. Geologists, foresters, petroleum engineers, and other professionals employed by resource developers often chose their careers because they had an interest in nature and wanted to work outdoors. Many of them spend their free time in outdoor pursuits and decorate their workspaces with pictures of wildlife and beautiful natural settings. But they cannot make environmental decisions explicitly in order to do the right thing. Companies may claim in their press releases and corporate communications programs that they are committed to doing the right thing with respect to the environment, but that is merely good corporate public relations.

Companies take pro-environmental or pro-conservation actions primarily because they make good business sense. These decisions must be compatible with the market discipline in which the company operates. These decisions must hold up under the critical tests of whether they will contribute positively to the firm's profitability or contribute positively to the firm's share price. Some companies will tolerate some short-term negative effects on profitability or share price if senior management can make a credible case that the long-term effect will be positive. But this is really the only criterion on which business decisions are evaluated.

Although market discipline affords a comparatively narrow scope for environmental goals of a corporation as a societal institution, there is possibly more scope than many environmentalists and environmental policy makers often assume. And, as discussed above, corporations make many decisions related to the natural landscape and have considerable resources at their disposal. Why, then, might resource-based companies be interested in participating in species and habitat conservation efforts? There are a number of reasons, all of them related to the market discipline which rules corporate behavior.

A forestry company might be concerned about potentially losing a major customer like retailer Home Depot, who might require that a forestry company demonstrate responsible conservation-oriented forestry practices. Home Depot, in turn, is concerned about their customers. Even if only a small minority of customers actually care about forestry practices, Home Depot is still concerned. Home Depot takes note because if they continue to buy from a supplier who is seen as operating in an environmentally irresponsible manner, they, as a retailer, will become a convenient high-profile target of an advocacy environmental group. The bad publicity from such an environmental campaign will hurt their sales as the issue becomes salient to more customers. In addition, this sort of campaign will introduce uncertainty into the company's retail business, which makes future corporate sales forecasts suspect. This uncertainty will, in turn, cause shareholders to sell the stock and depress the share price. Home Depot senior management and board of directors want to avoid this scenario and so put pressure on their suppliers to be seen as being a responsible forester. The forestry firm becomes committed to taking conservation action while not hurting their overall wood sales. These forestry firms have a strong incentive to find innovative ways that will allow them to contribute to conservation without hurting their business seriously. Participating in conservation workshops in order to learn more about conservation issues and altering their resource extraction plans is an attractive option and one worth the investment.

An oil and gas exploration and production company might have various development options in ecologically sensitive areas of North America. Unlike in the forestry example above, these companies are not particularly vulnerable to a large customer pressuring them to behave in a more environmentally responsible manner. If the oil and gas company is not a fully integrated company but an "upstream" exploration and production firm, their product will be mixed with that of other firms and will be sold as an undifferentiated commodity to end consumers. This makes it difficult for advocacy environmental groups to target a retailer selling their product. Furthermore, gasoline and heating fuels are seen by most consumers as necessities whose purchase provokes little thought; consumers have limited alternatives even if they do think about it. Thus, even integrated major oil companies that do have retail divisions have only limited exposure to boycott or public relations threats. (In the 1990s, Exxon was the target of such a boycott after the *Exxon Valdez* oil spill in Alaska and Shell was targeted by a boycott for their practices in Ogoniland, Nigeria). But all oil and gas firms do need to have their exploration and production plans approved by regulatory authorities in whichever jurisdiction they operate. Regulatory authorities must be in step with the public will or the governments they serve will be embarrassed and voted out of office at the next election. Throughout North America, publics, as voters, are quite concerned about protection of ecologically sensitive areas. It could be argued that publics are more concerned about environmental issues as voters than as consumers because as voters all bear the cost of environmental protection, while as consumers only the concerned bear the cost. As a consequence, regulatory agencies in many jurisdictions appear to be increasing their vigilance over the environmental aspects of resource development decisions.

Oil and gas exploration and production companies have learned that regulatory agencies are much more favorably disposed toward companies that proactively, voluntarily, and innovatively deal with environmental aspects of resource development. Being proactive and innovative with respect to environmental protection enhances a company's reputation with regulators, thereby reducing the time and cost of getting regulatory approval for projects and also reducing the probability that a project in which much time and expense has been invested will be rejected. Taking such a proactive stance could mean adjusting and minimizing landscape or species-impacting industrial activities or, in some cases, could mean avoiding industrial development in an ecologically sensitive area altogether. An example of the former is a hydroelectric company altering its water levels behind a dam to accommodate biological needs of fish stocks. The latter is the approach taken by one large, independent, fully integrated Canadian oil company in a recent case. When they learned from a biological science–based environmental organization how ecologically sensitive an area of prospective exploration and production was, they terminated plans for development and instead redirected their activities to another area that was less sensitive. This decision won the company a great deal of credibility with regulators, especially since they did it of their own volition. In sum, being environmentally proactive reduces regulatory approval costs and reduces risk of major project rejection. Both of these add to firm profitability. The second attribute also has a direct positive effect on share price, as a firm that loses major project approval decisions is a more risky investment proposition.

Companies operating in developing countries face a somewhat different set of business strategy parameters. Historically, resource-based companies from the developed countries saw the developing countries as a "free-for-all" where the societal rules of their home countries did not apply and where local rules were either nonexistent or not enforced. The records of these companies in the developing world were grim. But in the past few years things have been starting to change.

What is driving this change is that international environmental organizations are working with local groups in developing countries and, with the increased capabilities of rapid global telecommunications technology, are keeping pressure on multinational corporations wherever they operate in the world. Companies can no longer operate out of view of the world. Recently a North American mining company learned this when they were doing the preliminary work for developing a new mine in Bolivia. In Bolivia, they held meetings with local communities and, in order to minimize local opposition, talked about the relatively small mining operation they were planning. Meanwhile, head office personnel in

North America were hyping the huge new mine that was being planned in Bolivia to potential financial equity investors. One of the stock owners was an order of Roman Catholic brothers in Quebec who had missionary contacts in the communities in Bolivia. The order of brothers translated all the mining company's investment promotional statements into Spanish and Quechua (the local Andean indigenous language) and faxed them to the Bolivian communities. Company officials in the local communities were completely taken off guard when community leaders confronted them with the head-office information that conflicted with what the communities were being told. The company was thus forced to be more transparent and to show concern for the needs of the local community and the country (I. Thompson, personal communication with the author). This situation causes a potential chain of events. The company starts to look for mechanisms to move it beyond the adversarial and confrontational realm that developed. The community, though hopeful of good things coming as a consequence of resource development, has concerns about their environment and their health as well as sharing as a community in the benefits of resource development, but they only know how to oppose actions. Again, this company is now highly motivated to work with "reasonable" organizations to address local environmental and social concerns; and communities are anxious to participate in something that has the potential to move toward an integrated solution. Participating in and sponsoring a stakeholder-based conservation workshop with a science-based unit of the internationally respected IUCN becomes an attractive option to the mining company.

As oil companies look around the world for new reserves to develop, they are increasingly finding themselves in developing countries and in the middle of ecologically sensitive areas such as the Amazon. Developing-country governments want to develop the resources, as they need the foreign exchange generated to deal with national debts and human poverty alleviation. They do not, however, want to incur the wrath of the global environmental community and their local communities who oppose resource development. Multinational resource-based companies have an incentive to be innovative in their plans for developing resources consistent with local environmental and social exigencies. Companies that have demonstrated that they can develop resources while being sensitive to regional and local environmental and social concerns have a competitive advantage in gaining access to resource development opportunities in similar settings. For example, a company that has a good record of bringing oil to production with minimal environmental impact and social benefits in Amazonian Ecuador will be a favored bidder for developing oil resources in Amazonian Bolivia or Peru. The current high prices of oil and natural gas and the paucity of new reserves available in North America makes access to new reserves strategically critical for such companies; positioning one's firm as being a responsible developer of oil and gas in ecologically sensitive Less Developed Regions gives a company a distinct advantage in gaining access to new resources.

Being innovative with respect to environmental and social issues also minimizes the likelihood of a firm finding itself at the center of an international public relations fiasco. Being targeted by international environmental activist groups and finding one's company name on front pages of newspapers can be costly to firms in terms of their market valuation. A Canadian oil company operating in Sudan recently was the target of high-profile human rights activist organizations. The company's president estimated that this experience cost the firm millions of dollars in market valuation (due to the company's share price declining) as shareholders sold their shares because of the uncertainty introduced by the campaign and the resultant increase in risk of the stock as an investment (Allwright and Vredenburg 2002).

Because resource-based firms are strategically dependent on access to resources in ecologically sensitive areas (all remaining oil and gas, mineral, and timber resources are located in such areas) and the avoidance of environmental public relations disasters, these companies have a strong incentive to find new ways of arriving at integrative solutions to the problem of developing resources in a way that minimizes effects on species and habitat. Participating in and funding conservation workshops facilitated by a science-based conservation unit of the internationally reputable IUCN presents a way to do this.

Getting Industrial Developers Involved

Industry involvement in workshops, when it has happened at all, has been an explicit effort on the part of staff or Network individuals. There are no natural connections into the conservation world that can be easily tapped. Our first experience with resource-based companies who had a stake in the local area that was a focus of a workshop predated the Network experiment. It occurred in early 1994 in Southeast Asia. United States–based California Energy Company had commercial interests in developing thermal energy in both the Philippines and Indonesia.

Through the efforts of a CBSG Steering Committee member who knew the president of California Energy because they both served on the board of directors for the committee member's zoo, California Energy became the financial sponsor of a PHVA workshop on the tamarao in the Philippines and the gibbon in Indonesia.

This was during an earlier phase of my work in this area and I had the opportunity to participate in both the Philippine and Indonesian workshops. In the Philippines, California Energy sent their American expatriate public relations manager to the workshop venue at the University of the Philippines in Los Banos, a few hours drive north of Manila, to deliver a speech on behalf of the company at the workshop's opening ceremony. After her opening remarks, I spent some time interviewing her on her perceived role in the workshop and the role of California Energy in the Philippines. It was clear that California Energy saw that funding the conservation workshop was a charitable donation in a country that was important to their business. In order to be successful in a particular country, whether the United States or the Philippines, a company had to be a responsible corporate citizen and make charitable donations relevant to the communities in which the firm operated. This was clearly part of California Energy's policy wherever they operated and it was perceived to smooth community and government relations in the countries of operation.

When I suggested that perhaps she should stay at the workshop and participate, she protested that she knew nothing about the workshop topic and was merely a public relations person. Pursuing this line of inquiry further, I asked whether there were field operations managers in Manila who may have relevant information and may be able to participate from the company's perspective. These questions were met with slight confusion and a reiteration that she was sent as a public relations person and that's all. The experience in Indonesia was much the same, with California Energy seeing itself as a progressive American company making local charitable donations in the interest of being a responsible corporate citizen.

What intrigued me, however, was what an opportunity the conservation community was missing in not integrating industrial managers from such progressive companies into the conservation workshop process. This company certainly appeared to be motivated to invest in the local country in a broader sense; it appeared to have made the link from its business to local conservation interests. And yet there was this large gap. Scientists, government officials, and local community leaders were meeting to assess the conservation status of an important species and its habitat and devising plans for future conservation efforts in a workshop funded by a company important to the future of the regional landscape, without actively involving this company. The plans agreed to at the workshop could be seriously compromised by a single decision made by an operations manager at California Energy regarding location of industrial infrastructure or routing of an access road or pipeline.

The reason for this disconnect with the conservation process seems to be the differing cultures discussed above, the societally oriented science and government sector and the profit-oriented market-disciplined corporate sector. California Energy was a progressive company that understood that it had to play a broader role in society through charitable donations if it was to be profitable in international business in the long run. However, it did not see its role as becoming involved operationally in local environmental issues. Getting such companies that appeared motivated to become operationally involved became one of the missions of the Network effort.

Although efforts were made to incorporate industrial players in some of the first Network workshops in Uganda and Brazil, little progress resulted as we tried to work through CBSG's existing conservation community networks in each species' range country and continent. The conservation community just did not have the contacts and relationships with industry. As with the Network's experience with including social scientists in the workshop process, one cannot just invite industrial stakeholders in a regional landscape to come to a workshop on conservation. They simply will not come: they don't see the problem defined in a way that appears relevant to them, and they see the workshop as a gathering of scientists with whom they have little in common. What we learned from our attempts is that in order to succeed in getting industry to a conservation workshop there must be some longer-term connection between industry and the conservation community. In other words, the California Energy–sponsored workshops in the Philippines and Indonesia were truly missed opportunities. California Energy had personal connections to CBSG (through CBSG's Steering Committee member) that helped to engender trust in CBSG's scientific, nonadversarial approach. California Energy was sufficiently motivated to financially support a conservation effort. What we needed were situations such as these where a prior relationship with the conservation community had already developed trust and industry willingness to be involved.

The Network decided that we would try to experiment with at least one, and maybe two, workshops where we had a reasonable chance of involving industry stakeholders. The criterion that had emerged from our prior experience was that such potential participants should have a significant industrial stake in a regional landscape and should have prior relationships with the conservation community, CBSG, or the Network. The grizzly bear conservation domain in western Canada seemed to provide the necessary prior conditions (see chapter 10). The Eastern Slopes Grizzly Bear Project (ESGBP), led by bear biologist Steve Herrero, was a large conservation research effort that was funded in large part by industry, namely the oil and gas, forestry, tourism, and ranching industries. Herrero and his colleague Mike Gibeau had made strides in gaining the confidence of industry and involving them in a project steering committee that periodically met to hear about the results of the grizzly bear biology research. Herrero had also served as consultant to both industry and government, which helped engender good relations with these two groups. Gibeau was a national park warden who had grown up on a cattle ranch and who had taken leave to complete his doctorate. The ESGBP work formed a central part of his dissertation.

As both a Network member and a colleague at the University of Calgary, I was in a position to work with both of them in order to effect a workshop that would include industrial stakeholders. The first step in this direction was a presentation by Network members made in a public seminar series held at the university's business school. Some eighty people from the academic community, the national parks and Alberta provincial parks, the Alberta Energy and Utilities Board, various environmental organizations, the Canadian Association of Petroleum Producers, representatives from several other oil and gas companies, and the Calgary Zoo attended this presentation. There was a great deal of interest in the process and the presentation was followed by lively discussion. The Network members had a long dinner afterwards with Herrero and the Calgary Zoo representatives. Some historic conflicts between CBSG and IUCN's Bear Specialist Group, of which Herrero had been co-chair, were revisited, addressed, and put aside. A decision was made to move toward holding a comprehensive stakeholder-based PHVA workshop on the grizzly bear. The ESGBP would be the local organizer and host, and CBSG and Network members would facilitate the workshop. Steve Herrero asked to have me serve as co-organizer of the workshop, to reflect the broader purpose, and to facilitate communication between the Calgary-based ESGBP and CBSG and the Network, as I was a closely affiliated with both communities.

In order to build stakeholder commitment to involvement in the workshop, I attended the next semiannual meeting of the ESGBP Steering Committee in Calgary in order to make a presentation on the stakeholder-based PHVA workshop objectives and process. My presentation was the last item on the agenda, after several presentations on grizzly bear research findings. The presentation was well received and seemed to offer a natural extension of a role already played by the industry and government stakeholders in the ESGBP. The group appeared to be open to the idea of PHVA workshops in part because of the strong endorsement of several leading grizzly bear scientists and in part because, as a professor of management at the University of Calgary, I had credibility with the industrial stakeholders. We had little time for discussion, so I was invited back to the next meeting to be held six months later in Canmore. At this meeting we had a general discussion and endorsement of the entire steering committee. Many of us gathered socially afterwards in a Canmore bar and talked more about what we were planning.

Next we had to ensure that representatives of all the various groups holding a stake in the grizzly bear habitat area would be at the workshop and would come prepared with appropriate information. In order to underline the multidisciplinary and multi-stakeholder nature of the workshop, the invitation letter sent out was signed by both Steve Herrero and myself—Herrero as director of the ESGBP and professor of environmental science and me as member of the CBSG Network and professor of management. It was emphasized in the letter that everyone was expected to remain and participate in the workshop and that field operations–level individuals were needed, rather than corporate head office or public relations managers. In addition, efforts were made to entice individuals to the workshop who were believed to have expertise and information relevant to the workshop discussions. Because we personally knew most of these individuals, personal appeals were made to many with such expertise.

In the end, a number of industry managers did attend and participate in the workshop. All industry managers who participated were the ones who had already been active in the ESGBP and had already sat through a number of grizzly bear research presentations. These industry representatives were also the ones locally known individually and by their companies as leaders in proactively managing environmental issues. Notable among them were representatives from Husky Oil and Shell Canada from the oil and gas sector and Spray Lakes from the forestry sector. There were also some notable absences. The Rocky Mountain ski resorts, although financial contributors to the ESGBP, did not send a representative. The expert on the Rocky Mountain tourism industry based at the University of Calgary's business school declined the invitation to attend. In a conversation in response to my personal attempt to persuade a colleague I had known for years to participate, he argued that it was his perception that the workshop would just turn into another platform for environmentalist grizzly bear enthusiasts to again "gang up" on the beleaguered tourism industry, and he wanted no part of this. Other tourism experts in the university's geography department also declined to attend, while researchers from the environmental science and the biology departments attended in impressive numbers. A professional staff member of the Alberta Energy and Utility Board (AEUB), the regulatory agency, failed to attend. This individual is a former graduate student on whose committee I served, is a member of the ESGBP Steering Committee, understands the workshop issues both from ESGBP exposure and from his own graduate research, and has volumes of relevant landscape-level permit data at his disposal. In personal conversations after the workshop, he claimed that the AEUB was under budget and staff constraints and could not allow him to attend the workshop (it took place forty minutes drive west of the AEUB Calgary office in the spartan and low-cost Rocky Mountain community facility). One cannot help but speculate that there was concern on the AEUB's part about being involved in such an uncontrollable process as the workshop.

At the workshop itself, those industry managers who did attend played a participative role. An attempt was made in the working groups to do an analysis at the landscape level of current and planned industrial activity. Some of the interesting outcomes of their participation were the occasional expressions of surprise by, for example, the forestry manager. "You want data on what we are planning in that area? That's no problem, I can get you that." Unfortunately, despite attempts by the organizers and Network members to get them to bring data, managers usually did not have detailed landscape data with them at the workshop but promised to get it to the biologists later. There was also some interesting learning that went on by the industry managers. Upon learning of a certain localized area's unusually good suitability as bear habitat there were remarks such as, "Yes, that's slated as a cut-block but that's not a problem. We can leave a wide swathe there and cut down here instead." As we experienced in some of the other workshops, the VORTEX modeling played a central role in the workshop and largely left the industry data and discussions outside of the core. If it had been easier to "plug-in" the output from the group working with industry data, it might have integrated them more centrally into the workshop process. The key appeared to be geographic landscape maps. Regardless of industry, operations managers worked in a world of maps and that's where they could connect to the biological habitat information discussed by the biologists.

The managers also harbored some mistrust of the modeling. In interviews some eighteen months later (Stevens and Benning 2000), they expressed concern that the assumptions based on limited data available in the workshop were possibly biased to make the grizzly bear population appear more vulnerable than it was. This troubled them, as this bias would unnecessarily increase the cost of environmental mitigation for their companies. One could also speculate that this perception of bias on the part of industry managers was an artifact of their not having real input into the modeling. Had they been able to relatively accurately feed their landscape-based industrial infrastructure and extraction activities maps into the VORTEX model's habitat reduction and habitat quality portals, they may have been more confident in the model's predictions. Eighteen months later they also reported that they had not used the workshop report as a policy tool because they felt the data used led to disputable scenarios on the future viability of grizzly bear populations and habitat. Again, one could speculate that had their data been better utilized in the model, the outcome with respect to using the workshop report as a policy tool might have been different.

Our final effort to bring industrial stakeholders into the workshop process was in Ecuador in a PHVA workshop that occurred after the Network project ended. The Ecuador PHVA was not planned as one of the Network initiatives, but it presented itself as an opportunity to test some of our ideas in regards to both involving industry in the process and involving Indigenous peoples. One of the frustrations of the grizzly bear exercise had been that the industry participants still felt like strangers invited to someone else's party. The core discussions at that workshop had surrounded the biological modeling into which they had little input. Those who came and stayed to participate did so because of the good relationship they already had with the ESGBP, which their organizations were helping to fund. Unlike the California Energy Company in the Philippines and Indonesia, these company representatives understood how their landscape-based activities were also important to the equation, but somehow we were unable to significantly link their activities to the equation.

In Ecuador we intended to address this head on. We were dealing with the Indigenous participation issue directly by inviting the chief Indigenous leader to meet with the organizing group and also to view the video made of the Papua New Guinea workshop, where Indigenous people were a significant element. We would do the same thing with industry. I already had good relations with Alberta Energy Company and its predecessor in Ecuador, Pacalta Resources Inc., because I had been academic chairman of the University of Calgary's Quito-based Energy and Environment Project for Latin America for five years. I proposed a CBSG facilitated conservation workshop for the Cuyabeno Reserve region of northern Amazonian Ecuador to Fundación ÑanPaz, the quasi-independent sustainable development foundation funded by Alberta Energy. Fundación ÑanPaz, since its inception several years earlier, had been quite successful in its mandate to deal with community development in the region in which Alberta Energy had its oil production operations. Environment was the other half of the foundation's mandate from Alberta Energy, and it was this half that was relatively weak, largely because the Fundación ÑanPaz executive director's personal background was in socioeconomic and community development, not the environment. Assessing species conservation and developing a stakeholder-based conservation plan for the Cuyabeno Reserve and surrounding area where Alberta Energy's operations were was seen as a way that Fundación Ñan-Paz and Alberta Energy could play a larger role in Ecuador. In a way this was not dissimilar to the situation of California Energy in the Philippines and Indonesia, except that Alberta Energy would not just be giving a check to CBSG to run a workshop but would themselves, through their foundation, be the local workshop organizers. The thinking was that if the activity were more central to the company, then they would be more motivated to play a larger part in it.

As Fundación ÑanPaz was not a conservation-oriented organization, a great deal of consultation with both me and with CBSG staff was undertaken in order to prepare for the workshop. Numerous meetings were held over more than a twelve month period, in Quito, in Calgary, and in White Oaks, Florida, with Fundación ÑanPaz staff. It had been decided that it would be most appropriate to make this initial workshop a Conservation Assessment and Management Planning (CAMP) workshop with perhaps a concurrent PHVA exercise of one critical species. This decision was based on the fact that relatively little was known about the status of the species in the Cuyabeno and the status of their habitat. It was also believed that a CAMP process might be more amenable to industry landscape-based infrastructure data. Key to making this workshop a success with industry, as well as in terms of Indigenous inclusion, would be to get the right people to the workshop (in terms of field operations managers), to get them to bring the appropriate information (in terms of maps showing their installations and plans), and to get them engaged in the workshop process in a more integrated fashion than in the grizzly bear workshop. To this end I made several appeals personally to Alberta Energy senior management in Quito and to their field operations engineers to get their commitment. I used the grizzly bear workshop, in which some of their Canadian industry colleagues had participated, as a testimonial to the importance of industry involvement and success. Various individuals assured me that both senior people from Alberta Energy's Ecuador division and field engineering people would participate in the workshop. The president of the Ecuador division of Alberta Energy also assured me that he would make every effort to persuade his counterpart at the state oil company, Petroecuador, to send his field engineering manager as well. Alberta Energy made a significant financial investment in this workshop, funding everything.

Just prior to the start of the Ecuador workshop, CBSG chair Ulysses Seal developed a serious health problem that prevented him from traveling to Ecuador. This presented two problems for the workshop. First, in Latin America, having the senior official there is critical when you are attempting to do something significantly new. It ensures that an exercise will receive the appropriate attention. Secondly, Seal was to be the modeler on the team, in order to model the one species selected for the PHVA analysis. An attempt was made to replace Seal with another modeler, but none were available on such short notice. We adjusted to Seal's absence by eliminating the PHVA modeling exercise and limiting the workshop to a CAMP exercise. Also, I made a more formal opening presentation on behalf of CBSG's Steering Committee, on which I serve, and played the role of an additional CBSG team member.

Both Alberta Energy and Petroecuador had field engineering people at the workshop. Alberta Energy also had their second in command in Ecuador at the workshop for part of the first day. As requested, they also brought field maps with drill sites, roads, pipelines, and future plans for further infrastructure indicated. Things went well for the first day as the workshop went through a needs exercise and a general discussion among the stakeholders. The second day, however, did not go so well in regards to the industry participants. As workshop participants started focusing on taxon data sheets, biologists and Indigenous people, as well as government ministry representatives, were engaged in the minutiae of species-based data to be entered. The industry representatives were left essentially with nothing to do, as they had little to contribute to species-level biological discussions. Because they had pressing matters to attend to back in the Quito office they started making plans to head back for the day. I tried to persuade them to stay, but with nothing for them to contribute to it was not an easy argument to make. I did manage to extract a promise that they would return that evening and participate in the next day's activities by telling them that we would be focusing on superimposing biological and industry maps. That evening, however, a phone message came through saying that other pressing matters had come up and the Alberta Energy operations manager would not be able to make it back for the next day. The Petroecuador manager did stay for the next day and tried to remain involved, but he was not able to contribute much because the process was not structured to accommodate his data. When he received a phone call about a small oil spill in the jungle he too was gone from the workshop for good. Looking back at the workshop, industry representatives' partial participation did help to make the biologists and the Indigenous people more aware of the oil industry's activities and willingness to work with them on conservation issues; but being underutilized in the process caused the industry representatives to leave the workshop.

Lessons Learned

Increasingly, progressive resource-based companies are motivated to become involved in a more integrated way in conservation of species and habitat. Through extensive development of personal relationships with industrial resource development companies on the part of a CBSG-linked broker, the appropriate representatives of these industrial stakeholders can be persuaded to participate in conservation workshops and bring the appropriate landscape-level industrial data.

One cannot assume that simply sending an invitation will result in industry participation. Obtaining participation requires a concerted preworkshop effort to explain the workshop objectives and process and the importance of industry involvement in attaining conservation goals. Appropriate bridging individuals are also important as brokers. These are individuals who understand and are sympathetic to the CBSG-style PHVA or CAMP workshop process and have a reasonable understanding of the scientific underpinnings of the process while also understanding and having a rapport with industry.

I personally played this role in Ecuador. But my task was helped by the fact that Fundación ÑanPaz and its executive director were organizationally already positioned as a bridge between the oil company and the environmental and local development communities. Oil company personnel were already comfortable with Fundación ÑanPaz and its activities, so participating in a workshop requested by the foundation was not a stretch.

In the case of the Alberta grizzly bear workshop, the process was made easier by the fact that the ESGBP, like Fundación ÑanPaz, was already playing a bridging role between conservation science and resource-based industries. Key scientists on the ESGBP Steering Committee are individuals who have extensive personal contact with resource industries. They are unique as scientists in that they are not remote "ivory tower" researchers. Perhaps they have gained plenty of experience interacting with industrial stakeholders because these researchers work with large carnivore species that have huge ranges in some of the most industrially developed wilderness habitat on the planet.

Even in the case of California Energy Company in the Philippines and Indonesia, which represented the least integrated involvement of an industrial stakeholder in our studies, there was someone who played the role of broker and there was a sort of bridging organization. The CBSG Steering Committee member and his organization were the trusted bridge that made it possible for the energy company to work with CBSG.

In sum, we have learned that it is possible to get appropriate industrial stakeholders to the workshop with the relevant data. It requires an understanding of what the motivations are for industrial players to be involved. It requires a protocol for involving an individual or organization playing a bridging role in order to connect the conservation and industrial communities that normally do not have natural links between them.

The second major lesson we have learned is somewhat more problematic in that we have not yet worked it out and tested our solution. This is the fact that getting industrial stakeholders to the workshop with the right information is only half of the task. The other half is getting them and keeping them engaged in the workshop process. This problem is of a similar type as obtaining participation and maintaining engagement of Indigenous people. In that case we have learned to take traditional ecological knowledge, as used by Indigenous people, and translate and incorporate it into Western science–based models. When we allow Indigenous people to speak and be listened to about their information and integrate it as we did in Papua New Guinea and in Ecuador, we have richer scientific models and we have support from Indigenous people for the conservation endeavor. We have yet to do the same with the industrial community.

, We have speculated based on our experience with resource-based industries that conservation-relevant industrial information exists in company geographic maps. If we can find a way to incorporate this map-based habitat impact data into our core simulation models (perhaps linking Geographic Information System data into the habitat size and habitat quality portals of the VORTEX model), we will have found a way to integrate industrial landscape information. Having their information heard and used in the core modeling exercise of a workshop will ensure that industrial participants stay and become engaged. It will also increase the probability that they will use the workshop outcomes for policy decision making, thereby contributing to the species and habitat conservation enterprise.

Conclusions

This chapter has argued that commercial resource industries can play a significant role in endangered species conservation. It describes how commercial forestry, mining, and oil and gas industries have historically not participated in conservation, but have instead been exploiters and despoilers of the environment with adversarial relationships with environmental nongovernmental organizations. This chapter argues that industry's role may be in flux as modern international public corporations headquartered in North America or Europe increasingly have strong market-based incentives to work with "constructive" (as opposed to adversarial activist) environmentalists in order to pursue their corporate interests of maintaining profits and market-share value. Although often these companies are not landowners in the traditional sense, in that they own only government-granted, constrained timber or mineral rights to the land, they are often the de facto landscape decision makers.

These companies initially showed their interest by agreeing to fund CBSG workshops in the areas of their industrial operations in the Philippines and Indonesia as good public relations. This initiative was expanded into more active participation in the Alberta grizzly bear workshop, through discussion and contribution to working groups, and in the Ecuador CAMP workshop, through contributing landscape-based maps of current industrial activity and future plans. The success, such as it was, resulted in all cases from personal contact with "bridging" individuals who had good prior relationships with both industry and conservationists in the local area. In all cases, getting industry to the table was a challenge but not an insurmountable one. Many forward-looking companies today have real market-based incentives to participate in stakeholder-driven conservation initiatives in the areas where they operate.

The real challenge was keeping industry representatives at the table once we had them. In both the Alberta grizzly bear case and the Ecuador CAMP case individuals from resource extraction companies who were in a position to make constructive contributions to the conservation enterprise were alienated by the process; representatives had planned to attend and participate for the full workshop duration, but when it seemed that there was no way to use their industry-based landscape data in the modeling or species inventory exercises, they left. Future efforts must find better ways for industrial landscape-based data to interact with and contribute to species-based conservation data. Only in this way will conservation risk assessment be able to integrate industrial landscape planning, and only in this way will industrial landscape planning be able to integrate conservation goals.

PART FIVE

Reflections on Consilience

❖❖ Chapter 17

On Building Bridges between Specializations

GAYL D. NESS

Working with this Network is only the most recent in a long line of experiences I have had in the general arena of building bridges between specializations. Such endeavors can go by many names and usually involve some form of interdisciplinary or multidisciplinary effort to bring different scientific disciplines together. The past century has seen the development of many of these new interdisciplinary activities. Some have generated new scientific specializations, such as molecular biology, social psychology, economic psychology, bioengineering, and most recently a whole series of interdisciplinary initiatives to deal with the new information technology. In this Network, the concern has been to bring social and biological scientists together to address problems of biodiversity, especially species and habitat protection.

But more than science is involved. There is a parallel problem in bringing together the different specializations of governance to promote economic development or to address specific social problems. Agencies of health and education often need to be linked with those of agriculture, irrigation, road building, and market development if governments are to be successful in raising human productivity and welfare. Addressing persistent poverty and racial discrimination in the more developed countries is thought to require bringing together such specializations as education, urban planning, community development and banking. Thus I prefer the term "bridge building" rather than "inter-" or "multidisciplinary" because it includes the broader activity of linking different specializations in both science- and action-oriented organization.

Recognizing the Power of Specialization

It seems to me that one of the first requirements of successful bridge building is to recognize the importance and power of specialization. In sciences, disciplinary specialization enhances our powers of theory building and our capacities for observation. Enhanced powers of observation are achieved, for example, by building better telescopes or microscopes, or by developing more precise concepts for social, economic, or psychological accounting. But these specializations necessarily place blinders on us, obstructing from view other important things.

There are two diametrically opposed tendencies emerging around the powers and limitations of specialization. The first is a resistance to broadening perspective, and here two complementary forces are at work. One is the more psychological tendency often referred to as "the law of the instrument." People who are experts in social surveys find that all problems are best addressed through doing a survey. Modelers may distrust predictions based on intuition or experience because the assumptions are not made explicit; they prefer both the elegance and the explicit assumptions that go into modeling. In addition, and perhaps more important, is the tendency to "organize" work around specializations. These organizations then take on something of a political nature, with recognized boundaries, rules of admission, patterns of communication, such as professional journals whose editors have powers to define "appropriate" activities, and professional associations. The organizational phenomena imply real powers to reward and punish, thus helping to keep a specialization more or less focused on a relatively narrow range of activities.

But there is another and opposite tendency as well. Humans are good problem solvers and they seem to delight in learning new things. Thus, addressing a problem can lead to the search for new tools for a solution; and learning new tools is often a delight in itself, largely because it opens our eyes to things we have otherwise not noticed. Moreover, there are organizations created to pursue problems and activities beyond a specific specialization. Multidisciplinary research institutes provide a good example: some of the most innovative work in the field of population-environment dynamics came out of the International Institute for Applied Systems Analysis, where much cross-disciplinary effort is accorded to solving problems (Lutz 1994).

Overcoming the limitations of specializations while retaining their powers requires extended "organizational work." This work is what our

Network has been pursuing for the past three years. Reflecting on what we have done, and how this fits with my other experiences in bridge building, leads me to make a number of informal observations. Specialization is necessary, but so is building bridges between specializations.

A Focused Task

One of the first extensive and successful experiences of bridge building I encountered was one I observed as a postdoctoral fellow in Malaysia from 1961 to 1964. This was the country's Rural Development Program, which began about 1959, shortly after the country achieved independence in 1957 (Ness 1967). The program succeeded in bringing together the different specialized agencies of government to promote extensive infrastructure construction. Deputy Prime Minister Tun Abdul Razak headed the program, investing it with immense power. At district, state, and national levels, he created "operations rooms" for rural development complete with large (2 foot by 2 foot) Red Books. Each book contained a map of the relevant area with a series of overlays on which development projects were identified: schools, water systems, mosques, roads, and irrigation projects. Representatives of all government agencies at the respective levels were to meet weekly in the operations room to draw up plans for the projects and to review their progress. One reason for the success here was that the process of physical infrastructure construction constituted a highly focused task for which the necessary specific steps were well-known.

Advantages

This has a parallel in our Network. The Population and Habitat Viability Assessment (PHVA) provides a highly focused task in which to incorporate social science perspectives and information with those of the biological sciences. Thus a highly focused task offers an advantage, facilitating bridge building between specializations.

Malaysian rural development was also successful due to political conditions that do not necessarily apply elsewhere, though it is important to specify these. The periodic face-to-face meetings of the various government agents proved a distinct advantage. With all agents in attendance, problems could be solved on the spot without waiting for the time-consuming business of moving files from one office to another. Perhaps more important, the deputy prime minister went on official tours, visiting the local operations rooms and being briefed by the staff. Where he found good progress, he could readily reward the work; he could loosen bottlenecks at the center with a phone call to the capital from the local office; he could punish lazy or recalcitrant officers for not doing their work. In these cases the rewards and punishments were based on precise and accurate information about the causes of progress and delays. Using power without accurate information often leads to demoralization of a staff fearing arbitrary action. In this case, however, the morale of the civil service was high because members knew that if they did their work, they would see immediate results and would also be rewarded.

Where power is not available, as in the work of the PHVA, other kinds of influence must be used. Although systematic evaluation is not available, it appears that both the acknowledged professional expertise that PHVA staff bring to the workshops and the workshop design to effectively manage group dynamics do provide the influence needed. Power or influence is needed to achieve effective bridge building; but that power or influence must be based on accurate information concerning both the resistance to and support for bridge building.

Disadvantages

There also may be a disadvantage to the focused task. This, too, is evident in the PHVA process. The PHVA is a well-developed and powerful process with a large and significant constituency. It can work without the social sciences and if social scientists and their necessary data are not available the process can go ahead nonetheless. This may hinder or discourage social scientists from investing themselves in the process. It may also make biological scientists impatient with the bridge-building activities, leading them to press on regardless of the lack of effective collaboration.

At this point it is difficult to say whether the net effect of the focused activity is positive or negative. My own intuitive sense is that it is positive, but this also depends in substantial part on the openness of the PHVA scientists and their own desires to include the social sciences. There is, however, another advantage to the specific linkage with the PHVA workshop that deserves separate discussion.

Validation

One of the decided advantages of our Network experiment in bridge building derives from the ongoing nature of the PHVA. This provides for quick and ready empirical testing and validation of ideas. Network discussions were often wide ranging, drawing on a great variety of current and historical examples as a basis for proposing specific lines of action. At each discussion, there were already a number of PHVA workshops in the planning stage. We could select an upcoming workshop with specific characteristics and set up a process of building social science inputs into the PHVA. Work in the field followed within a short time. By the time we had our next Network discussion there was usually evidence from the field that could be used to assess the ideas on which we planned the next specialization-linking activity.

The six workshops described in part 3 of this book provide the basis for some generalizations on what works, and why, in our attempt to bring social and biological specializations together. The Network has provided opportunities to empirically test ideas about how to build bridges between biological and social sciences.

Lack of social science data, even the basic demographic data that are usually the most available and most readily quantifiable, has been a major weakness in the workshops. Where local populations have been available (e.g., Papua New Guinea, chapter 9), they have been most useful in providing both the biological and social data needed for effective PHVA.

Resources

Here the metaphor becomes literal. Building bridges requires resources. When Malaysia began its Rural Development Program, creating district level committees with operations rooms and Red Books and demanding that the different specialized units of government work together, it gave each local committee M$50,000 to start work immediately on projects they considered important. (At the current exchange rate of M$3 to US$1, that meant only about US$17,000. But with a per capita gross national product of just under US$300, those funds were substantial.) In subsequent briefings for state and federal officials, telephone calls from powerful people could release state or federal funds to the district officers. These resources were of great importance. They gave district officers and their counterparts in all agencies the confidence that they could obtain resources for necessary projects. In effect, it showed that interagency activity could pay off.

Some years later, in 1979, I was involved in evaluating the U.S. Agency for International Development (USAID) inputs into the Indonesian Family Planning Program (BKKBN) (Ness, Heiby, and Pillsbury 1979). One of our discoveries was the "Fast Money Moving Mechanism" that

USAID and BKKBN had developed together to promote local-level initiative in family planning programming. USAID had sufficient funds to provide provinces and districts with resources to try out innovative projects. Proposals flowed into Jakarta from provincial and district offices for funds. USAID could review them and pass them on in quick order, but the Indonesian director of BKKBN showed USAID how they could assure that the funds would actually go quickly to the districts and provinces. He advised USAID to write into each contract that "these funds must be delivered in full to the district recipient within four working days." Without this legal provision, headquarters would have dolled out one-quarter of the funds and demanded full accounting before the next quarter would be dolled out. With the director's suggested legal stipulation, the funds flowed quickly to the local areas where innovative projects were generated, often bringing together very different groups, individuals, and government agencies. The result was a rapid expansion of the program, rapid increases in contraceptive use, and rapid declines in fertility.

In 1989, the University of Michigan Population Environment Dynamics Project began. A generous grant from the MacArthur Foundation enabled us to organize a biweekly evening research seminar on global change. Within a year we had a regular attendance of fifty to seventy faculty and graduate students from every discipline from atmospheric science to zoology. With the MacArthur resources we were able to offer summer grants for feasibility research projects. These required direct participation from faculty and graduate students from more than one discipline. In two years, we generated more than a score of projects, some of which developed into much larger research projects that could attract national government and foundation funding.

These are only simple examples of the basic point. Building bridges requires resources. If social scientists are to work with biologists or biologists are to work with social scientists, resources must be found to support the group being recruited. Required resources might be very modest, such as travel and maintenance funds to attend meetings. They might also be more substantial, including salaries, materials, and research costs such as demographic or biological surveys. However modest or large, resources must be found for specific bridge-building activities.

In part, the reason for the necessity of resources lies in the point made in the discussion regarding the power of specialization above. Disciplines are *organized*, like states, guilds, or unions, with powers to identify citizenship, exclude foreigners, and to reward and punish members for their work. If a building contractor wants plumbers, electricians, and carpenters to work on a house, he must pay them for their time. The same is true in building interdisciplinary bridges. Just as there is no free lunch, there is no (or little) free interdisciplinary collaboration. Interdisciplinary bridge building requires resources designated specifically for the interdisciplinary work.

Drift

Genetic drift refers to the natural process of new genetic characteristics changing over time. Thus a new strain of rice may well lose some of its valued characteristics over time and will have to be replaced with another strain. There is a parallel organizational process in bridge building between specializations. Of the many successful cases I have observed over time (e.g., Malaysia's Rural Development Program; Taiwan's JCRR; Comilla's—now Bangladesh's—then East Pakistan Academy for Rural Development), none has survived unchanged for long. Each has shown a tendency to drift from an innovative bridging activity to a somewhat more specialized activity.

The reasons lie in things organizational sociologists have observed and assessed for the better part of this century. Philip Selznick (1949) noted that as organizations move along through time they tend to become "infused with value." They begin as simple, rationally designed *instruments* to do specific kinds of work and to achieve specific kinds of goals. But in time, they become something of an end in themselves, an entity valued for itself rather than for what it does. Processes originally established to direct attention to specific problems and to seek solutions to those problems become routine activities done simply because "this is the way we do things." Nothing lasts: one should be modest in expectations and note that sustaining effective, well-traveled bridges requires constant attention to innovation.

Location Specificity Again

Chapter 13 noted the importance of location specificity in population-environment dynamics. This importance has been amply demonstrated in the workshop descriptions as well as in the other analytical chapters. We have sufficient evidence that drawing together different specializations must be done on the ground where a specific problem lies.

Social science information can be generated for the nation-state and often for lower-level units as well (e.g., states or provinces). Such information is often available from international organizations that collect and publish data. The United Nations agencies and the U.S. Department of Commerce provide excellent time-series data sets that include the kinds of demographic and social data usually needed for a PHVA. But the scale is wrong. PHVAs work on very specific settings. Often these are rather small, as in the Indonesian or Thai gibbon cases (see chapter 14) or in the small habitats of the Brazilian muriqui (chapter 7). They may also cross administrative boundaries, as in the mountain gorilla case (chapter 6). The habitat of the gorillas is a relatively small area that crosses the boundaries of Uganda, Rwanda, and the Democratic Republic of the Congo. Although there are aggregate data on population, wealth, and welfare for each country, they cannot be used to tell us what is happening in the gorilla habitat. In the case of the Indian rhino (chapter 14), neither aggregate Indian nor state (West Bengal) data would show the migration streams that were important for Jaldapara. Overall Indian national or state migration streams are small, but those streams, small at a state level, overwhelmed Jaldapara and threatened the protected area. If social science data are to be included in PHVAs, these data must pertain to the site of the PHVA. Location specificity is the rule.

Communities or Communities of Interest

One of my earliest experiences in social science work came as I examined the growth of the cooperative movement, based on the famous Rochdale cooperatives of mid-nineteenth century England. I examined the growth of cooperatives in the United States, especially in the north central states and around the San Francisco Bay Area. Then I spent a year in Denmark reading and interviewing to understand how these organizations emerged in Scandinavia at the end of the nineteenth century and how they fared in the modern world. Later, in Malaysia, I studied the (quite unsuccessful) history of British colonial attempts to foster the growth of cooperatives (Ness 1961). I also discovered the largely unsuccessful efforts at "Community Development" in both British and independent India. Finally, I was made aware of the highly successful efforts of the (then) East Pakistan (now Bangladesh) Academy for Rural Development, in Comilla.

In all cases, successes in these new organizations lay in working with distinct communities of interest: small Danish or Bengali farmers, Scandinavian urban workers, or the new "skilled factory workers" in the English textile mills. Unsuccessful efforts were those where an outside power tried to tie together an entire village community. In some Indian villages this resulted in building latrines or fences, the easiest thing for the villagers to do in order to get the outside government off their backs. In other villages Community Development was a way for the wealthy villagers to get the poor to work for them without compensation; projects in these cases were usually roads or facilities that benefited the rich.

One of the apparent motivations behind our Network activities is to make the PHVA process more successful in actually promoting the protection and survival of a given species. It is recognized that success depends on bringing to the table the full range of stakeholders whose activities affect species survival. Gathering all the stakeholders together is essentially an attempt to build a community of interest around species survival. Working only with the professional biological conservationists leaves out important elements of the communities that affect survival. But if different stakeholders are to be brought to the table, there must also be social scientists who have experience dealing directly with these other interests (e.g., anthropologists and sociologists for Indigenous peoples; management and economic scientists to deal with industrial and commercial interests). What is required is bringing together people of diverse interests to build a new community of interest. Diverse scientific perspectives are needed to build new communities of interest around the issues of species survival.

A Bottom Line?

Is there a bottom line in this experiment? If so, I suspect it resides somewhere in the use of interdisciplinary activity to solve problems. In this Network we have been concerned with one fundamental problem: can we find ways to manage a local environment so as to save a specific species from extinction? The question we face is whether or not interdisciplinary activity is necessary to solve the problem of species survival. From the Network experiment, we can only say that we are not sure. On the one hand we believe that human activities—population growth, economic development, land usage, patterns of production and consumption, and national and international conservation policies, among other things— greatly affect location specific attempts to protect a habitat and a species in that habitat. But this remains something of a belief. I know of no systematic assessment of conservation activities that can attribute their success or failure to the linking of biological and social sciences. The PHVA workshop process by itself has shown many successes in producing effective management plans for species conservation. (It is important to note, however, that to date [2000] the PHVA workshop process has not yet been subjected to systematic evaluation. Although it has had scores of activities, in scores of countries, dealing with many different species and habitats, its outcomes have yet to be evaluated. We do not know how many or which of the workshops has been followed by species protection or extinction, nor are we able to connect protection or extinction to any specific conditions.)

It is possible, however, that our belief of success is ill founded. It may well be that the PHVA workshop process in itself, without any social science input, is quite capable of solving the problem of species protection. It may be that the additional cost of attempting to build bridges between the social and biological sciences will show little or no advantage in solving the problem of species protection.

In effect, the Network experiment has not been designed as a critical experiment, whose results would answer the bottom-line question of utility or value added. My sense is that after these few years of working together, we really do not know whether or not collaboration has been useful. We have certainly enjoyed the professional and personal exchanges, and I should imagine that we have all individually learned a great deal. Whether or not that has directly promoted the business of species conservation, I do not know, and I do not think our experiment allows us to answer this question. It remains, however, a basic question that should be asked of all attempts to build bridges between specializations.

I suspect that many (most or all?) of the Network participants will gain advantages. Certainly individual academic careers will be enhanced by the publications that will inevitably ensue. It may also be that the fundraising efforts of the PHVA process will be enhanced, though these seem now so well developed and so effective that it is difficult to see how they could be enhanced by being able to claim greater social science capacities for the process.

It does seem to me, however, that the Network experiment has been a positive one. It may also have far greater symbiotic effects than we can now trace. Humans delight in solving problems and also in learning things. It is always difficult to find the direct connection between new learning, new problem solving, and any specific major breakthrough in knowledge and action. The point here is that we simply do not know. But that knowledge constitutes what I would call the bottom line. Building bridges between social and biological sciences for species conservation will only be useful if it contributes to species conservation. That is the bottom line.

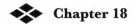

Chapter 18

Metamodels as a Tool for Risk Assessment

PHILIP S. MILLER AND ROBERT C. LACY

Models can be very powerful tools for integrating knowledge across scientific disciplines. However, they can also paradoxically serve as barriers to understanding when the models represent specialized knowledge of only one or two specialties. As the Network has evolved, we have come to realize that many of our early struggles with transdisciplinary activity revolved around the difficulties of getting disparate models to effectively communicate. Within a given discipline, the method with which we represent our knowledge of a system defines our model of the problem or issue at hand. In order to incorporate knowledge from another discipline into our understanding, we must find ways to get that model to talk to our own discipline's model.

A "model" is any representation of a part of our world (Britt 1997). Because a model is, by definition, an abstract depiction that uses various codes to which we have assigned specific qualitative or quantitative meaning, models that are developed independently for different purposes will usually be incompatible until someone develops a common dictionary that translates concepts and terminologies between the individual approaches. The problem of communicating between models is even more complex when the concepts being represented are themselves not overlapping. In that case, the concepts of the individual component models must be linked through the use of translational models (figure 18-1).

The term "modeler" is a label often applied to technical specialists, as though there were a cohesive field of modeling that operates largely independently of the means through which the data that help to construct the model are gathered. In reality, however, there are many types of models, and different modelers may have very different uses for these

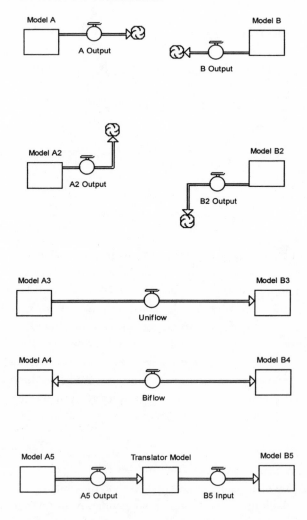

Figure 18-1. Representations, using the STELLA systems modeling package, of possible connections between models of two interacting systems. In the first case, Model A and Model B are producing outputs that are expressed in different terms and therefore have not been translated into useful joint information. In the second case, the outputs of the two models have been transformed in an attempt to reach common expressions, but the new outputs have instead diverged further from a common ground. In the third case, the output of Model A has been successfully transformed to provide input that can be used by Model B. In the fourth case, the output from each model has been altered to provide input into the other. In the last case, a Translator is used that models the connections between the outputs produced by Model A and the inputs required by Model B.

constructs (Morrison and Morgan 1999). For example, models can be qualitative and heuristic diagrams or maps of the connections among entities, ideas, or processes. In this context, they can serve as very useful tools to organize and describe our thinking about a topic. Importantly, they then act as means of expressing ourselves to others and can be indispensable in this role even if they are informal and incomplete. Additionally, models can help us extract fundamental patterns or principles about our world. In this context, they encapsulate the specialized knowledge and framework of a discipline, are used to develop a deeper understanding of that specialty, and are considered best when they are elegant (having few components), general, and formal. Models can also be used to make detailed, quantitative predictions or projections. In this context, they are used to convert a series of input data through an engine that describes relationships and into an output that makes a forecast. Such models are best when they are sufficiently detailed and complete to yield accurate and precise results.

We found that considerable frustration occurs when people use multiple models (or sometimes even the same model!) for these very different purposes. For example, a heuristic model may enlighten one person about connections to be explored, while providing no useful tool for someone else needing instead to forecast a likely future trajectory. On the other hand, a highly specific and detailed model may provide desired forecasts, while obscuring general properties of the system. We have had the experience of being told by a colleague using a model to develop generalized theories that we should not mess up his elegant model by forcing him to incorporate data from the real world. We wanted a case-specific prediction, while the theorist wanted to find the general properties that emerged from his model-driven world. The VORTEX population viability assessment (PVA) model we describe in chapter 3 is a highly detailed model of wildlife population dynamics designed primarily to explore numerical outcomes of individual scenarios and to compare the characteristics of alternative scenarios developed as part of a larger analysis of model sensitivity. VORTEX does not readily reveal general truths about the processes it represents, and it does not routinely provide a good mechanism for describing and using transdisciplinary linkages required to achieve more effective biodiversity conservation. For those purposes, other models and tools are necessary.

Even when models are employed for diverse purposes, a number of common features come into play to help us express and explore our understanding of a given system. Perhaps most importantly, models force us to explicitly state our assumptions about the general characteristics of a system at the very least and, if appropriate, the detailed mechanisms that govern its function. In a model we specify the system's components, connections, inputs, and outputs. Moreover, models often provide an invaluable visualization of our understanding of the system. Through this visual representation, models can provide a structure for combining system components suggested by various collaborators or, perhaps, by the data itself. In this way, models can help us to assess complex systems using methodologies that would be impossible in an unstructured analysis. Finally, by providing the framework onto which other disciplinary experts can contribute their specific knowledge, models developed collaboratively can be extremely powerful tools to bridge disciplines. It is in this way that models most effectively act as true transdisciplinary mediators (Morgan and Morrison 1999).

A Chronology of the Network Metamodeling Experience

In mid-1998, the Network began developing general heuristic models of the process by which disparate social and biological data types could be translated into input within a wildlife population risk assessment framework. These first attempts were very simple, largely verbal in nature, and focused on the required physical communication linkages between experts in diverse domains such as human population demography, anthropology, agricultural economics, geography, wildlife biology, and ecological risk assessment. An example of these early diagrammatic explorations of the process is described in some detail in chapter 3.

Once this general protocol was in place, we were able to generate more graphical models of the types of linkages between human and wildlife population processes. Figure 18-2 shows a relatively simple system in which data on human demographic processes—including survival, reproductive rate, and migration (dispersal) patterns—can be used collectively to describe the growth dynamics of a human population system. Moreover, relatively simple descriptors such as sanitation regimes, socioeconomic status, and health protocols can serve as modifiers to these central demographic processes. When these basic variables and their modifiers are identified and quantified, one can generate an expected future population growth rate. Armed with knowledge of the general biological and social structure of a given human population, one can assess the current status of household size and number and, by extension, the nature and extent of land demands as a function of socioeconomic condition and dispersal patterns across the landscape. Using the types of translational processes mentioned in the previous paragraph and in chapter 3, these activities can finally be translated in terms of their more direct impacts on wildlife populations or more indirect impacts on wildlife habitat. More importantly, however, we can use the aforementioned human population demographic data to develop projections of how human population size is expected to change over time and, by extension, how those populations will change the nature and extent to which they interact with the landscape. These projections can then be used to gain insight into how local habitats will change in character over time, which can be directly applied to an assessment of future wildlife population persistence.

While appearing rather complex at first glance, figure 18-2 takes a very narrow view of the ways in which humans and their activities on the landscape impact the persistence of wildlife populations. Many broader social processes are not accounted for in this figure. For example, how does a nation's governance structure affect household economic opportunities, the rate of urbanization, or the expansion of technological capacity? In what ways do random fluctuations or long-term deterministic trends in nonlocal economic markets affect a region's agricultural or industrial

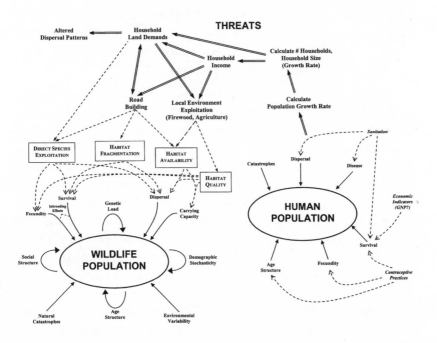

Figure 18-2. Diagrammatic model of human population growth dynamics, generalized land-use patterns, and their points of impact on wildlife population biology.

capability? How do these factors in turn influence social and political instability, ultimately leading to a major change in natural resource utilization patterns, or perhaps an increased likelihood of major cultural conflict? Questions such as these are of fundamental concern if we want to improve our ability to understand wildlife population risk assessment in the context of the future behavior of expanding human populations.

Towards that end, we expanded our thinking beyond the simpler example shown in figure 18-2 in an attempt to develop a more global picture of these complex relationships. This larger view is shown in figure 18-3 (note that this is essentially a graphical depiction of the information presented in box 3-1 and associated text in chapter 3). While certainly not exhaustive in its complexity, we think this heuristic diagrammatic tool serves as a very useful vehicle for three major reasons:

1. It can help to identify those broad human population processes and activities that are primarily responsible for the endangerment of specific wildlife populations under study.
2. It can graphically illustrate the nature of the relationships between humans and wildlife.
3. By illustrating these relationships in a relatively simple form, it can be used in specific instances as a tool to engage relevant stakeholder groups and demonstrate the utility of their information and expertise to the wildlife population risk assessment process.

This figure, commonly referred to within our Network as the "mega-bubble diagram," has become our central heuristic model of the ways in which human and wildlife populations interact and, all too often, come into conflict. However, despite our recognition of its value for defining broad relationships, we quickly understood that the diagram would always retain a high level of abstraction and, therefore, would be of comparatively little use in helping us to quantify the nature and magnitude of specific mechanisms by which wildlife populations are placed at risk. Each individual analysis employing the expanded process would be required, in effect, to generate its own detailed bubble diagram of specific human processes and subsequent threats as in figure 18-2. The abbreviated case studies discussed below (and presented in much greater detail in other chapters in this book) attempt to illustrate our efforts at making these detailed linkages and quantifying them within our own population viability modeling environment, VORTEX (see chapter 3). In particular, we emphasize the dependence of successful model-based integration on the existence and availability of both human and wildlife population data.

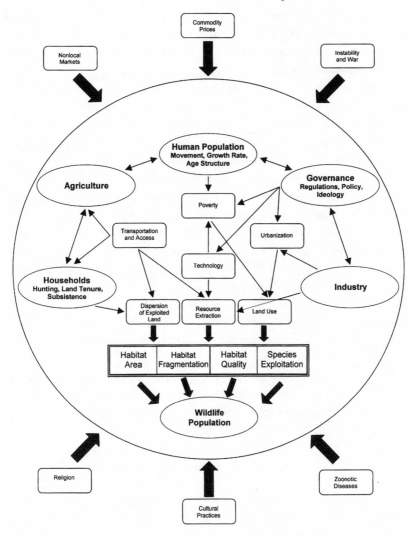

Figure 18-3. Heuristic model of human population processes and activities and the pathways through which they could link into traditional models of wildlife population viability.

Mountain Gorilla PHVA

As discussed in chapter 6, substantial field data on the demography and ecology of the mountain gorilla (e.g., Weber and Vedder 1983; Watts 1991; Robbins 1995) were available for detailed analysis and basic model construction at the workshop (Werikhe et al. 1998). However, based on recent historical events reviewed in Prunier (1997), the frequent outbreak of severe political and civil unrest in eastern Africa is a major concern for mountain gorilla conservation. This is especially true for the population of mountain gorillas inhabiting the Virunga volcanoes region straddling the borders of Uganda, Rwanda, and the Democratic Republic of the Congo.

Population and Habitat Viability Assessment (PHVA) workshop participants were encouraged to use some type of variant of the protocol outlined in figure 18-2 to discuss the ways in which regional war would impact mountain gorillas. Effects of such an event are thought to include the potential for large-scale loss of suitable gorilla habitat, an increase in the direct take of gorillas through poaching or accidental shooting or shelling, and decreased reproductive output resulting from the considerable stress put on affected gorilla groups. It is important to note that, while only a small percentage (about 3–5 percent) of available mountain gorilla habitat in the Virunga region was known to be directly destroyed by people fleeing the 1994 Rwandan civil war, it is possible (and perhaps even likely) that a larger proportion of the total gorilla habitat would be rendered unavailable due to the proximity of gorillas to large concentrations of refugees.

Based on these discussions, the participants put together a set of additional VORTEX scenarios in an attempt to simulate armed conflict and its potential for impact on mountain gorilla population viability. In all scenarios, we assumed that such a conflict would last for about ten years and a war in the area would break out about every thirty years. Four primary war types and scenarios were envisioned (illustrated in figure 18-4):

Scenario 1: During the war, the average proportion of adult female gorillas successfully breeding would be reduced by 10 percent from the baseline peace-time value. In other words, where we assumed that 31.3 percent of females bred in a year without war (based directly on field data collected over a twenty-year period), only 28.2 percent would successfully breed under conditions of war. Moreover, an additional 5 percent annual mortality was imposed on infants and adults (both sexes) during war years. When a given war event was over, breeding and mortality rates would return to their baseline values. Finally, each war event would include a cumulative and permanent 25 percent reduction in habitat carrying capacity. This reduction would be direct, in the form of targeted habitat destruction resulting from collection of forest materials for cooking and heating, and indirect, in the form of reduced gorilla habitat availability resulting from closer proximity to the newly displaced refugees.

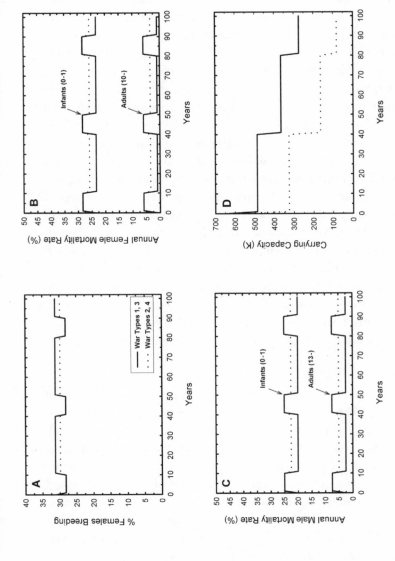

Figure 18-4. Simulated effects of war in the Virunga volcanoes region, eastern Africa, on the local mountain gorilla population. Specific demographic or environmental variables affected are (A) proportion of females breeding in a given year; (B) annual female mortality; (C) annual male mortality; and (D) Virunga habitat carrying capacity. (Source: Werikhe et al. 1998.)

Scenario 2: This scenario is similar to the previous one except that, following a given war event, the simulated reductions in breeding and mortality rates are not eliminated but rather show only a 50 percent recovery to the baseline values. This was intended to simulate a more perilous set of ecological conditions facing gorillas during the periods between wars.

Scenario 3: This situation is identical to scenario 1 but with a 50 percent reduction in carrying capacity during and following each war event.

Scenario 4: This situation is identical to scenario 2 but with a 50 percent reduction in carrying capacity during and following each war event.

It is important to realize that most of the numerical details summarized in these war scenarios are based on best guesstimates on the part of the workshop participants involved in model development. These estimates are necessary because the data needed to precisely quantify the social and ecological characteristics of a major civil uprising in this region, and its demographic impact on local mountain gorilla populations, simply do not exist. As a result, our projections of the impact of war on mountain gorilla population persistence did not reach the level of sophistication to which we originally aspired. Nevertheless, to our knowledge this is one of the first attempts at directly quantifying the expected demographic and ecological impacts of specific human activities on wildlife populations in the context of population viability analysis (chapter 3 discusses model results). Although the results may be highly speculative at this time, it would be even more fanciful to presume that wars in this region will not have impacts on the mountain gorilla populations.

Tree Kangaroo PHVA

More than 95 percent of the land in Papua New Guinea (PNG), and wildlife living there, are privately owned by the people of the country. Consequently, our conservation assessment workshop (Bonaccorso et al. 1999) would succeed only if we had strong participation by local landowners (see chapter 9). Moreover, nearly all of the relevant quantitative information on species and habitat utilization was not published but existed only in verbal form within the local communities. The Network participants realized, therefore, that a concerted interview process involving translation of information (as in figure 18-2) would be necessary for a successful synthesis and improved population risk assessment.

Researchers currently studying PNG tree kangaroos identify hunting by local villagers for the food and pet trade as a primary threat to the future viability of the group of species as a whole. In order to compile the information necessary to evaluate the impact of this hunting threat, the Network put together a team of human demographers and social scientists to conduct detailed interviews with nearly a dozen landowners from across the country (see chapters 9 and 14 for a more detailed discussion of these methodologies). The interview team attempted to estimate the annual total rate of animal extraction for a given tree kangaroo population in a particular area, given information on human population densities and hunting practices. This rate of removal could then be added to the estimated baseline mortality to assess the direct impacts of human hunting on tree kangaroo population viability.

The models developed from this information demonstrated that current hunting rates can indeed be a major force influencing the future viability of tree kangaroo populations in PNG, particularly when those populations are quite small. In addition, the workshop participants were able to observe the serious demographic consequences of preferential hunting of adult females over males in this polygynous species, an unexpected result for those engaged in hunting throughout the country. Despite our interest in developing increasingly detailed models of wildlife population viability, it is important to remember that neither sophisticated demographic models nor comprehensive data sets from the field are always necessary to generate insights that can be vital to the conservation decision-making process.

At the same time, it is worthwhile to point out that two major pieces of data were missing that prevented us from using the integrative process to its fullest potential. We made an attempt before the workshop to engage a geographer at the local university in Lae, PNG, who had mapped human population distribution and density across the country. Our hope was that this information would allow us to estimate the total human population size in and around species-specific tree kangaroo habitat. If this estimate were to be combined with an estimate of the total number of tree kangaroos hunted per village annually, we would have an estimate of the total number of animals extracted per year. In addition, estimates from field researchers on the total number of tree kangaroos inhabiting individual forest patches would allow us to translate the raw number of animals killed by hunters into a mortality rate—the form of demographic data required for input into a PVA model like VORTEX.

Unfortunately, we were unable to effectively distill the raw data on human population distribution down to the required total population size during the workshop. As a result, despite the fact that we had valuable quantitative data from the perspective of the activities of human populations, we could not come up with complete estimates of the number of tree kangaroos harvested per year. We were therefore compelled to put forward our best guesstimates of local human population size in order to derive estimates of annual tree kangaroo harvest. To further complicate the situation, we could not effectively merge these estimates with similar data from the wildlife population perspective because the latter data type was largely unknown. Tree kangaroos are elusive and hard to spot in the dense jungles of New Guinea (at least for scientific researchers), making the study of their population dynamics difficult indeed. As a result, we had virtually no information on the numbers of tree kangaroos inhabiting a particular patch of forest. Therefore, even if we had precise figures on the total numbers of tree kangaroos harvested per year, we could not effectively translate that number into an estimated rate of population mortality because we lacked data on total tree kangaroo population size. We were once again forced to make general assumptions about population size, providing us with a more general means to move forward in our calculations but with significantly less precision than we had originally hoped.

This example provided the Network with a clear illustration of the true complexity involved in integrating diverse sets of data into a PVA context. We had many elements of a successful integration process in place, but a few critical pieces of data were missing that stifled the overall synthesis. While some of these missing elements could be pulled together at future workshops through more effective preworkshop preparation, others may be missing simply because of the difficulty in field method collection and cannot be obtained simply through a more detailed literature search.

Grizzly Bear PHVA

The focus of this PHVA workshop (see chapter 10) was to develop broad plans for the conservation of grizzly bears in the central Rocky Mountain ecosystem of Canada, including the area in and around Banff National Park (Herrero, Miller, and Seal 2000). Demographic projections of bear population dynamics were initially made on the assumption that future estimates of bear mortality would persist at about the rate observed in recent years. This assumption of rather constant long-term demographic rates is the underlying basis characterizing most population viability analyses. Yet it is unlikely that mortality of grizzlies will remain constant into the future. Virtually all of the recorded bear deaths in the region are human caused; spatial analysis of the locations of dead bears revealed that nearly all were within a few hundred meters of roads or hiking trails

(McClellan et al. 1999). As the human population in this economically prosperous region is expected to increase by as much as 4 percent per year over at least the next decade, the frequency of human–bear encounters is likewise anticipated to increase alarmingly. Therefore, the Network had an opportunity and obligation to integrate information on human population growth into projections of grizzly bear population viability as humans expand into more and more area.

Insight into the utility of data integration within the PVA context was provided most effectively in this workshop by the addition of complex spatial models of human-induced alterations of grizzly bear habitat. Extensive Geographic Information System (GIS) data exist for the Central Rockies ecosystem, including not only standard vegetation classifications but also spatial distributions of human land-use patterns such as mean road density. Since data on bear demographics and habitat use have been collected over the past twenty-five years (Gibeau and Herrero 1999), a significant opportunity existed for a productive synthesis of these two sets of information. However, to date, these data had not been synthesized in the context of a quantitative grizzly bear population risk assessment.

During the course of the PHVA workshop, a group of participants began their analysis by constructing a descriptive model of the nature and consequences of human–grizzly bear conflict in the Eastern Slopes region (figure 18-5). The team recognized the complex nature of interactions between factors. For example, closure of roads to reduce the frequency of human–bear contact may lead to increased antagonism in certain sectors of the human population, thereby leading to an increased lethality of contact. With this descriptive model in place, the group then developed an algorithm for using GIS and animal telemetry data to predict grizzly bear mortality risk as a function of selected map variables. Logistic regression techniques similar to those described in Mace et al. (1999) were used to derive an expression for the relative probability that an animal would die in a specific area defined by a collection of map features, namely elevation, political jurisdiction, and mean road density. Figure 18-6 depicts relative bear mortality as a function of elevation and political jurisdiction. Preliminary analyses indicate that the highest probability of grizzly bear mortality in British Columbia and Alberta occurs at lower elevations; and overall mortality is greatest at low elevations and high mean road density. While perhaps intuitive, these results nevertheless demonstrate the power of combining GIS and demographic data to derive functional relationships between the spatial characteristics of a given habitat, use of the habitat by humans, and the population dynamics of wildlife

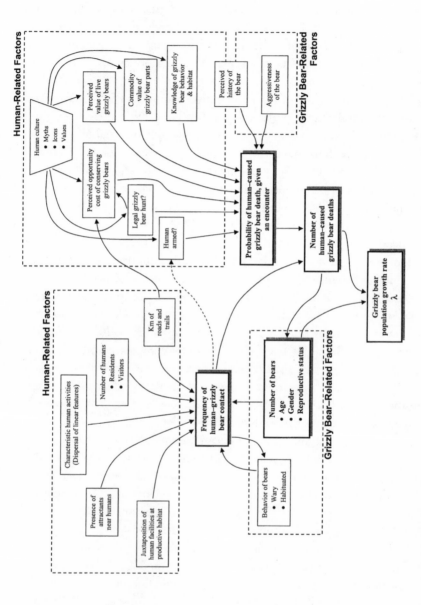

Figure 18-5. Descriptive model of the nature and consequences of human–grizzly bear conflict in the Canadian Eastern Slopes region. (Source: Herrero, Miller, and Seal 2000.)

using that habitat. To date, many in the Network feel that this has been our most successful attempt at quantitative synthesis of data describing the interactive dynamics of both human and wildlife populations.

We can perhaps extend this technique to more precisely evaluate future trends in human activities on the landscape and their impacts on wildlife. Given predictive data on how the landscape might change over time—primarily due to human use—one can easily incorporate future changes into the logistic model and the GIS maps. For example, if we estimate that road density will increase by 20 percent in the next decade, this process can identify where that will most likely occur, the types of habitat it will influence, and consequently how the risk of grizzly bear mortality will change through time and space. This function can then be entered into a population viability model to more realistically simulate metapopulation dynamics in a landscape increasingly modified by human disturbance.

As the Network experiment has evolved, we have met with greater levels of success in our search to develop and apply a variety of tools to integrate information. We have a long way to go, however, and our efforts to date represent only a first step in the right direction.

Where Are We Now and Where Are We Heading?

In our quest to improve the means by which discipline-specific models can effectively communicate, we did not make much progress when we tried to start with detailed technical solutions (i.e., making direct links between parameters in one model and those of another). For relatively simple cases, such as using a human demographic projection to predict a proportional change in the human-induced mortality of grizzly bears, the communication among models could have been encoded directly into the model structure. However, doing so would have forced cumbersome additional complexity into a merged model without any obvious gain of understanding. Indeed, the combined model may have obscured the relationships.

As described above, we are now working with an approach that we feel will be a more useful way to gain understanding across disciplines. We begin not with a precise, complete, predictive model, but instead with a general heuristic model or map of the component subsystems we wish to consider (such as figure 18-2). At this stage of an analysis, we are working to understand which models need to be considered and what information needs to flow between them. Generalized modeling tools such as STELLA (High Performance Systems, Inc; *www.hps-inc.com*) or VENSIM (Ventana

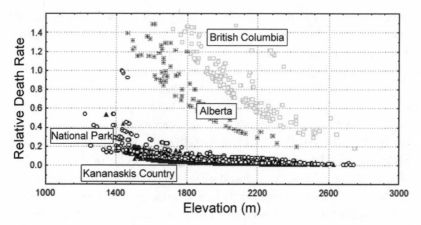

Figure 18-6. Preliminary analysis of relative mortality rate among grizzly bears of the Canadian Eastern Slopes region as a function of landscape elevation and political jurisdiction. "National Park" and "Kananaskis County" data are collected within protected areas, while "British Columbia" and "Alberta" are outside such protection. Analytical methodology outlined in Mace et al. (1999) and Herrero, Miller, and Seal (2000).

Systems Inc; *www.vensim.com*) can greatly facilitate the graphical construction of these model linkages.

After the landscape of talking models has been mapped, the next stage is to identify the specific information flows that are needed. In some cases, as when human population growth was translated directly into animal mortality, the task of entering results from one model system into the receiving model is relatively simple. In other cases, as when we converted patterns of war into patterns of refugee flow and then into patterns of disease outbreak among animal populations, intervening models are needed to complete the relationships that connect the distant original models. A new Network member is currently developing a sophisticated metamodel of measles introduction from human trackers into a simulated mountain gorilla population (Hungerford et al. 2003). The model includes information on measles epidemiology in humans and gorillas, tracker behavior with respect to the dynamics of contact with gorillas, effects of human public health programs on the spread of the disease among trackers, and detailed information on gorilla vaccination protocols (figure 18-7). Only after we have reached this level of specificity and understanding of the components and connections among the systems can we begin to undertake the detailed technical job of mathematically connecting the models.

We have, to date, completed this process of making models communicate in some rather simple cases. Specifically, several Network researchers have created an epidemiological modeling program to simulate infectious disease in wildlife populations and, ultimately, the role that disease can play in influencing population viability (Lacy et al. forthcoming). Currently titled OUTBREAK, the program will soon be one component of a metamodel through its linkage to VORTEX. This linkage can be

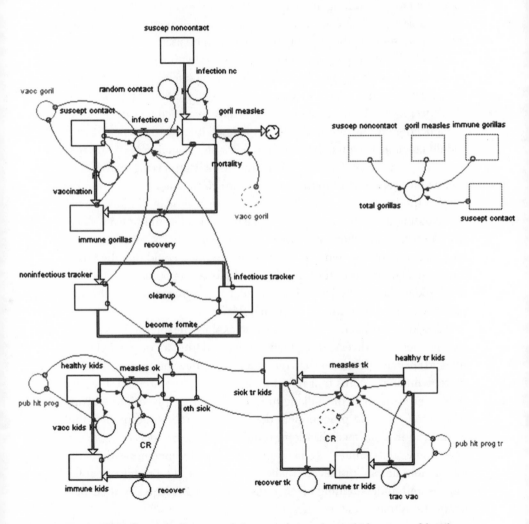

Figure 18-7. STELLA diagram of the complex and coupled system of local human trackers, isolated mountain gorilla populations, measles epidemiology, and transmission dynamics. (Source: Hungerford et al. forthcoming.)

accomplished through an "open-data" protocol, where specification of input for each component submodel and its descriptors of the state of the system are stored in a form that is accessible to other programs. Other computer-based methodologies are then employed to specify and operationalize all dynamic interactions among the models. In our linked disease–PVA metamodel, a VORTEX user will call on OUTBREAK to simulate the dynamics and impacts of disease within each year of the VORTEX simulation. Working in the alternative interface, an OUTBREAK user will be able to ask VORTEX to provide the model of population changes. The data describing the state of the system can be used or changed from either side. For example, susceptibility to disease (within OUTBREAK) can be a function of the genes of individuals, mating interactions, or population density (each modeled in VORTEX), while mortality or reproduction (in VORTEX) can be functions of disease state (modeled by OUTBREAK).

Conceptually, however, this approach of progressive specification should facilitate even much more complex systems analyses. We are now exploring the use of GIS as a mechanism for transferring spatial data among models—at least in those cases in which the information from the various models are all spatially structured. In a similar fashion, we are investigating the creation of an interface that will allow multiple instances of VORTEX to simultaneously model interacting species, as in predator-prey or competitor systems. Individual metamodels could then be linked to create a larger entity such as an individual-based population epidemiology metamodel linked to a spatial model of landscape change through GIS and a rule-based model of animal dispersal.

Thus, our approach to using models to explore the interactions among the knowledge of diverse disciplines has evolved from an original plan for merged megamodels to our current concept of developing metamodels, thereby linking systems that retain their original structure and integrity. A simpler framework of "talking models" has now evolved into "talking modelers," with the practitioners defining the larger-scale models that connect individual specializations, rather than supermodels being developed that would define—as emergent properties of their complexity—the interactions among displaced smaller-scale models. The result is that the models and understanding of each discipline are enriched by their new connections to other systems.

We are continuing to explore this path—the Network described throughout this book has been recently expanded by the addition of many more collaborators—while focusing on a more specific subset of biodiversity problems (i.e., wildlife disease, impacts of roads and industrial infrastructure on the landscape, localized pollution and harvest of wildlife for local consumption). Interestingly, the Network itself has become a "Meta-Network," funded through the U.S. National Science Foundation's Biocomplexity Initiative. Our primary goal in this expanding research project is to improve science-based capabilities for decision making by scientists, managers, and policy makers addressing environmental problems. To accomplish this, our diverse team of collaborators will improve our capacity to work as effective interdisciplinary teams, create modeling and process tools to facilitate the study of biocomplexity, apply these approaches to study complex interactions among coupled human and natural systems, build educational tools that bring to students these interdisciplinary approaches, and develop decision-making tools and methods based on these processes.

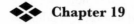 **Chapter 19**

Far from Land: Further Explorations in Consilience

FRANCES R. WESTLEY, PHILIP S. MILLER,
AND ROBERT C. LACY

I would not give a fig for the simplicity this side of complexity,
but I would give my life for the simplicity on the other side of complexity.

Oliver Wendell Holmes

The research in this book was designed to allow the scientists in the Network access to an important, ongoing set of collaborations around a critical conservation issue, that of saving endangered species. It was also designed to allow us to test out the current theories about transdisciplinary research and action. As a group we learned a great deal, some of it through our experiments with the Population and Habitat Viability Assessment (PHVA) process and some of it through our attempts to work with each other.

While it is widely recognized among social scientists that interorganizational and interdisciplinary collaboration is essential for resolving complex "domain" problems (Trist 1983; MacNeill, Winsemius, and Yakushiji 1991), we began this project in the understanding that there was much still to be learned about the limits and possibilities of successful collaboration. We were struck early by the challenges of trying to experiment in real time, with workshops that were intended to result in policy and action around particular species. One of the clearest of all messages from the accumulated research on successful management of endangered spaces and species is to "involve all stakeholders and involve them early" (Yaffee et al. 1996, 35). However, if a wider variety of stakeholders, such as local landowners or Indigenous peoples, grassroots organizations, and industrial actors, comes to the table, how do convenors keep the process from becoming so conflict ridden or political that the chances for positive action are reduced? And how can a process that is designed for scientific accuracy and risk evaluation maintain that validity while continuing to add new and different kinds of conceptual structures?

The creation of an interdisciplinary research team represented an important step in that direction by creating an intensive exchange between social and natural scientists around such tools and processes. The group became something of a lab for exploring the limits of interdisciplinary research and stakeholder involvement. Tensions were experienced between the social and natural scientists as well as between those with more action- versus more research-oriented perspectives. Some of these tensions were resolved by conceptual breakthroughs but others remained unresolved.

In this chapter, we summarize the learnings of this project, both from the point of view of collaborative interdisciplinary process and of modeling or data integration. These learnings are drawn both from our participation in the PHVA processes and in our five Network meetings over the three-year period. We conclude by discussing future directions for experiments on consilience.

Process Learnings

Early on, the majority of the Network's time was spent on the question of how to get the right people and information to the PHVA. An important tool developed by the Network, our bubble diagram (chapter 18, figure 18-3), was constructed to help identify and incorporate the necessary people and information into the workshop process. The idea was to better define the connections and relationships between human populations and behavior, household economics and resource use, industry, and government and wildlife populations and then to use the diagram to help different disciplinary and stakeholder groups to identify their potential contribution to the deliberations.

We initially thought that the Conservation Breeding Specialist Group (CBSG) and the Network should be responsible for getting the right people from these various domains to the table, but as evidenced in this books' case studies, this proved extremely difficult. For example, at the mountain gorilla PHVA in Uganda (chapter 6), we had plenty of biological information about the species and we had human demographic data for Uganda, but we were unsuccessful at getting local information and local stakeholders into the process. The Network experiment at the tree kangaroo workshop in Papua New Guinea (PNG) was far more successful in this regard (chapter 9). We had village-level human population information and local landowners participating, but their presence was coincident to, rather than as a result of, our efforts. The workshop organizer in PNG had a long-term relationship with several nongovernmental organizations working there and it was through them that the local people and the localized information were brought to the workshop.

Even more problematic was that we did not have the necessary local contacts and networks established nor did we understand how to frame the workshops in a way to make them relevant to social scientists. We wrestled with how to convey to social scientists the value of their involvement. We learned fairly quickly that the species focus of the PHVA was potentially alienating, even to those in our immediate Network. We were faced with having to define the problem differently if we hoped to get more social science involvement, but this had implications for and threats to the integrity of the CBSG workshop process. Risks included the trade-off between inclusion and precision and the fact that CBSG had been established with a particular orientation toward species-based conservation activities. Additionally, biological scientists—the most established network in the CBSG system—were at times resistant to premature inclusion of additional stakeholders if the scientists did not feel their own data was in order (see chapter 7); and biological scientists may be inherently suspicious of social scientists (i.e., natural scientists think social scientists are there to study them; see chapter 6).

Over time the challenge of including social scientists in the workshops came to be identified by Network members as the issue of "problem definition" and can perhaps be seen as a key dynamic in the entry stage of collaborations. Who defines the nature of the problem, the scale of analysis (genetic, landscape, ecosystem), and the level of complexity (deterministic, stochastic, chaotic)? Whoever dominates may be based more on implicit power than on explicit task goals. As noted, PHVA design was initially sensitive to possible power issues between biologists and wildlife managers and sought to equalize power between the groups in the interest of better information integration. A learning of the broadened Network processes is that interdisciplinary hierarchies exist as well, with quantitative and deductive disciplines favored over qualitative, inductive, or abductive ones. Equalizing this balance was difficult due to a negative reinforcing loop: to attract more social scientists means to change the problem definition, but to change the problem definition requires a critical mass of social scientists. Other problems such as unfamiliar concepts and vocabularies, or confusion over different meanings of words (e.g., regime), paled beside the problem of getting these people to the table. Once there, the challenge was to give each participant's contribution equal weight and to set aside territoriality in order to develop a common language. Our Network experiments have illustrated clearly that effective communication across disciplines may in fact require not only this setting but a reframing of the central question as well.

From a practical viewpoint (i.e, the organization of future expanded PHVA processes), this experience made clear that it cannot be CBSG's (or any single organization's) responsibility alone to get the right people and information to the table. A small organization like CBSG cannot expect to be able to identify local participants or gain access to local data and knowledge, particularly in a country where the organization has not worked before.

Partnerships of more long-standing duration may be critical for sustained conservation action (see chapter 16). In this case, we envision that CBSG workshops could operate as joint ventures with partners who have specific local knowledge and can follow up on the workshop recommendations, including those that are relevant to local community participation in ongoing conservation actions. Involving a local partner focused on those other stakeholders (local communities, indigenous groups, industrial firms) from the early planning stages would allow the problem in question to be more widely defined so as to engage all stakeholders. This would have the added advantage of strengthening the likelihood of policy outcomes in species' range countries. Ultimately, as the social scientists in the Network group clearly believed (see part 4), policy change is necessary to support lasting protection for species. While CBSG itself can act as a catalyst, the PHVA process must be supported by in-country organizations if long-term change is to happen.

From a more conceptual viewpoint (i.e., the self-reflection of Network members on their own processes), this experience made clear that there is no "comfort zone" for experiments in consilience. As Stacey (1996) has indicated, complex problems such as conservation require people to work in a zone that is far from agreement and also far from clarity (figure 19-1). While the natural inclination (and in many ways the healthy one in terms of disciplinary rigor) is to try to move problems back into zones that are better mapped conceptually and methodologically, in transdisciplinary groups this generally results in one disciplinary group being alienated (generally those associated with softer, more social science–based or more practical methodologies). A great deal of time needs to be spent in

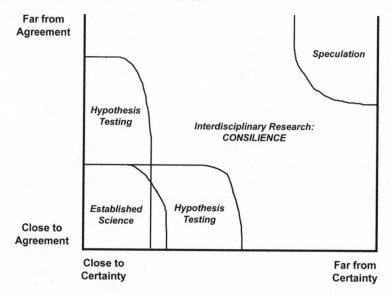

Figure 19-1. The Stacey Complexity Matrix. (Adapted from Stacey 1996 by B. Zimmerman.)

the zone that is far from land (where it is hard to get any a disciplinary-based sense of bearings) before enough common language and perspectives, or at least bridging language and perspectives, can be built.

If convergence is sought too early, covert or overt power dynamics tend to come into play, with the "harder" sciences exerting a greater claim to truth. The danger here is that others in the group, such as the social scientists or the managers, may withdraw. The Network social scientists had moments of alienation and questioning of the whole project, linked to what was felt to be an overly narrow focus on species alone. Trust and good communications sufficient to tolerate the expression of honest conflict helped . . . but in the process of struggling with this central challenge to consilience, we also came to realize that good will and good process, although necessary, were not sufficient.

A deeper problem lies in the fact that as a society of specialists we have literally diverged in terms of the ways we structure and embed our concepts and our data. Even when data from different disciplines has a bearing on a similar problem and we can get specialists to recognize this and agree to work together, there are major obstacles in getting the data or the models to talk to each other.

Learnings from Modeling

Many of the key breakthroughs of the Network project, in our opinion, were in recognizing both the limits and the possibilities of integrating knowledge through a modeling process. In the workshop process, attempts to integrate the knowledge and data from the working groups into the VORTEX model produced the following insights:

- Human demographic models generally dealt with data sets at too large a scale to be easily integrated into VORTEX as data input. Interpretation of the models in the light of local patterns and motivations was necessary. For example, human population density at a park border might or might not result in increased mortality of species. The threat might be direct or it might be through habitat erosion. In order to make the threat meaningful in terms of VORTEX, these differences are important.
- The potential for human diseases to cross the species barrier and vice versa, particularly in the case of primates, had implications for understanding situations where humans and wild primates are brought into close contact, either through wars (refugee camps on park borders) or through ecotourism. However, detailed epidemiological models that integrate human and nonhuman disease transmission are rare.
- War is a threat to species and to habitat. However, social science models of war are not geared to elements such as frequency of deforestation or environmental pollution, all elements of importance to species risk assessment.
- Social scientists understand that human cultural attitudes toward a particular species make up one of the variables that influence the way in which the species is treated (hunted for food, hunted as a threat). However, it is difficult to translate these understandings directly into such variables as rate of harvesting or amount of habitat fragmentation that can be meaningfully entered into a program like VORTEX.
- Human economic activities also have direct and indirect impact on species survival, but the problem again is to translate these into specific numerical threats to species or to habitat. However, this seems one of the most promising avenues for future integration, given the evidence at the muriqui and grizzly bear workshops that specific human impacts on the landscape could be discussed in enough detail in working groups to come up with numerical estimates (see chapters 7 and 10). An additional

problem concerning the impacts of economic activities seemed to be the relatively low incentives (other than external regulatory pressure) available to bring industry stakeholders to the table and the difficulty in making their role obvious once they are there.

The pattern across all these learnings is the challenge of a discipline-based world finding means to integrate not just the people, but the knowledge they possess. It was alarming to realize the extent to which we had built up knowledge structures that were incommensurate while all bearing on a central problem to be solved. One learning of this project has been that while most of the literature on interorganizational and intersectoral collaboration has focused on the human process barriers (power, mistrust, building common ground), rational barriers are at least as important. We need to find ways to translate, easily and effectively, from one discipline's knowledge base to another.

Future Experiments in Consilience

The environment consists of biological, physical, and human domains. Traditionally, disciplines focus on one of these (or a small subset of one), although some fields such as ecology often look at interactions between two. Analytical, computational and social tools for exploring the dynamics between the biological, physical, and human domains are not nearly as well developed as are tools for addressing within-domain problems. Each domain can be further partitioned among scales of space, time, degree of reductionism, quantification, and generality versus specificity. Disciplines within each domain tend to focus on processes at one end of these scales, and these differences in scales of analysis further hinder study of the complex environmental issues that lie at the intersection of domains and across scales.

Our experiences convinced us of the need to increase our understanding of both content and process in biocomplexity. Without knowing how to get models—and modelers—to effectively communicate, how to work across disciplines, and how to engage stakeholders in deliberations, we cannot, in a timely fashion, assist managers and policy makers facing complex environmental problems. One important future direction for experiments in consilience must be aimed at further developing and refining the conceptual and theoretical models, modeling capabilities, and process tools necessary to carry out biodiversity risk assessments.

The Open-Data Approach to Developing Metamodels

As the result of the Network project we have already begun pursuing "open-data" metamodels for understanding interacting systems. Description of our work is publicly available on the Web at *www2.netcom.com/~biocomplexity*. In an open-data metamodel, the specifications of the input parameters for each component submodel and its descriptors of the system are stored on the computer system in a form that is accessible to other programs. Each program has access to the data structures of each other program, using and possibly changing the values generated by other programs. To allow dynamic interaction among models, each model provides function calls within a dynamic link library (DLL) for initialization, incrementing time steps, and closing. In this way, any one model, or a higher-level interface program, can provide the interface through which users view the overall system. To capture the advantages of open-data models requires that data specifications be widely available so that potential users and contributors have access.

We hypothesize that open-data metamodels will promote the study of interactions by making the components modeled by each discipline available for use by others. We will be asking at each stage:

- How far can we push the open-data approach to integrate models from diverse fields?
- What challenges are encountered as we link progressively divergent models?
- Do linkages lead to understanding within fields that was not apparent before?
- Do metamodels reveal emergent properties of the interactions between systems?

Scenario Development and Scenario Testing

While the open-data approach allows communication among models that work on partly shared sets of data, our experiences through this project have led us to realize that at some stage the divergence of data types or scales of analysis will be such that exchange of data will no longer be possible or sufficient. Often, many components are not yet understood at a level that allows quantitative modeling. Yet, much of the challenge of biocomplexity lies in integrating knowledge between quantitative and qualitative models, between reductionist and holistic perspectives, or even between value systems and biophysical processes, as in the case of integrating traditional ecological knowledge and Western science.

The integration of knowledge of divergent types can be accomplished by developing scenarios that encapsulate the understanding at one (higher, more qualitative, holistic, or causal) level in a way that defines the processes that need to be represented and examined in the lower (more mechanistic) models. These models are then used to test scenarios that represent alternative conceptions of the higher-level view. This strengthens the analyses at each level, challenging reductionist models to generate higher-level phenomena and testing the more holistic hypothesis for consistency with presumed underlying processes.

Scenario planning is a widely used methodology in the private sector and increasingly in ecosystem assessments (Gallopin 2002). Known as "the gentle art of reperceiving," scenario planning is intended not so much as a means of predicting the future as a means of "questioning assumptions about how the world works ... liberating people's insights" (Schwartz 1991, 9) and directing attention to unfolding alternatives. Excellent scenario planning is a labor-intensive process involving (1) a lengthy, wide-spectrum data-gathering phase that explores all relevant information for a focal issue; (2) the selection of scenario logics; (3) the development of integrated scenarios; and (4) identifying indicators that allow for early detection of the unfolding implications for the focal issue.

As such, scenario planning makes an ideal complement to more focused, reductionist models like VORTEX in that it allows participants to use their judgment of probabilities and sense of system interconnectedness to make predictions, which in turn can be translated into VORTEX inputs.

Implications for Action

Introducing scenario planning in the PHVA process may demand some redesign of that process, in particular:

- Earlier gathering of data about the human institutional and governance context, which will frame any activities of species conservation. This greater lead time will also allow the development of the relationships that this project has indicated are important, ensuring that both the right stakeholders and the right information arrive at the workshop.

- More time spent in the workshop on the analysis stage, for all groups, including those dealing with "soft" or nonquantitative information. Such groups might also need greater support in both the divergence and convergence stage 2 (see chapter 4 and figure 4-2) that deals with analysis. System mapping tools may play a role here.

- Return to the central role of modeling for all workshops, as this introduces a discipline of exploring data integration, irrespective of barriers of discipline, scale, or knowledge structure.
- Support for the role of the modeler. As data integration appears to be one of the greatest barriers to consilience, the modeler must be personally and professionally prepared to take on the central responsibility for innovating around a new workshop process element. Added to the other roles of teacher and small-group leader, this key role can be easily overburdened. A variety of solutions might be proposed to deal with these problems, including pre-workshop research into the challenges of data integration, additional modelers at the workshop, or relieving the modeler of small-group responsibility.

Ultimately, the Network experience once again proved the robust and resilient nature of the PHVA process. Building on the base that Ulysses Seal and CBSG has designed and perfected, it remains one of the best current models in existence for combining action and science and for experimenting with consilience. Researching and writing this book has led us to believe that this process can be strengthened further by building partnerships, which will allow greater leverage on conservation policy as well as strengthen the workshop process, and by continuing to unravel the mysteries of reintegrating our knowledge across conceptual, scale, and disciplinary barriers. Last but not least, we must continue to refine this and other conservation processes to make best use of the conservation world's greatest resource: the judgment and wisdom of all those who care about saving our endangered spaces and the species therein.

References

Agrawal, A., and C. C. Gibson. 1999. "Enchantment and Disenchantment: The Role of Community in Natural Resource Conservation." *World Development* 27 (4): 629–649.

Akçakaya, H. R., and L. R. Ginzburg. 1991. "Ecological Risk Analysis for Single and Multiple Populations." In *Species Conservation: A Population-Biological Approach*, ed. A. Seitz and V. Loeschke, 73–87. Basel, Switzerland: Birkhaüser Verlag.

Alcorn, J. B., ed. 1993. *Papua New Guinea Conservation Needs Assessment*. Vol. 1. Boroko, PNG: Government of Papua New Guinea Department of Environment and Conservation.

Algonquin Wolf Advisory Group (AWAG). 2000. *Report to the Honourable John C. Snobelen, Minister of Natural Resources*. Peterborough, Ontario: AWAG.

Allwright, D., and H. Vredenburg. 2002. "Deep, Heavy, and Ugly: Talisman Energy (Sudan) Inc." Calgary, Albert: TCPL International Institute for Resource Industries and Sustainability Studies (TC-IRIS) Case Series, June.

Alpert, P. 1993. "Conserving Biodiversity in Cameroon." *Ambio* 22 (1): 144–149.

———. "Applying Ecological Research at Integrated Conservation and Development Projects." *Ecological Applications* 5 (4): 877–860.

———. 1996. "Integrated Conservation and Development Projects: Examples from Africa." *BioScience* 46 (11): 845–855.

Alvarez, K. 1993. *The Twilight of the Panther*. Sarasota, Fla.: Myakka River Publishing.

Andersen, K. E. 1995. "Institutional Flaws of Collective Forest Management." *Ambio* 24 (6): 349–353.

Araya, B., D. Garland, G. Espinoza, A. Sanhuesa, R. C. Lacy, A. Teare, and S. Ellis, eds. 1999. *Taller análisis de la viabilidad del hábitat y de la población del pingüino Humboldt (Spheniscus humboldti)*. Borrador del informe. Apple Valley, Minn.: Conservation Breeding Specialist Group (SSC/IUCN).

Asian Rhino Specialist Group (AsRSG). 1996. Report of a meeting of the IUCN SSC's AsRSG, Sandakan, Malaysia, 29 November–1 December 1995.

Assembly of First Nations and the Inuit Circumpolar Conference. 1994. "A Preliminary Research Prospectus." In *Traditional Ecological Knowledge and Modern Environmental Assessment*, ed. B. Sadler and P. Boothroyd. Vancouver: Center for Human Settlements, University of British Columbia.

Baden, J. A., and D. Leal. 1990. *The Yellowstone Primer: Land and Resource Management in the Greater Yellowstone Ecosystem*. San Francisco: Pacific Research Institute for Public Policy.

Badola, R. 1998. "Attitudes of Local People towards Conservation and Alternatives to Forest Resources: A Case Study from the Lower Himalayas." *Biodiversity and Conservation* 7 (10): 1245–1259.

Ballou, J. D., R. C. Lacy, D. Kleiman, A. Rylands, and S. Ellis, eds. 1998. *Leontopithecus II. The Second Population and Habitat Viability Assessment for Lion Tamarins (Leontopithecus).* Apple Valley, Minn.: Conservation Breeding Specialist Group (SSC/IUCN).

Barrett, C. B., and P. Arcese. 1995. "Are Integrated Conservation-Development Projects (ICDPs) Sustainable? On the Conservation of Large Mammals in Sub-Saharan Africa." *World Development* 27 (7): 1073–1084.

Battiste, M., and J. S. Y. Henderson. 2000. *Protecting Indigenous Knowledge and Heritage.* Saskatoon, Sask.: Purich.

Baydack, R. K., H. Campa, and J. B. Haufler, eds. 1999. *Practical Approaches to the Conservation of Biological Diversity.* Washington, D.C.: Island Press.

Beissinger, S. R. and M. I. Westphal. 1998. "On the Use of Demographic Models of Population Viability in Endangered Species Management." *Journal of Wildlife Management* 62: 821–841.

Beissinger, S. R., and D. R. McCullough, eds. 2002. *Population Viability Analysis.* Chicago: University of Chicago Press.

Beltran, J., ed. 2000. *Indigenous and Traditional Peoples and Protected Areas: Principles, Guidelines and Case Studies.* Gland, Switzerland: IUCN and WWF International.

Berkes, F. 1993. "Traditional Ecological Knowledge in Perspective." In *Traditional Ecological Knowledge: Concepts and Cases,* ed. Julian T. Inglis. Ottawa: Canadian Museum of Nature and International Development Research Centre.

———, ed. 1989. *Common Property Resources, Ecology, and Community-Based Sustainable Development.* London: Bellhaven.

Birch, L. C. 1948. "The Intrinsic Rate of Natural Increase of an Insect Population." *Journal of Animal Ecology* 17: 15–26.

Biswas, A. K., and H. C. Tortajada-Quiroz. 1996. "Environmental Impacts of the Rwandan Refugees on Zaire." *Ambio* 25 (6): 403–408.

Bonaccorso, F., P. Clark, P. S. Miller, and O. Byers, eds. 1999. *Conservation Assessment and Management Plan for the Tree Kangaroos of Papua New Guinea and Population Habitat Viability Analysis Assessment for Matschie's Tree Kangaroo: Final Report.* Apple Valley, Minn.: Conservation Breeding Specialist Group (SSC/IUCN)

Borrini-Feyerabend, G. 1996. *Collaborative Management of Protected Areas: Tailoring the Approach to the Context.* Gland, Switzerland: IUCN.

Boyce, M. S. 1992. "Population Viability Analysis." *Annual Review of Ecology and Systematics* 23: 481–506.

Britt, D. W. 1997. *A Conceptual Introduction to Modeling.* Mahwah, N.J.: Lawrence Erlbaum and Associates.

Brook, B. W., J. J. O'Grady, M. A. Burgman, H. R. Akçakaya, and R. Frankham. 2000. "Predictive Accuracy of Population Viability Analysis in Conservation Biology." *Nature* 404: 385–387.

Brook, B. W., J. R. Cannon, C. Mirande, R. C. Lacy, and R. Frankham. 1999. "Comparison of the Population Viability Analysis Packages GAPPS, INMAT, RAMAS and VORTEX for the Whooping Crane *(Grus americana).*" *Animal Conservation* 2: 23–31.

Brotton, J., and G. Wall. 1997. *The Bathurst Caribou Herd in a Changing Climate.* Ottawa: Environment Canada, Climate Change Digest.

Brown, L. D. 1980. "Planned Change in Underorganized Systems." In *Systems Theory for Organization Development*, ed. T. Cummings, 181–203. New York: John Wiley and Sons.

———. 1991. "Bridging Organizations and Sustainable Development." *Human Relations* 44 (8): 807–831.

Brown, L. D., and D. Ashman. 1995. "Intersectorial Problem Solving and Social Capital Formation," Paper presented to the Academy of Management Special Conference, Cleveland, Ohio.

———. 1996. "Participation, Social Capital and Intersectoral Problem-Solving: African and Asian Cases." *World Development* 24, 9: 1467–1479.

Brunnee, J., and S. J. Toope. 1997. "Environmental Security and Freshwater Resources: Ecosystem Regime Building." *American Journal of International Law* 91: 26–59.

Brunner, R. D., and T. W. Clark. 1997. "A Practice-Based Approach to Ecosystem Management." *Conservation Biology* 11 (1): 48–58.

Buck, S. J. 1998. *The Global Commons. An Introduction*. Washington, D.C: Island Press.

Burgman, M., S. Ferson, and H. R. Akçakaya. 1993. *Risk Assessment in Conservation Biology*. New York: Chapman and Hall.

Buss, M., and M. de Almeida. 1997. *A Review of Wolf and Coyote Status and Policy in Ontario*. Peterborough: Ontario Ministry of Natural Resources, December.

Caldwell, J. 1979. "Toward a Restatement of Demographic Transition Theory." *Population and Development Review* 2–3/4: 321–366.

Cajete, G. 2000. *Native Science, Natural Laws of Interdependence*. Santa Fe: Clear Light Publishers.

Canada. Supreme Court. 1997. *Delgamuukw v. British Columbia*. File No. 23799, June 16, 17, December 11, under General Principles, Item No. 84.

Caro, T. M. and M. K. Laurenson. 1994. "Ecological and Genetic Factors in Conservation: A Cautionary Tale." *Science* 263: 485–486.

Case, D. D. 1990. *The Community's Toolbox: The Idea, Methods and Tools for Participatory Assessment, Monitoring and Evaluation in Community Forestry*. Community Forestry Manual 2. Rome: Food and Agriculture Organization of the United Nations. The complete document is available online at *www.fao.org/docrep/X5307E/X5307E00.htm*.

Caughley, G. 1994. "Directions in Conservation Biology." *Journal of Animal Ecology* 63: 215–244.

CBSG. 1993. *Futures Search Report*. Apple Valley, Minn.: Conservation Breeding Specialist Group (SSC/IUCN).

Central Rockies Ecosystem Interagency Liaison Group (CREILG). 1995. *Atlas of the Central Rockies Ecosystem*. Calgary, Komex International, Ltd.

Chambers, R. 1992. *Rural Appraisal: Rapid, Relaxed and Participatory*. Discussion Paper 311. Brighton, England: Institute of Development Studies.

———. 1994. "Participatory Rural Appraisal (PRA): Analysis of Experience." *World Development* 22 (9): 1253–1268.

Chapin, F. S. III, E. S. Zavaleta, V. T. Eviner, R. L. Naylor, P. M. Vitousek, H. L. Reynolds, D. U. Hooper, S. Lavorel, O. E. Sala, S. E. Hobbie, M. C. Mack, and S. Díaz. 2000. "Consequences of Changing Biodiversity." *Nature* 405: 234–242.

Cialdini, R. B. 1984. *Influence*. New York: William Morrow and Company.

Clark, T. W. 1989. *Conservation Biology of the Black-Footed Ferret*. Philadelphia, Penn.: Wildlife Preservation Trust International.

————. 1999. "Interdisciplinary Problem Solving: Net Steps in the Greater Yellowstone Ecosystem." *Policy Sciences* 32 (4): 393–414.

Clark, T. W., P. C. Paquet, and A. Peyton Curlee. 1996. "Introduction to Special Section: Large Carnivore Conservation in the Rocky Mountains of the United States and Canada." *Conservation Biology* 10 (4): 936–939.

Clark, T. W., R. M. Warneke, and G. G. George. 1990. "Management and Conservation of Small Populations." In *Management and Conservation of Small Populations,* ed. T. W. Clark and J. H. Seebeck, 1–18. Brookfield, Ill.,: Chicago Zoological Society.

Clark, W. C. 1989. "Managing Planet Earth." *Scientific American* 261 (3): 46–57.

Clayton, L., M. Keeling, and E. J. Milner-Gulland. 1997. "Bringing Home the Bacon: A Spatial Model of Wild Pig Hunting in Sulawesi, Indonesia." *Ecological Applications* 7: 642–652.

Collings, P., G. Wenzel, and R. G. Condon. 1998. "Modern Food Sharing Networks and Community Integration in the Central Canadian Arctic." *Arctic* 51 (4): 301–314.

Coomes, O. T. 1995. "A Century of Rain Forest Use in Western Amazonia: Lessons for Extraction-Based Conservation of Tropical Forest Resources." *Forest and Conservation History* 39: 108–120.

Coomes, O. T., and B. L. Barham. 1997. "Rain Forest Extraction and Conservation in Amazonia." *The Geographical Journal* 163: 180–188.

Cooperrider, D. L. 1990. "Positive Image, Positive Action: The Affirmative Basis of Organizing." In *Appreciative Management and Leadership: The Power of Positive Thought and Action in Organizations,* ed. S. Srivastva and D. L. Cooperrider, 91–125. San Francisco: Jossey-Bass.

Cooperrider, D. L., and W. Pasmore. 1991. "Global Social Change: A New Agenda for the Social Sciences." *Human Relations* 44 (10): 1037–1055.

COSEWIC. 2000. *Canadian Species at Risk.* Committee on the Status of Endangered Wildlife in Canada. Ottawa: Canadian Wildlife Service. COSEWIC maintains a Web site at *www.cosewic.gc.ca.*

Cox, P., and T. Elmquist. 1997. "Ecocolonialism and Indigenous-Controlled Rainforest Preserves in Samoa." *Ambio* 26 (2): 84–89.

Cruikshank, J. 1981. "Legend and Landscape: Convergence of Oral and Scientific Tradition in the Yukon Territory." *Arctic Anthropology* 18 (2): 67–93.

————. 1998. *The Social Life of Stories.* Lincoln: University of Nebraska Press.

Daily, H., and P. Erhlich. 1999. "Managing Earth's Ecosystems: An Interdisciplinary Challenge." *Ecosystems* 2: 277–280.

Daly, M., and M. Wilson. 1988. *Homocide.* New York: Aldine DeGruyter.

De Boer, W. F., and D. S. Baquete. 1998. "Natural Resource Use, Crop Damage and Attitudes of Rural People in the Vicinity of the Maputo Elephant Reserve, Mozambique." *Environmental Conservation* 25 (3): 208–218.

Decher, J. 1997. "Conservation, Small Mammals, and the Future of Sacred Groves in West Africa." *Biodiversity and Conservation* 6 (7): 1007–1026.

Deevey, E. S. Jr. 1947. "Life Tables for Natural Populations of Animals." *Quarterly Review of Biology* 22: 283–314.

Dene Cultural Institute. 1994. "Traditional Ecological Knowledge and Environmental Impact Assessment." In *Traditional Ecological Knowledge and Modern Environmental Assessment,* ed. B. Sadler and P. Boothroyd. Vancouver: Center for Human Settlements, University of British Columbia.

Department of Indian Affairs and Northern Development (DIAND). 1987. *The Western Arctic Land Claim: The Inuvialuit Final Agreement.* Ottawa.

———. 1993. Agreement between the Inuit of the Nunavut Settlement Area and Her Majesty the Queen in the Right of Canada. Ottawa.

Duerden, F., and R. G. Kuhn. 1998. "Scale, Context and Application of Traditional Knowledge in the Canadian North." *Polar Record* 34 (188): 31–38.

Durant, S. M., and G. M. Mace. 1994. "Species Differences and Population Structure in Population Viability Analyses," In *Creative Conservation: Interactive Management of Wild and Captive Animals,* ed. P. G. S. Olney, G. M. Mace and A. T. C. Feistner, 67–91. London: Chapman & Hall.

Ewins, P., M. de Almeida, P. S. Miller, and O. Byers, eds. 2000. *Population and Habitat Viability Assessment Workshop for the Wolves of Algonquin Park: Final Report.* Apple Valley, Minn.: Conservation Breeding Specialist Group (SSC/IUCN).

Fiallo, E. A. and S. K. Jacobson. 1995. "Local Communities and Protected Areas: Attitudes of Rural Residents towards Conservation and Machalilla National Park, Ecuador." *Environmental Conservation* 22 (3): 241–249.

Fisher, R. A. 1958. *The Genetical Theory of Natural Selection.* 2d ed. New York: Dover.

Flannery, T. F. 1995. *Mammals of New Guinea.* Ithaca, N.Y.: Cornell University Press.

Flannery, T. F., R. Martin, and A. Szalay. 1996. *Tree Kangaroos: A Curious Natural History.* Melbourne, Australia: Reed Books.

Forbes, G. 1994. "Wolf-Ungulate Relationships in Algonquin Provincial Park." Ph.D. dissertation, Department of Geography, University of Waterloo.

Forbes, G., and J. Theberge. 1996. "Cross-Boundary Management of Algonquin Park Wolves." *Conservation Biology* 10 (4): 1091–1097.

Foreman, D., J. Davis, D. Johns, R. Noss, and M. E. Soulé. 1992. *The Wildlands Project: Plotting a North American Wilderness Recovery Strategy.* Canton, N.Y.: Cenozoic Society.

Franklin, I. R. 1980. "Evolutionary Change in Small Populations." In *Conservation Biology: An Ecological-Evolutionary Perspective,* ed. M. E. Soulé and B. A. Wilcox, 135–149. Sunderland, Mass.: Sinauer Associates.

Freeman, M. M. R. 1985. "Appeal to Tradition: Different Perspectives on Wildlife Management." In *Native Power: The Quest for Autonomy and Nationhood of Aboriginal Peoples,* ed. J. Brosted et al., 265–282. Oslo: Universitetsforlaget.

———. 1995. "The Nature and Utility of Traditional Ecological Knowledge." In *Consuming Canada,* ed. C. Gaffield and P. Gaffield, CBSG. 1993. *Futures Search Report.* Apple Valley, Minn.: Conservation Breeding Specialist Group (SSC/IUCN). 39–47. Toronto: Copp Clark Ltd.

The Futures Group. 1993. DEMPROJ, *A Demographic Projection Model for Development Planning, Version 3.* Washington D.C.: The Futures Group.

Gadjil, M., F. Berkes, and C. Folks. 1993. "Indigenous Knowledge for Biodiversity Conservation." *Ambio* 22 (2–3): 151–156.

Gagné, W. C., and J. L. Gressitt. 1982. "Conservation in New Guinea." In *Monograph Biologicae,* Vol. 42, ed. J. L. Gressitt, 945–966.

Gallopin, G. C. 2002. "Scenarios, Surprises and Branch Points." In *Panarchy,* ed. L. Gunderson and C. S. Holling, 361–392. Washington, D.C.: Island Press,.

Ghimire, K. B. 1994. "Parks and People: Livelihood Issues in National Parks in Thailand and Madagascar." *Development and Change* 25 (1): 195–229.

Gibbon, G., ed. 1998. *Archeology of Prehistoric North America, An Encyclopedia.* New York: Garland Publishers.

Gibeau, M. L., and S. Herrero. 1999. *Eastern Slopes Grizzly Bear Project: Progress Report for 1998.* Prepared for the ESGBP Steering Committee, Calgary, Alberta.

Gibson, C. C., and S. A. Marks. 1995. "Transforming Rural Hunters into Conservationists: An Assessment of Community-Based Wildlife Programs in Africa." *World Development* 23 (6): 941–957.

Giddens, A. 1984. *The Construction of Society: Outlining of the Theory of Structuration.* Oxford: Policy Press.

Gilpin, M. E. and M. E. Soulé. 1986. "Minimum Viable Populations: Processes of Extinction." In *Conservation Biology: The Science of Scarcity and Diversity,* ed. M. E. Soulé, 19–24. Sunderland, Mass.: Sinauer Associates.

GNWT Department of Culture and Communications. 1991. *Report of the Traditional Knowledge Working Group.* Yellowknife.

Goodman, D. 1987. "The Demography of Chance Extinction." In *Viable Populations for Conservation,* ed. M. E. Soulé, 11–34. Cambridge: Cambridge University Press.

Gouldner, A. 1976. *Dialectics of Ideology and Technology.* London: Macmillan.

Granovettor, M. 1985. "Economic Action and Social Structure: the Problem of Embeddedness." *American Journal of Sociology* 91: 481–510.

Gray, B. 1989. *Collaborating: Finding Common Ground for Multiparty Problems.* San Francisco: Jossey-Bass.

Gray, B., and D. J. Wood. 1991. "Collaborative Alliances: Moving from Practice to Theory." *Journal of Applied Behavioral Science* 27 (1): 3–22.

Groom, M. J. 1998. "Allee Effects Limit Population Viability of an Annual Plant." *American Naturalist* 151: 487–496.

Groom, M. J., and M. A. Pascual. 1998. "The Analysis of Population Persistence: An Outlook on the Practice of Viability Analysis." In *Conservation Biology,* 2d ed., ed. P. L. Fiedler and P. M. Kareiva, 4–27. New York: Chapman and Hall.

Grove, J. M., and W. R. Burch. 1997. "A Social Ecology Approach and Application of Urban Ecosystem and Landscape Analysis: A Case Study of Baltimore." *Urban Ecosystems* 1: 259–275.

Groves, C. P. 1982. "The Systematics of Tree Kangaroo *(Dendrolagus; Marsupialia, Macropodidae).*" *Australian Mammalogy* 5: 157–186.

———. 2001. *Primate Taxonomy.* Washington, D.C.: Smithsonian Institute Press.

Grumbine, E. R. 1994. "What is Ecosystem Management?" *Conservation Biology* 8 (1): 27–38.

———. 1997. "Reflections on 'What is Ecosystem Management?'" *Conservation Biology* 11 (1): 41–47.

Guha, R. 1997. "The Authoritarian Biologist and the Arrogance of Anti-Humanism: Wildlife Conservation in the Third World." *The Ecologist* 27 (1): 14–20.

Gunderson, L., C. S. Holling, and S. Light, eds. 1995. *Barriers and Bridges to Renewal of Ecosystems and Institutions.* New York: Columbia University Press.

Gunn, A., U. S. Seal, and P. S. Miller, eds. 1998. *Population and Habitat Viability Assessment Workshop for Peary Caribou and Arctic-Island Caribou (Rangifer tarandus).* Apple Valley, Minn.: Conservation Breeding Specialist Group (SSC/IUCN).

Haas, P. M. 1992. "Introduction: Epistemic Communities and International Policy Coordination." *International Organization* 46 (1): 1–35.

Hanna, S. S., C. Folke, and K. G. Mäler, eds. 1996. *Rights to Nature. Ecological, Economic, Cultural, and Political Principles of Institutions for the Environment.* Washington, D.C.: Island Press.

Harcourt, A. H. 1995. "Population Viability Estimates: Theory and Practice for a Wild Gorilla Population." *Conservation Biology* 9: 134–142.

Hardy, C., and N. Phillips. 1998. "Strategies of Engagement: Lessons from the Critical Examination of Collaboration in an Interorganizational Domain." *Organization Science* (March/April): 217–230.

Harrison, P., and F. Pearce. 2000. *AAAS Atlas of Population & Environment.* Berkeley: University of California Press.

Hart, T., and J. Hart. 1997. "Conservation and Civil Strife: Two Perspectives from Central Africa." *Conservation Biology* 11: 308–309.

Hartup, B. K. 1994. "Community Conservation in Belize: Demography, Resource Use, and Attitudes of Participating Landowners." *Biological Conservation* 69: 235–241.

Harvey, L. D. 1995. "Creating a Global Warming Implementation Regime." *Global Environmental Change* 5 (5): 415–432.

Hedrick, P. W., R. C. Lacy, F. W. Allendorf, and M. E. Soulé. 1996. "Directions in Conservation Biology: Comments on Caughley." *Conservation Biology* 10: 1312–1320.

Herrero, S., P. S. Miller, and U. S. Seal, eds. 2000. *Population and Habitat Viability Assessment (PHVA) Workshop for the Grizzly Bear of the Central Rockies Ecosystem (Ursus arctos).* Apple Valley, Minn.: Conservation Breeding Specialist Group (SSC/IUCN).

Hirsch, A., L. G. Dias, L. de O. Martins, R. F. Campos, E. C. Landau, and N. A. T. Resende. Forthcoming. *BDGEOPRIM—Database of Geo-Referenced Localities for Neotropical Primates.* Neotropical Primates. See also *www.icb.ufmg.br/~primatas/home_bdgeoprim.htm.*

Hoare, R. E., and J. T. Du Toit. 1999. "Co-Existence between People and Elephants in African Savannas." *Conservation Biology* 13 (3): 633–639.

Holling, C. S., and G. K. Meffe. 1996. "Command and Control and the Pathology of Natural Resource Management." *Conservation Biology* 10 (2): 328–337.

Horowitz, L. S. 1998. "Integrating Indigenous Resource Management with Wildlife Conservation: A Case Study of Batang Ai National Park, Sarawak, Malaysia." *Human Ecology* 26 (3): 371–403.

Hough, J. L.1993. "Why Burn the Bush? Social Approaches to Bush-Fire Management in West African National Parks." *Biological Conservation* 65, 23–28.

Hudson, W. E., ed. 1991. *Landscape Linkages and Biodiversity.* Washington, D.C.: Island Press.

Hungerford, L., M. Cranfield, L. Gaffikin, A. Mudakikwa, and J. B. Nizey. Forthcoming. "Simulation Modeling as a Decision Aid for Wildlife Disease Management: Measles in Gorillas."

Hutchins, M., G. M Smith, D. C. Mead, S. Elbin, and J. Steenberg. 1991. "Social Behavior of Matschie's Tree Kangaroo *(Dendrolagus matschiei)* and Its Implications for Captive Management." *Zoo Biology* 10: 147–164.

IUCN. 1996. *1996 IUCN Red List of Threatened Animals.* Gland, Switzerland: IUCN.

———. 2000. *IUCN Red List Categories.* Gland, Switzerland: IUCN.

Janoff, S., and M. Weisbord. 2000. *Future Search: An Action Guide to Finding Common Ground in Organizations and Communities.* San Francisco: Berett-Khoehler.

Jessop, B. 1997. "Capitalism and Its Future: Remarks in Regulation, Government and Governance." *Review of International Political Economy* 4 (3): 561–581.

Kaner, S. 1996. *Facilitator's Guide to Participative Decision Making.* Gabriola Island, British Columbia: New Society Publishers.

Kanter, R. M. 1989. "Becoming PALS: Pooling, Allying and Linking across Companies." *Academy of Management Executive* 33: 183–193.

Kaus, A. 1993. "Environmental Perceptions and Social Relations in the Mapimi Biosphere Reserve." *Conservation Biology* 7 (2): 398–406.

Kellert, S. R. 1985. "Social and Perceptual Factors in Endangered Species Management." *Journal of Wildlife Management* 49: 526–528.

Kellert, S. R., and E. O. Wilson. 1993. *The Biophilia Hypotheses.* Washington, D.C.: Island Press.

Kelly, B. T., P. S. Miller, and U. S. Seal, eds. 1999. *Population Habitat Viability Assessment Workshop for the Red Wolf (Canis rufus): Final Report.* Apple Valley, Minn.: Conservation Breeding Specialist Group (SSC/IUCN).

Kennedy, M. 1992. *Australasian Marsupials and Monotremes: An Action Plan for Their Conservation.* Gland, Switzerland: IUCN.

Keohane, R. O., and E. Ostrom. 1994. "Introduction [to special issue on 'Local Commons and Global Interdependence: Heterogenity and Cooperation in Two Domains']." *Journal of Theoretical Politics* 6 (4): 403–428.

Khan, J. A. 1995. "Conservation and Management of Gir Lion Sanctuary and National Park, Gujarat, India." *Conservation Biology* 73: 183–188.

Kiley-Worthington, M. 1997. "Wildlife Conservation, Food Production and 'development': Can They Be Integrated? Ecological Agriculture and Elephant Conservation in Africa." *Environmental Values* 6 (4): 455–470.

Kjos, C., O. Byers, P. S. Miller, J. Borovansky and U. S. Seal, eds. 1998. *Population and Habitat Viability Assessment Workshop for the Winged Mapleleaf Mussel (Quadrula fragosa).* Apple Valley, Minn.: Conservation Breeding Specialist Group (SSC/IUCN).

Knight, J. 1992. *Institutions and Social Conflict.* Cambridge: Cambridge University Press.

Kostoff, R. N. 2002. "Overcoming Specialization." *BioScience* 52 (10): 937–941.

Kothari, A., S. Suri, and N. Singh. 1995. "People and Protected Areas: Rethinking Conservation in India." *The Ecologist* 25 (5): 188–194.

Krippendorff, K. 1998. "On the Essential Contexts of Artifacts or on the Proposition that Design is Making Sense of Things." In *The Idea of Design,* ed. V. Margolin and R. Buchanan, 156–186. Cambridge: MIT Press.

Lacy, R. C. 1993a. "Vortex: A Computer Simulation Model for Population Viability Analysis." *Wildlife Research* 20: 45–65.

———. 1993b. "Impacts of Inbreeding in Natural and Captive Populations of Vertebrates: Implications for Conservation." *Perspectives in Biology and Medicine* 36: 480–496.

———. 1993/1994. "What is Population (and Habitat) Viability Analysis?" *Primate Conservation* 14/15: 27–33.

———. 1997. "Importance of Genetic Variation to the Viability of Mammalian Populations." *Journal of Mammalogy* 78: 320–335.

———. 2000a. "Considering Threats to the Viability of Small Populations." *Ecological Bulletins* 48: 39–51.

———. 2000b. "Structure of the Vortex Simulation Model for Population Viability Analysis." *Ecological Bulletins* 48: 191–203.

Lacy, R. C., A. M. Petric, and M. Warneke. 1993. "Inbreeding and Outbreeding Depression in Captive Populations of Wild Species." In *The Natural History of Inbreeding and Outbreeding,* ed. N. W. Thornhill, 352–374. Chicago: University of Chicago Press.

Lacy, R. C., P. S. Miller, J. P. Pollak, R. P. Bright, and J. D. Ballou. 2003. "Outbreak: A Stochastic Simulation Model of Infectious Disease Processes in Wildlife Populations." Unpublished manuscript.

Lande, R., and G. F. Barrowclough. 1987. "Effective Population Size, Genetic Variation, and Their Use in Population Management." In *Viable Populations for Conservation,* ed. M. E. Soulé, 87–123. Cambridge: Cambridge University Press.

Lanjouw, A., G. Cummings, and J. Miller. 1996. "Gorilla Conservation Problems and Activities in North Kivu, Eastern Zaire (February 1996)." *African Primates* 1 (2): 44–46.

Leach, M., R. Mearns, and I. Scoones. 1999. "Environmental Entitlements: Dynamics and Institutions in Community-Based Natural Resource Management." *World Development* 27 (2): 225–247.

Legare, A. 1997. "The Government of Nunavut (1999): A Prospective Analysis." In *First Nations in Canada: Perspectives on Opportunity, Empowerment and Self-Determination,* ed. R. J. Ponting, 404–431. Toronto: McGraw-Hill Ryerson Ltd.

Lertzman, D. 1999. "Planning between Cultural Paradigms: Traditional Knowledge and the Transition to Ecological Sustainability" Ph.D. dissertation, School of Community and Regional Planning, University of British Columbia.

———. 2002. "Rediscovering Rites of Passage: Education, Transformation and the Transition to Sustainability." *Conservation Ecology* 5 (2): 30.

Lindenmayer, D. B. 1996. *Wildlife and Woodchips: Leadbeater's Possum—a Test Case for Sustainable Forestry.* Sydney: University of New South Wales Press.

Lindenmayer, D. B., and H. P. Possingham. 1994. *The Risk of Extinction: Ranking Management Options for Leadbeater's Possum Using Population Viability Analysis.* Canberra: Centre for Resource and Environmental Studies, the Australian National University.

Lindenmayer, D. B., M. A. Burgman, H. R. Akçakaya, R. C. Lacy, and H. P. Possingham. 1995. "A Review of the Generic Computer Programs ALEX, RAMAS/space and VORTEX for Modeling the Viability of Wildlife Populations." *Ecological Modelling* 82: 161–174.

Lindenmayer, D. B., R. C. Lacy, and M. L. Pope. 2000. "Testing a Simulation Model for Population Viability Analysis." *Ecological Applications* 10: 580–597.

Livi-Bacci, M. 1997. *A Concise History of World Population.* Trans. C. Ipsen. 2nd ed. Oxford: Blackwell Publishers.

Lutz, W. 1994. *Population, Development and Environment: Understanding Their Interactions in Mauritius.* Berlin: Springer-Verlag.

Lynch, O., and A. Marat. 1993. "A Review and Analysis of National Laws and Policies Concerning Customary Owner's Rights and the Conservation and Sustainable Development of Forests and Other Biological Resources." In *Papua New Guinea Conservation Needs Assessment,* Vol. 1, ed. J. B. Alcorn, 7–29. Boroko, PNG: Government of Papua New Guinea Department of Environment and Conservation.

Mace, G. M., and R. Lande. 1991. "Assessing Extinction Threats: Toward a Reevaluation of IUCN Threatened Species Categories." *Conservation Biology* 5: 148–157.

Mace, G. M., N. Collar, J. Cooke, K. Gaston, J. Ginsberg, N. Leader Williams, M. Maunder, and E. J. Milner-Gulland. 1992. "The Development of New Criteria for Listing Species on the IUCN Red List." *Species* 19: 16–22.

Mace, R. D., J. S. Waller, T. L. Manley, K. Ake, and W. T. Wittinger. 1999. "Landscape Evaluation of Grizzly Bear Habitat in Western Montana." *Conservation Biology* 13: 367–377.

MacNeill, J., P. Winsemius, and T. Yakushiji. 1991. *Beyond Interdependence: The Meshing of the World's Economy and the Earth's Ecology.* New York: Oxford University Press.

Maehr, D. S., and J. A. Cox. 1995. "Landscape Features and Panthers in Florida." *Conservation Biology* 9: 1008–1019.

Maehr, D. S., R. C. Lacy, E. D. Land, O. L. Bass, and T. S. Hoctor. 2002. "Evolution of Population Viability Assessment for the Florida Panther: A Multiperspective Approach." In *Population Viability Analysis,* ed. S. Beissinger and D. McCullough, 284–311. Chicago: University of Chicago Press.

Manansang, J., D. Siswomartono, T. Soehartono, U. S. Seal, P. S. Miller, and S. Ellis. eds. 1997. *Marine Turtles of Indonesia: Population Viability and Conservation Assessment and Management Workshop.* Apple Valley, Minn.: Conservation Breeding Specialist Group (SSC/IUCN).

Matamoros, Y., G. Wong, and U. S. Seal, eds. 1995. *Taller de evaluación de viabilidad de población y hábitat de Saimiri oerstedi citrinellus.* Apple Valley, Minn.: Fundación pro Zoológicos (FUNDAZOO) and Conservation Breeding Specialist Group (SSC/IUCN).

Matzke, G. E. and N. Nabane. 1995. "African Wildlife Conservation, Utilization and Community Empowerment: Zambezi Developments Continue." *Ambio* 24 (5): 318–319.

Maxwell, B. 1997. *Responding to Global Climate Change in Canada's Arctic.* Vol. 2 of *Canada Country Study: Climate Impacts and Adaptation.* Downsview, Ontario: Environment Canada, Atmospheric Environment Service.

McCann, K. S. 2000. "The Diversity-Stability Debate." *Nature* 405: 228–233.

McClellan, B. N., F. W. Hovey, R. D. Mace, J. G. Woods, D. W. Carney, M. L. Gibeau, W. L. Wakkinen, and W. F. Kasworm. 1999. "Rates and Causes of Grizzly Bear Mortality in the Interior Mountains of British Columbia, Alberta, Montana, Washington, and Idaho." *Journal of Wildlife Management* 63: 911–920.

McDevitt, T. M. 1998. *World Population Profile: 1998.* U.S. Bureau of the Census, Report WP/98. Washington, D.C.: U.S. Government Printing Office. Available online at *www.census.gov/ipc/www/wp98.html.*

McGrady-Steed, J., P. M. Harris, and P. J. Morin. 1997. "Biodiversity Regulates Ecosystem Predictability." *Nature* 390: 162–165.

McNeely, J. A. 1993. "Economic Incentives for Conserving Biodiversity: Lessens for Africa." *Ambio* 22 (2–3): 144–150.

Mesqueda, C. G., and N. L. Wiener. 1996. "Human Collective Aggression: A Behavioral Ecological Approach." *Ethology and Sociobiology* 17: 247–262.

———. 1999. "Male Age Composition and Severity of Conflicts." *Politics and the Life Sciences* 18 (2): 181–189.

Michael, D. N. 1993. "Governing by Learning: Boundaries, Myths and Metaphors." *Futures* 25 (1): 81–89.

Miller, D. 1993. "The Architecture of Simplicity." *Academy of Management Review* 18 (1): 116–138.

Miller, P. S., and R. C. Lacy. 1999. VORTEX: *A Stochastic Simulation of the Extinction Process. Version 8 User's Manual.* Apple Valley, Minn.: Conservation Breeding Specialist Group (SSC/IUCN).

Mills, L. S., C. Baldwin, M. J. Wisdom, J. Citta, D. J. Mattson, and K. Murphy. 1996. "Factors Leading to Different Viability Predictions for a Grizzly Bear Data Set." *Conservation Biology* 10: 863–873.

Molur, S., R. Sukumar, U. S. Seal, and S. Walker, ed. 1995. *Population and Habitat Viability Assessment for the Great Indian One-Horned Rhinoceros.* Coimbatore: CBSG India (SSC/IUCN).

Morgan, M. S., and M. Morrison, eds. 1999. *Models as Mediators.* Cambridge: Cambridge University Press.

Morrison, M., and M. S. Morgan. 1999. "Models as Mediating Instruments." In *Models as Mediators,* ed. M. S. Morgan and M. Morrison, 10–37. Cambridge: Cambridge University Press.

Muchaal, P. K., and G. Ngandjui. 1999. "Impact of Village Hunting on Wildlife Populations in the Western Dja Reserve, Cameroon." *Conservation Biology* 13 (2): 385–396.

Nadasdy, P. 1999. "The Politics of TEK: Power and the 'Integration' of Knowledge." *Arctic Anthropology* 36 (1–2): 1–18.

Naeem, S., and S. Li. 1997. "Biodiversity Enhances Ecosystem Reliability." *Nature* 390: 507–509.

Naiman, R. J. 1999. "A Perspective on Interdisciplinary Science." *Ecosystems* 2: 292–295.

Ness, G. D. 1961. "The Cooperative Movement and Industrial Capitalism in England and Denmark." *Berkeley Journal of Sociology* 6 (1): 1–14.

———. 1967. Bureaucracy and Rural Development in Malaysia. Berkeley: University of California Press.

———. 1993. "The Long View: Population Environment Dynamics in Historical Perspective." In *Population Environment Dynamics: Ideas and Observations*, ed. G. D. Ness, W. Drake, and S. Brechin. Ann Arbor: University of Michigan Press.

Ness, G. D., and H. Ando. 1984. *The Land Is Shrinking: Population Planning in Asia.* Baltimore: Johns Hopkins University Press.

Ness, G. D., and M. Golay, 1997. *Population and Strategies for National Sustainable Development.* London: Earthscan Publications Ltd.

Ness, G., J. Heiby, and B. Pillsbury. 1979. *AIDS' Role in Indonesian Family Planning: USAID Program Evaluation Report No. 2.* Washington D.C.: U.S. Agency for International Development.

Nettle, D., and S. Romaine. 2000. *Vanishing Voices: The Extinction of the World's Languages.* New York: Oxford University Press.

Neumann, R. P. 1997. "Primitive Ideas: Protected Buffer Zones and the Politics of Land in Africa." *Development and Change* 28 (3): 559–582.

Newman, B., H. Irwin, K. Lowe, A. Mostwill, S. Smith, and J. Jones. 1992. "Southern Appalachian Wildlands Proposal." *Wild Earth* (special issue): 46–60.

Newmark, W. D., N. L. Leonard, H. L. Sariko, and D. M. Gamassa. 1993. "Conservation Attitudes of Local People Living Adjacent to Five Protected Areas in Tanzania." *Biological Conservation* 63: 177–183.

Nicolson, C. R., A. Starfield, G. Kofinas, J. Kruse. 2002. "Ten Heuristics for Interdisciplinary Modeling Projects." *Ecosystems* 5: 376–384.

Nyhus, P. J., F. R. Westley, R. C. Lacy, and P. S. Miller. 2002. "A Role for Natural Resource Social Science in Biodiversity Risk Assessment." *Society & Natural Resources* 10: 923–932.

Oli, M. K., I. R. Taylor, and M. E. Rogers. 1994. "Snow Leopard *(Panthero uncia)* Predation on Livestock: An Assessment of Local Perceptions." *Biological Conservation* 68: 63–68.

Ontario Negotiation Bulletin. 1994. *The Algonquins of Golden Lake Negotiations.* Issue No. 6 (October). Pembroke: Ontario Native Affairs Secretariat.

———. 1998. *The Algonquin Land Claim Negotiations.* Issue No. 7 (March). Pembroke: Ontario Native Affairs Secretariat.

Ostrom, E. 1990. *Governing the Commons: The Evolution of Institutions for Collective Action.* Cambridge: Cambridge University Press.

Ostrom, E., J. Burger, C. B. Field, R. B. Norgaard, and D. Policansky. 1999. "Revisting the Commons: Local Lessons, Global Challenges." *Science* 248: 278–282.

Paijmans, K. ed. 1976. *New Guinea Vegetation.* Canberra, Australia: National University Press.

Pandey, S., and M. P. Wells. 1997. "Ecodevelopment Planning at India's Great Himalayan National Park for Biodiversity Conservation and Participatory Rural Development." *Biodiversity and Conservation* 6 (9): 1277–1292.

Pasquero, J. 1991. "Supraorganizational Collaboration: the Canadian Environmental Experiment." *Journal of Applied Behavioral Science* 27 (1): 38–64.

Pearce, F. 1996. "Soldiers Lay Waste to Africa's Oldest Park." *New Scientist* 3 December: 4.

Pearlstine, L. G., L. A. Brandt, W. M. Kitchens, and F. J. Mazzotti. 1995. "Impacts of Citrus Development on Habitats of Southwest Florida." *Conservation Biology* 9: 1020–1032.

Perrings, C. A., K. G. Mäler, C. Folke, C. S. Holling and B. O. Jansson, eds. 1995a. *Biodiversity Conservation: Problems and Policies.* Dordrecht, The Netherlands: Kluwer Academic Publishers.

———. 1995b. *Biodiversity Loss: Economic and Ecological Issues.* Cambridge: Cambridge University Press.

Peterson, D., and J. Goodall. 2000. *Visions of Caliban: On Chimpanzees and People.* Athens: University of Georgia Press.

Peterson, K. 1992. "Basic Program Design for Aboriginal Involvement in Environmental Assessment." Presentation to Ontario Hydro's Board of Directors, Ottawa, June.

———. Forthcoming. *A Planning Framework for Collaboration and Meaningful Participation of Aboriginal Peoples in Environmental Decision Making Practice.* Ph.D. dissertation. Faculty of Environmental Design, University of Calgary.

Peterson, K., and C. Westley-Esquimaux. 1992. "Dancing in the Dark . . . A New Era of Consultation." Presentation to IAIA conference, Washington, D.C., November.

Pimlott, D. H., J. A. Shannon, and G. B. Kolenosky. 1969. *The Ecology of the Timber Wolf in Algonquin Provincial Park.* Toronto: Ontario Department of Lands and Forests.

Plumptre, A. J., J. B. Bizumuremyi, F. Uwimana, and J. D. Ndaruhebeye. 1997. "The Effects of the Rwandan Civil War on Poaching of Ungulates in the Parc National des Volcans." *Oryx* 31: 265–273.

Population Action International (PAI). 2000. *Nature's Place.* Washington, D.C.: PAI.

Prickett, S. T. A., W. R. Burch, Jr., and J. M. Grove. 1999. "Interdisciplinary Research: Maintaining the Constructive Impulse in a Culture of Criticism." *Ecosystems* 2: 302–307.

Primm, S. A., and T. W. Clark. 1996. "The Greater Yellowstone Policy Debate: What Is the Policy Problem?" *Policy Sciences* 29 (2): 137–166.

Prunier, G. 1997. *The Rwanda Crisis: History of a Genocide.* New York: Columbia University Press.

Purvis, A., and A. Hector. 2000. "Getting the Measure of Biodiversity." *Nature* 405: 212–219.

Quinn, N., and J. Inglis. 2000. *A Summary of Information Related to the Movement and Mortality of Wolves and Deer around Algonquin Park.* Peterborough: Ontario Parks.

Ralls, K., J. D. Ballou, and A. R. Templeton. 1988. "Estimates of Lethal Equivalents and the Cost of Inbreeding in Mammals." *Conservation Biology* 2: 85–93.

Rapport, D. 2000. "Transdisciplinarity: An Approach to Problem-Solving in a Complex World." In *Transdisciplinarity: Recreating Integrated Knowledge,* ed. M. Somerville and D. Rapport, 135–144. Oxford: EOLSS, Publishers Co.

Reed, J. M., L. S. Mills, J. B. Dunning, E. S. Menges, K. S. McKelvey, R. Frye, S. Beissinger, M. C. Anstett, and P. S. Miller. 2002. "Use and Emerging Issues in Population Viability Analysis." *Conservation Biology* 16 (1): 7–19.

Renew 2000. Annual Report No. 10. Recovery of Nationally Endangered Wildlife in Canada. Ottawa: Canadian Wildlife Service.

Resource Assessment Commission. 1992. *Forest and Timber Inquiry.* Vol. 1. Canberra: Australian Government Printing Office.

Richards, M. 1996. "Protected Areas, People and Incentives in the Search for Sustainable Forest Conservation in Honduras." *Environmental Conservation* 23 (3): 207–217.

Robbins, M. M. 1995. "A Demographic Analysis of Male Life History and Social Structure of Mountain Gorillas." *Behaviour* 132: 21–47.

Robertson, A. 1960. "A Theory of Limits in Artificial Selection." *Proceedings of the Royal Society of London* 153B: 234–249.

Rodon, T. 1998. "Co-Management and Self-Determination in Nunavut." *Polar Geography* 22 (2): 119–135.

Sandelands, L., and V. Srivastan. 1993. "The Problem of Experience in the Study of Organizations." *Organization Studies* 14 (1): 1–22.

Scheffer, M., C. Brock, and F. R. Westley. 2000. "Socio-Economic Mechanisms Analysis." *Ecosystems* 3: 451–471.

Schwartz, P. 1996. *The Art of the Longview: Paths to Strategic Insight for Yourself and Your Company.* New York: Doubleday.

Scientific Panel for Sustainable Forest Practices in Clayoquot Sound. 1995. *Report 3, First Nations' Perspectives Relating to Forest Practices in Clayoquot Sound.* Victoria: Cortex Consultants Inc.

Scott, P. J., J. A. Burton, and R. Fitter. 1987. "Red Data Books: The Historical Background." In *The Road to Extinction,* ed. R. Fitter and M. Fitter, 1–5. Gland, Switzerland: IUCN.

Seal, U. S. 1992. *Genetic Management Strategies and Population Viability of the Florida Panther (Felis concolor coryi).* Report of a workshop to the U.S. Fish and Wildlife Service. Apple Valley, Minn.: Captive Breeding Specialist Group (SSC/IUCN).

Seal, U. S., and R. C. Lacy. 1989. *Florida Panther Population Viability Analysis.* Report to the U.S. Fish and Wildlife Service. Apple Valley, Minn.: Captive Breeding Specialist Group (SSC/IUCN).

Sekhar, N. U. 1998. "Crop and Livestock Depredation Caused by Wild Animals in Protected Areas: the Case of Sariska Tiger Reserve, Rajasthan, India." *Environmental Conservation* 25(2): 160–171.

Selznick, P. 1949. *TVA and the Grass Roots: A Study in the Sociology of Formal Organizations.* Berkeley: University of California Press.

Shaffer, M. L. 1981. "Minimum Population Sizes for Species Conservation." *BioScience* 31: 131–134.

———. 1990. "Population Viability Analysis." *Conservation Biology* 4: 39–40.

Shyamsundar, P. 1996. "Constraints on Socio-Buffering around the Mantadia National Park in Madagascar." *Environmental Conservation* 23 (1): 67–73.

Sibanda, B. M. C. 1995. "Wildlife, Conservation and the Tonga in Omay." *Land Use Policy* 12 (1): 69–85.

Simberloff, D. A. 1986. "The Proximate Causes of Extinction." In *Patterns and Processes in the History of Life,* ed. D. M. Raup and D. Jablonski, 259–276. Berlin: Springer-Verlag.

———. 1988. "The Contribution of Population and Community Biology to Conservation Science." *Annual Review of Ecology and Systematics* 19: 473–511.

Smith, L. T. 2001. *Decolonizing Methodologies: Research and Indigenous Peoples.* New York: Zed Books Ltd.

Somerville, M., and D. Rapport. 2000. *Transdisciplinarity: Recreating Integrated Knowledge.* Oxford: EOLSS Publishers Co.

Soulé, M. E., ed. 1987. *Viable Populations for Conservation.* Cambridge: Cambridge University Press.

Soulé, M. E., M. Gilpin, W. Conway, and T. Foose. 1986. "The Millennium Ark: How Long a Voyage, How Many Staterooms, How Many Passengers?" *Zoo Biology* 5: 101–113.

Southgate, D., and H. L. Clark. 1993. "Can Conservation Projects Save Biodiversity in South America?" *Ambio* 22 (2–3): 163–166.

Sproule-Jones, M. 1989. "Multiple Rules and the 'Nesting' of Public Policies." *Journal of Theoretical Politics* 1 (4): 459–477.

Stacey, R. 1996. *Strategic Management and Organization Dynamics.* London: Pitman Publishing.

Stanford, C. B. 2001. "The Subspecies Concept in Primatology: The Case of Mountain Gorillas." *Primates* 42: 309–319.

Starfield, A. M., and A. L. Bleloch. 1986. *Building Models for Conservation and Wildlife Management.* New York: Macmillan.

Stevens, S., ed. 1997. *Conservation through Cultural Survival: Indigenous Peoples and Protected Areas.* Washington, D.C.: Island Press.

Stevens, S., and M. Benning. 2000. "Evaluating a Collaborative Approach towards Grizzly Bear Conservation in the Central Canadian Rockies Ecosystem." Course: MGST 797.04, University of Calgary.

Stiles, D. 1994. "Tribals and Trade: A Strategy for Cultural and Ecological Survival." *Ambio* 23 (2): 106–111.

Stover, J., and S. Kirmeyer. 1997. DEMPROJ: *A Computer Program for Making Population Projections. Version 4 User's Manual.* Washington, D.C.: The Futures Group International.

Strier, K., and G. Fonseca. 1996. "The Endangered Muriqui in Brazil's Atlantic Forest." Paper presented at the meeting of the International Primatological Society, Madison, WI. August 1996.

Supriatna, J., R. Tilson, K. J. Gurmaya, J. Manansang, W. Wardojo, A. Sriyanto, A. Teare, K. Castle, and U. S. Seal, eds. 1994. *Javan Gibbon and Javan Langur Population and Habitat Viability Analysis Report.* Apple Valley, Minn.: Captive Breeding Specialist Group (SSC/IUCN).

Swanson, T. 1999. "Why Is There a Biodiversity Convention?: The International Interest in Centralized Development Planning." *International Affairs* 75 (2): 307–331.

Tacconi, L. 1997a. "Property Rights and Participatory Biodiversity Conservation: Lessons from Malekula Island, Vanuatu." *Land Use Policy* 14 (2): 151–161.

———. 1997b. "An Ecological Economic Approach to Forest and Biodiversity Conservation: The Case of Vanuatu." *World Development* 25 (12): 1995–2008.

Tambiah, S. J. 1990. *Magic, Science, Religion and the Scope of Rationality.* Cambridge: Cambridge University Press.

Theberge, J., and M. Theberge. 1997. *Wolf-Prey-Ecosystem Research: Status Report Algonquin-Frontenac.* Waterloo, Ontario: Faculty of Environmental Studies, University of Waterloo.

———. 1998. *Wolf Country: Eleven Years Tracking the Algonquin Wolves.* Toronto: McClellan and Stewart Inc.

Tilman, D. 2000. "Causes, Consequences, and Ethics of Biodiversity." *Nature* 405: 208–211.

Tilman, D., and J. A. Downing. 1994. "Biodiversity and Stability in Grasslands." *Nature* 367: 363–365.

Tilson, R., U. S. Seal, K. Soemarna, W. Ramono, E. Sumardja, S. Poniran, C. van Schaik, M. Leighton, H. Rijksen, and A. Eudey, eds. 1993. *Orangutan Population and Habitat Viability Analysis Report.* Apple Valley, Minn.: Captive Breeding Specialist Group (SSC/IUCN).

Trist, E. 1983. "Referent Organizations and the Development of Interorganizational Domains." *Human Relations* 36: 269–284.

Tudge, C. 1991. *The Last Animals in the Zoo.* London: Hutcheson.

Turner, B. L. II, W. C. Clark, R. W. Kates, J. F. Richards, J. T. Mathews, and W. B. Meyer, eds. 1990. *The Earth as Transformed by Human Action: Global and Regional Changes in the Biosphere over the Past 300 Years.* Cambridge: Cambridge University Press.

UNESCO. 1996. *Biosphere Reserves: The Seville Strategy and the Statutory Framework of the World Network.* Paris: UNESCO.

UNHCR. 1996. *Executive Committee of the High Commissioner's Programme. Standing Committee, 3rd Meeting.* Update on regional developments in Africa. 31 May.

United Nations (UN). 1966. *World Population Prospects, the 1966 Assessment.* New York: United Nations.

———. 1996. *World Population Prospects, the 1996 Revision.* New York: United Nations.

———. 1998. *Sex and Age Distributions, the 1998 Revision.* New York: United Nations.

———. 1999. *World Urbanization Prospects, the 1999 Revision.* New York: United Nations.

———. 2000. *World Population Prospects, the 2000 Revision.* New York: United Nations.

———. 2001. *World Population Prospects, the 2001 Revision.* New York: United Nations.

Usher, P. J. 1993. "The Beverly-Kaminuriak Caribou Management Board: An Experience in Co-Management)." In *Traditional Ecological Knowledge: Concepts and Cases,* ed. J. T. Inglis, 111–120. Ottawa: Canadian Museum of Nature and International Development Research Centre.

Vanclay, J. K. 1998. "FLORES: For Exploring Land Use Options in Forested Landscapes." *Agroforestry Forum* 9: 47–52.

Vandergeest, P. 1996. "Property Rights in Protected Areas: Obstacles to Community Involvement as a Solution in Thailand." *Environmental Conservation* 23 (3): 259–268.

Vincent, M., G. Slater, and P. Clark. 2000. *Recovery Plan for the Scott's Tree Kangaroo (Dendrolagus scottae) and Golden-Mantled Tree Kangaroo (Dendrolagus goodfellowi pulcherrimus).* Draft.

Vredenburg, H. 1986. "On the Theoretical Interpretation of a Multiple Request Influence Strategy in an Industrial Marketing Setting." *Developments in Marketing Science* 19: 436–441.

Vredenburg, H., and J. J. Marshall. 1988. "Extending the External Validity of the FITD Effect to the Industrial Marketplace." *Journal of the Academy of Marketing Science* 16 (2): 49–56.

———. 1991. "A Field Experimental Investigation of a Social Influence Application in Industrial Communication Strategy." *Journal of Direct Marketing* 5 (3): 7–18.

Vucetich, J., and P. Paquet. 2000. *The Demographic Population Viability of Algonquin Wolves.* Report prepared for the Algonquin Wolf Advisory Group.

Wahl, C. E. 1995. "Climate in Nunavut: A Strategy in Face of Change." Master's thesis, Environment and Resource Studies, University of Waterloo.

Wainwright, C., and W. Wehrmeyer. 1998. "Success in Integrating Conservation and Development?: A Study from Zambia." *World Development* 26 (6): 933–944.

Walker, S., and S. Molur. 1994. *Population and Habitat Viability Analysis (PHVA) Workshop for Indian/Nepali Rhinoceros.* Coimbatore: Zoo Outreach Organisation/CBSG India.

Wallis de Vries, M. F. 1995. "Large Herbivores and the Design of Large-Scale Nature Reserves in Western Europe." *Conservation Biology* 9 (1): 25–33.

Walters C. 1986. *Adaptive Management of Renewable Resources.* London: Collier-MacMillan.

Watts, D. P. 1991. "Strategies of Habitat Use by Mountain Gorillas." *Folia Primatologica* 56: 1–16.

Wear, D. 1999. "Challenges to Interdisciplinary Discourse." *Ecosystems* 2: 299–301.

Weber, A. W., and A. Vedder. 1983. "Population Dynamics of the Virunga Gorillas: 1959–1978." *Biological Conservation* 26: 341–366.

Weber, M. 1978. *Economy and Society.* Berkeley: University of California Press.

Weed, T. J. 1994. "Central America's 'Peace Parks' and Regional Conflict Resolution." *International Environmental Affairs* 6 (2):175–190.

Wells, M. P. 1996. "The Social Role of Protected Areas in the New South Africa." *Environmental Conservation* 23 (4): 322–331.

———. 1998. "Institutions and Incentives for Biodiversity Conservation." *Biodiversity and Conservation* 7 (6): 815–835.

Wells, M. P., and K. E. Brandon. 1992. *People and Parks: Linking Protected Area Management With Local Communities.* Washington, D.C.: The World Bank, World Wildlife Fund, and U.S. Agency for International Development.

———. 1993. "The Principles and Practice of Buffer Zones and Local Participation in Biodiversity Conservation." *Ambio* 22 (2–3): 157–162.

Wemmer, C. 1990. "Creative Collaboration." *Conservation and Research Center Newsletter* 2 (1): 1.

Werikhe, S., L. Macfie, N. Rosen, and P. S. Miller, eds. 1998. *Can the Mountain Gorilla Survive? Population and Habitat Viability Assessment Workshop for Gorilla gorilla beringei.* Apple Valley, Minn: Conservation Breeding Specialist Group (SSC/IUCN).

Westley, F. R. 1995. "Governing Design: The Management of Social Systems and Ecosystem Management." In *Barriers and Bridges to Renewal of Ecosystems and Institution,* ed. L. Gunderson, C. S. Holling, and S. Light, 391–427. New York: Columbia University Press.

———. 1999. "Not on Our Watch: The Biodiversity Crisis and Global Collaboration Response." In *Organizational Dimensions of Global Change,* ed. D. Cooperrider and J. Dutton, 88–113. Thousand Oaks, Cal.: Sage Publications.

Westley, F. R., and H. Vredenburg. 1991. "Strategic Bridging: The Alliances of Business and Environmentalists." *Journal of Applied Behavioral Science* 27 (1): 65–90.

———.1996. "Rethinking Sustainability: Criteria for Aligning Economic Practice with Environmental Protection." *Journal of Management Inquiry* 5 (2): 104–119.

———. 1997. "Interorganizational Collaboration and the Preservation of Biodiversity." *Organization Science* 8 (4): 381–403.

Wheatley, M. 1994. *Leadership and the New Science: Learning about Organization from an Orderly Universe.* San Francisco: Berett-Koehler.

White, C., and M. Scott-Brown. 1995. *Atlas of the Central Rockies Ecosystem: Towards an Ecologically Sustainable Landscape.* Calgary: Parks Canada and Komex International.

Wildlands League. 1997. *Kill the Myth, Not the Wolf: A Wolf Conservation Strategy for Algonquin Park and Region.* Toronto: Wildlands League.

Williams, J. S. 1994. *Population Dynamics in Participatory Rural Appraisal.* A manual to assist rural communities in population appraisal. Draft. Gland, Switzerland: IUCN.

Wilson E. O, ed.1986. *Biodiversity* Washington, D.C.: National Academy Press, National Academy of Sciences (U.S.), Smithsonian Institution.

———. 1989. "Threats to Biodiversity." *Scientific American* 261 (3): 108–117.

———. 1992. *The Diversity of Life.* New York: Norton.

———. 1998. *Consilience.* New York: Alfred A. Knopf.

Wilson, P. J. et al. 2000. "DNA Profiles of the Eastern Canadian Wolf and the Red Wolf Provide Evidence for a Common Evolutionary History Independent of the Gray Wolf." *Canadian Journal of Zoology* 78 (12): 2156–2166.

Wildlife Management Advisory Council (WMAC). 1988. *Inuvialuit Renewable Resource Conservation and Management Plan.* Yellowknife: Wildlife Management Advisory Council and Fisheries Joint Management Committee.

Woodley, S., and L. Sportza. 1996. *Ecosystem Stresses in Canada's National Parks: Guide and Questionnaire.* Revised version, typescript.

World Bank. 1993. *Country Profile, Uganda.* Washington, D.C.: The World Bank.

World Commission on Environment and Development (WCED). 1987. *Our Common Future.* Oxford, U.K.: Oxford University Press.

World Resources Institute (WRI). 1992. *Global Biodiversity Strategy: A Policy Maker's Guide.* Washington, D.C.: World Resource Institute, International Union for the Conservation of Nature and United National Environmental Protection.

Wright, R. G. 1999. "Wildlife Management in the National Parks: Questions in Search of Answers." *Ecological Applications* 9 (1): 30–36.

Yaffee, S. L., A. F. Phillips, I. C. Frentz, P. W. Hardy, S. M. Maleki, and B. E. Thorpe. 1996. *Ecosystem Management in the United States: An Assessment of Current Experience.* Washington, D.C.: Island Press.

Young, O. R. 1994. *International Governance: Protecting the Environment in a Stateless Society.* Ithaca: Cornell University Press.

———. 1997. "Arctic Governance: Bringing the High Latitudes in from the Cold." *International Environmental Affairs* 9 (1): 54–68.

Contributors

Jenna S. Borovansky has a bachelor of science in wildlife management and biology from the University of Wisconsin–Stevens Point and earned her master of environmental studies at Yale University's School of Forestry and Environmental Studies after working with CBSG for many years. She continues to use her conservation policy and ecosystem management expertise directing programs for a river conservation organization in the Pacific Northwest.

Onnie Byers earned her doctorate in reproductive physiology from the University of Minnesota and completed a postdoctoral fellowship at the Smithsonian Institution's National Zoo in Washington, D.C. She joined the CBSG staff in 1991 as a program officer. In addition to serving as a reproductive specialist in workshops conducted by CBSG and other conservation organizations, Byers is responsible for organization, design, and facilitation of CBSG's Conservation Assessment and Management Plan and PHVA workshops.

George Francis is a Distinguished Professor Emeritus in the Department of Environmental and Resource Studies at the University of Waterloo, Canada. Francis' academic background includes degrees in biology (ecology), economics and political science, and resource conservation (precursor to environmental studies). His main area of academic interest has been the implications of governance and management that arise from the adoption of "an ecosystem approach" to regional environmental and resource management issues. He has been involved in a number of university-based collaborative projects that investigated questions related to these issues in the context of the North American Great Lakes, he has had a long association with groups involved with conservation issues and biosphere reserves in Canada, and he has a number of academic publications and other reports associated with these various endeavors.

Robert C. Lacy is the population geneticist for the Department of Conservation Biology at Brookfield Zoo. In 2003, he also took on the role of chairing

CBSG. He was trained in quantitative genetics (B.A., M.A., Wesleyan University), evolutionary biology, ecology, and population genetics (Ph.D., Cornell University). He has published papers in evolutionary theory, genetics, population ecology, taxonomy, behavior, conservation, and wildlife management. Software he developed for modeling population dynamics is used by conservationists, wildlife managers, researchers, government agencies, NGOs, and universities throughout the world.

David A. Lertzman is an adjunct assistant professor in environmental management and sustainable development with the TCPL International Institute for Resource Industries and Sustainability Studies (TC-IRIS), Haskayne School of Business, University of Calgary; a research associate with the Arctic Institute of North America; and a private consultant. His research focus is on bridging traditional knowledge of Indigenous peoples and Western science in sustainable development.

Philip S. Miller is a population biologist and program officer with CBSG. His academic training includes degrees in chemistry (Purdue University, 1985), geology (Northwestern University, 1988), and zoology (Arizona State University, 1994). Since joining CBSG in late 1994, he has focused his research attention on the implementation and expansion of quantitative risk assessment methodologies for endangered species management. In addition, he continues to collaborate with Network members on the improved design of participatory workshop processes aimed at applying biological science to practical conservation decision making.

Gayl D. Ness is now professor emeritus of sociology at the University of Michigan. He received his doctorate in sociology from the University of California, Berkeley in 1961 and completed a postdoctoral fellowship 1961–64 in Southeast Asia. He has written extensively on Southeast Asia, population, environment, and development issues.

Philip J. Nyhus is an assistant professor of environmental studies in the Department of Geosciences at Franklin and Marshall College in Lancaster, Pennsylvania. He received his doctorate in land resources from the Institute for Environmental Studies at the University of Wisconsin. His research bridges the natural and social sciences to addresses human interactions with the environment. He has studied the human dimension of tiger and large-mammal conservation in Indonesia and the use of coupled human-natural systems to improve biodiversity risk assessment to inform conservation policy.

Emmanuel Raufflet has a doctorate in management (McGill University) and is an assistant professor in management at HEC Montreal. Emmanuel conducts research on institutional change in forest management in developing

areas and is interested in the role of organizations in the sustainable management of natural resources.

Harrie Vredenburg is professor of strategy and director of the TCPL
International Institute for Resource Industries and Sustainability Studies
(TC-IRIS) at the Haskayne School of Business, University of Calgary, where
he holds the Suncor Energy Chair in Competitive Strategy and Sustainable
Development. He received his doctorate in strategic management from the
Richard Ivey School of Business at the University of Western Ontario. His
research centers on competitive strategy and sustainable development and on
stakeholder-based strategic management.

Frances R. Westley is the James McGill Professor of Strategy, Faculty of
Management, McGill University. She received her doctorate in sociology from
McGill. Her research is on interorganizational, intersectoral, and interdisciplinary collaboration, particularly in the endangered spaces and species
domain. She has worked for many years assisting CBSG in developing its processes and products and has designed trainings and multi-stakeholder workshops around the globe.

John S. Williams, a demographer, specializes in population and environ-
ment programs. He is senior policy analyst at the Population Reference
Bureau in Washington, D.C., where he is responsible for program evaluation.
Williams received his doctorate from the Office of Population Research at
Princeton University and has served on the graduate faculty of the New
School for Social Research and the City University of New York. He is an
active member of IUCN's Species Survival Commission and has facilitated
endangered species workshops in Asia, Africa, Oceania, and Latin America.

Index